Undergraduate Texts in Mathematics

T0192067

For further volumes:
http://www.springer.com/series/666

Peter Petersen

Linear Algebra

 Springer

Peter Petersen
Department of Mathematics
University of California
Los Angeles, CA
USA

ISSN 0172-6056
ISBN 978-1-4899-9788-3   ISBN 978-1-4614-3612-6 (eBook)
DOI 10.1007/978-1-4614-3612-6
Springer New York Heidelberg Dordrecht London

Mathematics Subject Classification (2000): 15-01

Preface

This book covers the aspects of linear algebra that are included in most advanced undergraduate texts. All the usual topics from complex vector spaces, complex inner products the spectral theorem for normal operators, dual spaces, quotient spaces, the minimal polynomial, the Jordan canonical form, and the Frobenius (or rational) canonical form are explained. A chapter on determinants has been included as the last chapter, but they are not used in the text as a whole. A different approach to linear algebra that does not use determinants can be found in [Axler].

The expected prerequisites for this book would be a lower division course in matrix algebra. A good reference for this material is [Bretscher].

In the context of other books on linear algebra it is my feeling that this text is about on a par in difficulty with books such as [Axler, Curtis, Halmos, Hoffman-Kunze, Lang]. If you want to consider more challenging texts, I would suggest looking at the graduate level books [Greub, Roman, Serre].

Chapter 1 contains all of the basic material on abstract vector spaces and linear maps. The dimension formula for linear maps is the theoretical highlight. To facilitate some more concrete developments we cover matrix representations, change of basis, and Gauss elimination. Linear independence which is usually introduced much earlier in linear algebra only comes towards to the end of the chapter. But it is covered in great detail there. We have also included two sections on dual spaces and quotient spaces that can be skipped.

Chapter 2 is concerned with the theory of linear operators. Linear differential equations are used to motivate the introduction of eigenvalues and eigenvectors, but this motivation can be skipped. We then explain how Gauss elimination can be used to compute the eigenvalues as well as the eigenvectors of a matrix. This is used to understand the basics of how and when a linear operator on a finite-dimensional space is diagonalizable. We also introduce the minimal polynomial and use it to give the classic characterization of diagonalizable operators. In the later sections we give a fairly simple proof of the Cayley–Hamilton theorem and the cyclic subspace decomposition. This quickly leads to the Frobenius canonical form. This canonical form is our most general result on how to find a simple matrix representation for a linear map in case it is not diagonalizable. The antepenultimate section explains how the Frobenius canonical form implies the Jordan–Chevalley decomposition and the Jordan-Weierstrass canonical form. In the last section, we present a quick and elementary approach to the Smith normal form. This form allows us to calculate directly all of the similarity invariants of a matrix using basic row and column operations on matrices with polynomial entries.

Chapter 3 includes material on inner product spaces. The Cauchy–Schwarz inequality and its generalization to Bessel's inequality and how they tie in with orthogonal projections form the theoretical centerpiece of this chapter. Along the way, we cover standard facts about orthonormal bases and their existence through the Gram–Schmidt procedure as well as orthogonal complements and orthogonal projections. The chapter also contains the basic elements of adjoints of linear maps and some of its uses to orthogonal projections as this ties in nicely with orthonormal bases. We end the chapter with a treatment of matrix exponentials and systems of differential equations.

Chapter 4 covers quite a bit of ground on the theory of linear maps between inner product spaces. The most important result is of course the spectral theorem for self-adjoint operators. This theorem is used to establish the canonical forms for real and complex normal operators, which then gives the canonical form for unitary, orthogonal, and skew-adjoint operators. It should be pointed out that the proof of the spectral theorem does not depend on whether we use real or complex scalars nor does it rely on the characteristic or minimal polynomials. The reason for ignoring our earlier material on diagonalizability is that it is desirable to have a theory that more easily generalizes to infinite dimensions. The usual proofs that use the characteristic and minimal polynomials are relegated to the exercises. The last sections of the chapter cover the singular value decomposition, the polar decomposition, triangulability of complex linear operators (Schur's theorem), and quadratic forms.

Chapter 5 covers determinants. At this point, it might seem almost useless to introduce the determinant as we have covered the theory without needing it much. While not indispensable, the determinant is rather useful in giving a clean definition for the characteristic polynomial. It is also one of the most important invariants of a finite-dimensional operator. It has several nice properties and gives an excellent criterion for when an operator is invertible. It also comes in handy in giving a formula (Cramer's rule) for solutions to linear systems. Finally, we discuss its uses in the theory of linear differential equations, in particular in connection with the variation of parameters formula for the solution to inhomogeneous equations. We have taken the liberty of defining the determinant of a linear operator through the use of volume forms. Aside from showing that volume forms exist, this gives a rather nice way of proving all the properties of determinants without using permutations. It also has the added benefit of automatically giving the permutation formula for the determinant and hence showing that the sign of a permutation is well defined.

An * after a section heading means that the section is not necessary for the understanding of other sections without an *.

Let me offer a few suggestions for how to teach a course using this book. My assumption is that most courses are based on 150 min of instruction per week with a problem session or two added. I realize that some courses meet three times while others only two, so I will not suggest how much can be covered in a lecture.

First, let us suppose that you, like me, teach in the pedagogically impoverished quarter system: It should be possible to teach Chap. 1, Sects. 1.2–1.13 in 5 weeks, being a bit careful about what exactly is covered in Sects. 1.12 and 1.13. Then, spend 2 weeks on Chap. 2, Sects. 2.3–2.5, possibly omitting Sect. 2.4 covering the minimal polynomial if timing looks tight. Next spend 2 weeks on Chap. 3, Sects. 3.1–3.5, and finish the course by covering Chap. 4, Sect. 4.1 as well as Exercise 9 in Sect. 4.1. This finishes the course with a proof of the Spectral Theorem for self-adjoint operators, although not the proof I would recommend for a more serious treatment.

Next, let us suppose that you teach in a short semester system, as the ones at various private colleges and universities. You could then add 2 weeks of material by either covering the canonical forms from Chap. 2, Sects. 2.6–2.8 or alternately spend 2 weeks covering some of the theory of linear operators on inner product spaces from Chap. 4, Sects. 4.1–4.5. In case you have 15 weeks at your disposal, it might be possible to cover both of these topics rather than choosing between them.

Finally, should you have two quarters, like we sometimes do here at UCLA, then you can in all likelihood cover virtually the entire text. I would certainly recommend that you cover all of Chap. 4 and the canonical form sections in Chap. 2, Sects. 2.6–2.8, as well as the chapter on determinants. If time permits, it might even be possible to include Sects. 2.2, 3.7, 4.8, and 5.8 that cover differential equations.

This book has been used to teach a bridge course on linear algebra at UCLA as well as a regular quarter length course. The bridge course was funded by a VIGRE NSF grant, and its purpose was to ensure that incoming graduate students had really learned all of the linear algebra that we expect them to know when starting graduate school. The author would like to thank several UCLA students for suggesting various improvements to the text: Jeremy Brandman, Sam Chamberlain, Timothy Eller, Clark Grubb, Vanessa Idiarte, Yanina Landa, Bryant Mathews, Shervin Mosadeghi, and Danielle O'Donnol. I am also pleased to acknowledge NSF support from grants DMS 0204177 and 1006677.

I would also like to thank Springer-Verlag for their interest and involvement in this book as well as their suggestions for improvements.

Finally, I am immensely grateful to Joe Borzellino at Cal Poly San Luis Obispo who used the text several times at his institution and supplied me with numerous corrections.

Contents

Chapter 1
Basic Theory

In the first chapter, we are going to cover the definitions of vector spaces, linear maps, and subspaces. In addition, we are introducing several important concepts such as basis, dimension, direct sum, matrix representations of linear maps, and kernel and image for linear maps. We shall prove the dimension theorem for linear maps that relate the dimension of the domain to the dimensions of kernel and image. We give an account of Gauss elimination and how it ties in with the more abstract theory. This will be used to define and compute the characteristic polynomial in Chap. 2.

It is important to note that Sects. 1.13 and 1.12 contain alternate proofs of some of the important results in this chapter. As such, some people might want to go right to these sections after the discussion on isomorphism in Sect. 1.8 and then go back to the missed sections.

As induction is going to play a big role in many of the proofs, we have chosen to say a few things about that topic in the first section.

1.1 Induction and Well-Ordering*

A fundamental property of the *natural numbers*, i.e., the positive integers $\mathbb{N} = \{1, 2, 3, \ldots\}$, that will be used throughout the book is the fact that they are *well ordered*. This means that any nonempty subset $S \subset \mathbb{N}$ has a smallest element $s_{\min} \in S$ such that $s_{\min} \leq s$ for all $s \in S$. Using the natural ordering of the integers, rational numbers, or real numbers, we see that this property does not hold for those numbers. For example, the half-open interval $(0, \infty)$ does not have a smallest element.

In order to justify that the positive integers are well ordered, let $S \subset \mathbb{N}$ be nonempty and select $k \in S$. Starting with 1, we can check whether it belongs to S. If it does, then $s_{\min} = 1$. Otherwise, check whether 2 belongs to S. If $2 \in S$ and

P. Petersen, *Linear Algebra*, Undergraduate Texts in Mathematics,
DOI 10.1007/978-1-4614-3612-6_1, © Springer Science+Business Media New York 2012

$1 \notin S$, then we have $s_{\min} = 2$. Otherwise, we proceed to check whether 3 belongs to S. Continuing in this manner, we must eventually find $k_0 \le k$, such that $k_0 \in S$, but $1, 2, 3, \ldots, k_0 - 1 \notin S$. This is the desired minimum: $s_{\min} = k_0$.

We shall use the well-ordering of the natural numbers in several places in this text. A very interesting application is to the proof of the prime factorization theorem: any integer ≥ 2 is a product of prime numbers. The proof works the following way. Let $S \subset \mathbb{N}$ be the set of numbers which do not admit a prime factorization. If S is empty, we are finished; otherwise, S contains a smallest element $n = s_{\min} \in S$. If n has no divisors, then it is a prime number and hence has a prime factorization. Thus, n must have a divisor $p > 1$. Now write $n = p \cdot q$. Since $p, q < n$ both numbers must have a prime factorization. But then also $n = p \cdot q$ has a prime factorization. This contradicts that S is nonempty.

The second important idea that is tied to the natural numbers is that of *induction*. Sometimes, it is also called *mathematical induction* so as not to confuse it with the inductive method from science. The types of results that one can attempt to prove with induction always have a statement that needs to be verified for each number $n \in \mathbb{N}$. Some good examples are

1. $1 + 2 + 3 + \cdots + n = \frac{n(n+1)}{2}$.
2. Every integer ≥ 2 has a prime factorization.
3. Every polynomial has a root.

The first statement is pretty straightforward to understand. The second is a bit more complicated, and we also note that in fact, there is only a statement for each integer ≥ 2. This could be finessed by saying that each integer $n + 1$, $n \ge 1$ has a prime factorization. This, however, seems too pedantic and also introduces extra and irrelevant baggage by using addition. The third statement is obviously quite different from the other two. For one thing, it only stands a chance of being true if we also assume that the polynomials have degree ≥ 1. This gives us the idea of how this can be tied to the positive integers. The statement can be paraphrased as: Every polynomial of degree ≥ 1 has a root. Even then, we need to be more precise as $x^2 + 1$ does not have any real roots.

In order to explain how induction works abstractly, suppose that we have a statement $P(n)$ for each $n \in \mathbb{N}$. Each of the above statements can be used as an example of what $P(n)$ can be. The induction process now works by first ensuring that the anchor statement is valid. In other words, we first check that $P(1)$ is true. We then have to establish the *induction step*. This means that we need to show that if $P(n-1)$ is true, then $P(n)$ is also true. The assumption that $P(n-1)$ is true is called the *induction hypothesis*. If we can establish the validity of these two facts, then $P(n)$ must be true for all n. This follows from the well-ordering of the natural numbers. Namely, let $S = \{n : P(n) \text{ is false}\}$. If S is empty, we are finished, otherwise, S has a smallest element $k \in S$. Since $1 \notin S$, we know that $k > 1$. But this means that we know that $P(k-1)$ is true. The induction step then implies that $P(k)$ is true as well. This contradicts that S is nonempty.

Let us see if we can use this procedure on the above statements. For 1, we begin by checking that $1 = \frac{1(1+1)}{2}$. This is indeed true. Next, we assume that

$$1 + 2 + 3 + \cdots + (n-1) = \frac{(n-1)\,n}{2},$$

and we wish to show that

$$1 + 2 + 3 + \cdots + n = \frac{n\,(n+1)}{2}.$$

Using the induction hypothesis, we see that

$$
\begin{aligned}
(1 + 2 + 3 + \cdots + (n-1)) + n &= \frac{(n-1)\,n}{2} + n \\
&= \frac{(n-1)\,n + 2n}{2} \\
&= \frac{(n+1)\,n}{2}.
\end{aligned}
$$

Thus, we have shown that $P\,(n)$ is true provided $P\,(n-1)$ is true.

For 2, we note that two is a prime number and hence has a prime factorization. Next, we have to prove that n has a prime factorization if $(n-1)$ does. This, however, does not look like a very promising thing to show. In fact, we need a stronger form of induction to get this to work.

The induction step in the stronger version of induction is as follows: If $P\,(k)$ is true for all $k < n$, then $P\,(n)$ is also true. Thus, the induction hypothesis is much stronger as we assume that all statements prior to $P\,(n)$ are true. The proof that this form of induction works is virtually identical to the above justification.

Let us see how this stronger version can be used to establish the induction step for 2. Let $n \in \mathbb{N}$, and assume that all integers below n have a prime factorization. If n has no divisors other than 1 and n, it must be a prime number and we are finished. Otherwise, $n = p \cdot q$ where $p, q < n$. Whence, both p and q have prime factorizations by our induction hypothesis. This shows that also n has a prime factorization.

We already know that there is trouble with statement 3. Nevertheless, it is interesting to see how an induction proof might break down. First, we note that all polynomials of degree 1 look like $ax + b$ and hence have $-\frac{b}{a}$ as a root. This anchors the induction. To show that all polynomials of degree n have a root, we need to first decide which of the two induction hypotheses are needed. There really is not anything wrong by simply assuming that all polynomials of degree $< n$ have a root. In this way, we see that at least any polynomial of degree n that is the product of two polynomials of degree $< n$ must have a root. This leaves us with the so-called prime or irreducible polynomials of degree n, namely, those polynomials that are not divisible by polynomials of degree ≥ 1 and $< n$. Unfortunately, there is not much we can say about these polynomials. So induction does not seem to work well in this case. All is not lost however. A careful inspection of the "proof" of 3 can be modified to show that any polynomial has a prime factorization. This is studied further in Sect. 2.1.

The type of statement and induction argument that we will encounter most often in this text is definitely of the third type. That is to say, it certainly will never be of the very basic type seen in statement 1. Nor will it be as easy as in statement 2. In our cases, it will be necessary to first find the integer that is used for the induction, and even then, there will be a whole collection of statements associated with that integer. This is what is happening in the third statement. There, we first need to select the degree as our induction integer. Next, there are still infinitely many polynomials to consider when the degree is fixed. Finally, whether or not induction will work or is the "best" way of approaching the problem might actually be questionable.

The following statement is fairly typical of what we shall see: Every subspace of $\mathbb{R}^n$ admits a basis with $\leq n$ elements. The induction integer is the dimension n, and for each such integer, there are infinitely many subspaces to be checked. In this case, an induction proof will work, but it is also possible to prove the result without using induction.

1.2 Elementary Linear Algebra

Our first picture of what vectors are and what we can do with them comes from viewing them as geometric objects in the plane and space. Simply put, a vector is an arrow of some given length drawn in the plane. Such an arrow is also known as an oriented line segment. We agree that vectors that have the same length and orientation are equivalent no matter where they are based. Therefore, if we base them at the origin, then vectors are determined by their endpoints. Using a parallelogram, we can add such vectors (see Fig. 1.1). We can also multiply them by scalars. If the scalar is negative, we are changing the orientation. The size of the scalar determines how much we are scaling the vector, i.e., how much we are changing its length (see Fig. 1.2).

This geometric picture can also be taken to higher dimensions. The idea of scaling a vector does not change if it lies in space, nor does the idea of how to add vectors, as two vectors must lie either on a line or more generically in a plane. The problem comes when we wish to investigate these algebraic properties further. As an example, think about the associative law

$$(x + y) + z = x + (y + z).$$

Clearly, the proof of this identity changes geometrically from the plane to space. In fact, if the three vectors do not lie in a plane and therefore span a parallelepiped, then the sum of these three vectors regardless of the order in which they are added

Fig. 1.1 Vector addition

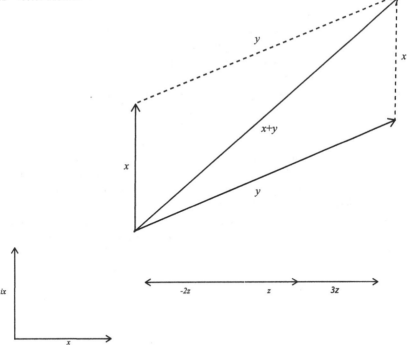

Fig. 1.2 Scalar multiplication

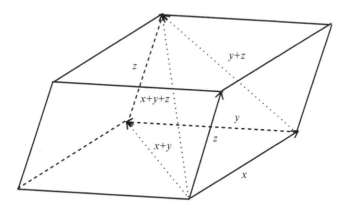

Fig. 1.3 Associativity

is the diagonal of this parallelepiped. The picture of what happens when the vectors lie in a plane is simply a projection of the three-dimensional picture on to the plane (see Fig. 1.3).

The purpose of linear algebra is to clarify these algebraic issues by looking at vectors in a less geometric fashion. This has the added benefit of also allowing other spaces that do not have geometric origins to be included in our discussion. The end result is a somewhat more abstract and less geometric theory, but it has turned out to be truly useful and foundational in almost all areas of mathematics, including geometry, not to mention the physical, natural, and social sciences.

Something quite different and interesting happens when we allow for complex scalars. This is seen in the plane itself which we can interpret as the set of complex numbers. Vectors still have the same geometric meaning, but we can also "scale" them by a number like $i = \sqrt{-1}$. The geometric picture of what happens when multiplying by i is that the vector's length is unchanged as $|i| = 1$, but it is rotated $90°$ (see Fig. 1.2). Thus it is not scaled in the usual sense of the word. However, when we define these notions below, one will not really see any algebraic difference in what is happening. It is worth pointing out that using complex scalars is not just something one does for the fun of it; it has turned out to be quite convenient and important to allow for this extra level of abstraction. This is true not just within mathematics itself. When looking at books on quantum mechanics, it quickly becomes clear that complex vector spaces are the "sine qua non"(without which nothing) of the subject.

1.3   Fields

The "scalars" or numbers used in linear algebra all lie in a *field*. A field is a set $\mathbb{F}$ of numbers, where one has both addition

$$\mathbb{F} \times \mathbb{F} \to \mathbb{F}$$

$$(\alpha, \beta) \mapsto \alpha + \beta$$

and multiplication

$$\mathbb{F} \times \mathbb{F} \to \mathbb{F}$$

$$(\alpha, \beta) \mapsto \alpha\beta.$$

Both operations are assumed associative, commutative, etc. We shall mainly be concerned with the real numbers $\mathbb{R}$ and complex numbers $\mathbb{C}$; some examples will be using the rational numbers $\mathbb{Q}$ as well. These three fields satisfy the axioms we list below.

Definition 1.3.1. A *field* $\mathbb{F}$ is a set whose elements are called numbers or when used in linear algebra *scalars*. The field contains two different elements 0 and 1, and we can add and multiply numbers. These operations satisfy

1. The associative law

$$\alpha + (\beta + \gamma) = (\alpha + \beta) + \gamma.$$

2. The commutative law

$$\alpha + \beta = \beta + \alpha.$$

3. Addition by 0:

$$\alpha + 0 = \alpha.$$

4. Existence of negative numbers: For each α, we can find $-\alpha$ so that

$$\alpha + (-\alpha) = 0.$$

5. The associative law:

$$\alpha\,(\beta\gamma) = (\alpha\beta)\,\gamma.$$

6. The commutative law:

$$\alpha\beta = \beta\alpha.$$

7. Multiplication by 1:

$$\alpha 1 = \alpha.$$

8. Existence of inverses: For each $\alpha \neq 0$, we can find α^{-1} so that

$$\alpha\alpha^{-1} = 1.$$

9. The distributive law:

$$\alpha\,(\beta + \gamma) = \alpha\beta + \alpha\gamma.$$

One can show that both 0 and 1 are uniquely defined and that the additive inverse $-\alpha$ as well as the multiplicative inverse α^{-1} is unique.

Occasionally, we shall also use that the field has *characteristic zero* this means that

$$n = \underbrace{\frac{n \text{ times}}{1 + \cdots + 1}} \neq 0$$

for all positive integers n. Fields such as $\mathbb{F}_2 = \{0, 1\}$ where $1 + 1 = 0$ clearly do not have characteristic zero. We make the assumption throughout the text that all fields have characteristic zero. In fact, there is little loss of generality in assuming that the fields we work are the usual number fields $\mathbb{Q}$, $\mathbb{R}$, and $\mathbb{C}$.

There are several important collections of numbers that are not fields:

$$\mathbb{N} = \{1, 2, 3, \ldots\}$$
$$\subset \mathbb{N}_0 = \{0, 1, 2, 3, \ldots\}$$
$$\subset \mathbb{Z} = \{0, \pm 1, \pm 2, \pm 3, \ldots\}$$
$$= \{0, 1, -1, 2, -2, 3, -3, \ldots\}.$$

1.4   Vector Spaces

Definition 1.4.1. A *vector space* consists of a set of vectors V and a field $\mathbb{F}$. The vectors can be added to yield another vector: if $x, y \in V$, then $x + y \in V$ or

$$V \times V \to V$$
$$(x, y) \mapsto x + y.$$

The scalars can be multiplied with the vectors to yield a new vector: if $\alpha \in \mathbb{F}$ and $x \in V$, then $\alpha x \in V$; in other words,

$$\mathbb{F} \times V \to V$$

$$(\alpha, x) \mapsto \alpha x.$$

The vector space contains a *zero vector* 0, also known as the *origin* of V. It is a bit confusing that we use the same symbol for $0 \in V$ and $0 \in \mathbb{F}$. It should always be obvious from the context which zero is used. We shall generally use the notation that scalars, i.e., elements of $\mathbb{F}$, are denoted by small Greek letters such as $\alpha, \beta, \gamma, \ldots$, while vectors are denoted by small roman letters such as $x, y, z, \ldots$. Addition and scalar multiplication must satisfy the following axioms:

1. The associative law:
$$(x + y) + z = x + (y + z).$$

2. The commutative law:
$$x + y = y + x.$$

3. Addition by 0:
$$x + 0 = x.$$

4. Existence of negative vectors: For each x, we can find $-x$ such that

$$x + (-x) = 0.$$

5. The associative law for multiplication by scalars:

$$\alpha (\beta x) = (\alpha \beta) x.$$

6. Multiplication by the unit scalar:

$$1x = x.$$

7. The distributive law when vectors are added:

$$\alpha (x + y) = \alpha x + \alpha y.$$

8. The distributive law when scalars are added:

$$(\alpha + \beta) x = \alpha x + \beta x.$$

Remark 1.4.2. We shall also allow scalars to be multiplied on the right of the vector:

$$x\alpha = \alpha x$$

The only slight issue with this definition is that we must ensure that associativity still holds. The key to that is that the field of scalars have the property that multiplication in commutative:

$$
\begin{aligned}
x\,(\alpha\beta) &= (\alpha\beta)\,x \\
&= (\beta\alpha)\,x \\
&= \beta\,(\alpha x) \\
&= (x\alpha)\,\beta
\end{aligned}
$$

These axioms lead to several "obvious" facts.

Proposition 1.4.3. *Let V be a vector space over a field $\mathbb{F}$. If $x \in V$ and $\alpha \in \mathbb{F}$, then:*

1. $0x = 0$.
2. $\alpha 0 = 0$.
3. $-1x = -x$.
4. *If $\alpha x = 0$, then either $\alpha = 0$ or $x = 0$.*

Proof. By the distributive law,

$$0x + 0x = (0 + 0)\,x = 0x.$$

This together with the associative law gives us

$$
\begin{aligned}
0x &= 0x + (0x - 0x) \\
&= (0x + 0x) - 0x \\
&= 0x - 0x \\
&= 0.
\end{aligned}
$$

The second identity is proved in the same manner.
For the third, consider

$$
\begin{aligned}
0 &= 0x \\
&= (1 - 1)\,x \\
&= 1x + (-1)\,x \\
&= x + (-1)\,x,
\end{aligned}
$$

adding $-x$ on both sides then yields

$$-x = (-1)\,x.$$

Finally, if $\alpha x = 0$ and $\alpha \neq 0$, then we have

$$\begin{aligned}
x &= \left(\alpha^{-1}\alpha\right) x \\
&= \alpha^{-1}\left(\alpha x\right) \\
&= \alpha^{-1}0 \\
&= 0.
\end{aligned}$$

$\square$

With these matters behind us, we can relax a bit and start adding, subtracting, and multiplying along the lines we are used to from matrix algebra and vector calculus.

Example 1.4.4. The simplest example of a vector space is the *trivial vector space* $V = \{0\}$ that contains only one point, the origin. The vector space operations and axioms are completely trivial as well in this case.

Here are some important examples of vectors spaces.

Example 1.4.5. The most important basic example is undoubtedly the Cartesian n-fold product of the field $\mathbb{F}$:

$$\mathbb{F}^n = \left\{ \begin{bmatrix} \alpha_1 \\ \vdots \\ \alpha_n \end{bmatrix} : \alpha_1, \ldots, \alpha_n \in \mathbb{F} \right\}$$

$$= \{(\alpha_1, \ldots, \alpha_n) : \alpha_1, \ldots, \alpha_n \in \mathbb{F}\}.$$

Note that the $n \times 1$ and the n-tuple ways of writing these vectors are equivalent. When writing vectors in a line of text, the n-tuple version is obviously more convenient. The column matrix version, however, conforms to various other natural choices, as we shall see, and carries some extra meaning for that reason. The ith entry α_i in the vector $x = (\alpha_1, \ldots, \alpha_n)$ is called the ith *coordinate* of x.

Vector addition is defined by adding the entries:

$$\begin{bmatrix} \alpha_1 \\ \vdots \\ \alpha_n \end{bmatrix} + \begin{bmatrix} \beta_1 \\ \vdots \\ \beta_n \end{bmatrix} = \begin{bmatrix} \alpha_1 + \beta_1 \\ \vdots \\ \alpha_n + \beta_n \end{bmatrix}$$

and likewise with scalar multiplication

$$\alpha \begin{bmatrix} \alpha_1 \\ \vdots \\ \alpha_n \end{bmatrix} = \begin{bmatrix} \alpha\alpha_1 \\ \vdots \\ \alpha\alpha_n \end{bmatrix},$$

The axioms are verified by using the axioms for the field $\mathbb{F}$.

Example 1.4.6. The *space of functions* whose domain is some fixed set S and whose values all lie in the field $\mathbb{F}$ is denoted by Func $(S, \mathbb{F}) = \{f : S \to \mathbb{F}\}$. Addition and scalar multiplication is defined by

$$(\alpha f)(x) = \alpha f(x),$$
$$(f_1 + f_2)(x) = f_1(x) + f_2(x).$$

And the axioms again follow from using the field axioms for $\mathbb{F}$.

In the special case where $S = \{1, \ldots, n\}$, it is worthwhile noting that

$$\text{Func}(\{1, \ldots, n\}, \mathbb{F}) = \mathbb{F}^n.$$

Thus, vectors in $\mathbb{F}^n$ can also be thought of as functions and can be graphed as either an arrow in space or as a histogram type function. The former is of course more geometric, but the latter certainly also has its advantages as collections of numbers in the form of $n \times 1$ matrices do not always look like vectors. In statistics, the histogram picture is obviously far more useful. The point here is that the way in which vectors are pictured might be psychologically important, but from an abstract mathematical perspective, there is no difference.

Example 1.4.7. The space of $n \times m$ matrices

$$\text{Mat}_{n \times m}(\mathbb{F}) = \left\{ \begin{bmatrix} \alpha_{11} & \cdots & \alpha_{1m} \\ \vdots & \ddots & \vdots \\ \alpha_{n1} & \cdots & \alpha_{nm} \end{bmatrix} : \alpha_{ij} \in \mathbb{F} \right\}$$
$$= \left\{ (\alpha_{ij}) : \alpha_{ij} \in \mathbb{F} \right\}.$$

$n \times m$ matrices are evidently just a different way of arranging vectors in $\mathbb{F}^{n \cdot m}$. This arrangement, as with the column version of vectors in $\mathbb{F}^n$, imbues these vectors with some extra meaning that will become evident as we proceed.

Example 1.4.8. There is a slightly more abstract vector space that we can construct out of a general set S and a vector space V. This is the set Map (S, V) of all maps from S to V. Scalar multiplication and addition are defined as follows:

$$(\alpha f)(x) = \alpha f(x),$$
$$(f_1 + f_2)(x) = f_1(x) + f_2(x).$$

The axioms now follow from V being a vector space.

The space of maps is in some sense the most general type of vector space as all other vector spaces are either of this type or *subspaces* of such function spaces.

Definition 1.4.9. A nonempty subset $M \subset V$ of a vector space V is said to be a *subspace* if it is closed under addition and scalar multiplication:

$$x, y \in M \Rightarrow x + y \in M,$$

$$\alpha \in \mathbb{F} \text{ and } x \in M \Rightarrow \alpha x \in M.$$

We also say that M is *closed* under vector addition and multiplication by scalars.

Note that since $M \neq \emptyset$, we can find $x \in M$; this means that $0 = 0 \cdot x \in M$. Thus, subspaces become vector spaces in their own right and this without any further checking of the axioms.

Example 1.4.10. The set of *polynomials* whose coefficients lie in the field $\mathbb{F}$

$$\mathbb{F}[t] = \left\{ p(t) = a_0 + a_1 t + \cdots + a_k t^k : k \in \mathbb{N}_0, a_0, a_1, \ldots, a_k \in \mathbb{F} \right\}$$

is also a vector space. If we think of polynomials as functions, then we imagine them as a subspace of Func $(\mathbb{F}, \mathbb{F})$. However, the fact that a polynomial is determined by its representation as a function depends on the fact that we have a field of characteristic zero! If, for instance, $\mathbb{F} = \{0, 1\}$, then the polynomial $t^2 + t$ vanishes when evaluated at both 0 and 1. Thus, this nontrivial polynomial is, when viewed as a function, the same as $p(t) = 0$.

We could also just record the coefficients. In that case, $\mathbb{F}[t]$ is a subspace of Func $(\mathbb{N}_0, \mathbb{F})$ and consists of those infinite tuples that are zero except at all but a finite number of places.

If

$$p(t) = a_0 + a_1 t + \cdots + a_n t^n \in \mathbb{F}[t],$$

then the largest integer $k \leq n$ such that $a_k \neq 0$ is called the degree of p. In other words,

$$p(t) = a_0 + a_1 t + \cdots + a_k t^k$$

and $a_k \neq 0$. We use the notation $\deg(p) = k$.

Example 1.4.11. The collection of formal power series

$$\mathbb{F}[[t]] = \left\{ a_0 + a_1 t + \cdots + a_k t^k + \cdots : a_0, a_1, \ldots, a_k, \ldots \in \mathbb{F} \right\}$$

$$= \left\{ \sum_{i=0}^{\infty} a_i t^i : a_i \in \mathbb{F}, i \in \mathbb{N}_0 \right\}$$

bears some resemblance to polynomials, but without further discussions on convergence or even whether this makes sense, we cannot interpret power series as lying in Func $(\mathbb{F}, \mathbb{F})$. If, however, we only think about recording the coefficients, then we see that $\mathbb{F}[[t]] = $ Func $(\mathbb{N}_0, \mathbb{F})$. The extra piece of information that both $\mathbb{F}[t]$ and $\mathbb{F}[[t]]$ carry with them, aside from being vector spaces, is that the elements can also be multiplied. This extra structure will be used in the case of $\mathbb{F}[t]$. Power series will not play an important role in the sequel. Finally, note that $\mathbb{F}[t]$, is a subspace of $\mathbb{F}[[t]]$.

Example 1.4.12. For two (or more) vector spaces V, W over the same field $\mathbb{F}$ we can form the (Cartesian) product

$$V \times W = \{(v, w) : v \in V \text{ and } w \in W\}.$$

Scalar multiplication and addition are defined by

$$\alpha (v, w) = (\alpha v, \alpha w),$$
$$(v_1, w_1) + (v_2, w_2) = (v_1 + v_2, w_1 + w_2).$$

Note that $V \times W$ is not in a natural way a subspace in a space of functions or maps.

Exercises

1. Find a subset $C \subset \mathbb{F}^2$ that is closed under scalar multiplication but not under addition of vectors.
2. Find a subset $A \subset \mathbb{C}^2$ that is closed under vector addition but not under multiplication by complex numbers.
3. Find a subset $Q \subset \mathbb{R}$ that is closed under addition but not scalar multiplication by real scalars.
4. Let $V = \mathbb{Z}$ be the set of integers with the usual addition as "vector addition." Show that it is not possible to define scalar multiplication by $\mathbb{Q}, \mathbb{R}$, or $\mathbb{C}$ so as to make it into a vector space.
5. Let V be a real vector space, i.e., a vector space were the scalars are $\mathbb{R}$. The *complexification* of V is defined as $V_{\mathbb{C}} = V \times V$. As in the construction of complex numbers, we agree to write $(v, w) \in V_{\mathbb{C}}$ as $v + iw$. Moreover, if $v \in V$, then it is convenient to use the shorthand notations $v = v + i0$ and $iv = 0 + iv$. Define complex scalar multiplication on $V_{\mathbb{C}}$ and show that it becomes a complex vector space.
6. Let V be a complex vector space i.e., a vector space were the scalars are $\mathbb{C}$. Define V^* as the complex vector space whose additive structure is that of V but where complex scalar multiplication is given by $\lambda * x = \bar{\lambda} x$. Show that V^* is a complex vector space.
7. Let P_n be the set of polynomials in $\mathbb{F}[t]$ of degree $\leq n$.

   (a) Show that P_n is a vector space.
   (b) Show that the space of polynomials of degree $n \geq 1$ is $P_n - P_{n-1}$ and does not form a subspace.
   (c) If $f(t) : \mathbb{F} \to \mathbb{F}$, show that $V = \{p(t) f(t) : p \in P_n\}$ is a subspace of Func $(\mathbb{F}, \mathbb{F})$.

8. Let $V = \mathbb{C}^{\times} = \mathbb{C} - \{0\}$. Define addition on V by $x \boxplus y = xy$. Define scalar multiplication by $\alpha \boxdot x = e^{\alpha} x$.

   (a) Show that if we use $0_V = 1$ and $-x = x^{-1}$, then the first four axioms for a vector space are satisfied.
   (b) Which of the scalar multiplication properties do not hold?

1.5   Bases

We are now going to introduce one of the most important concepts in linear algebra. Let V be a vector space over $\mathbb{F}$.

Definition 1.5.1. Our first construction is to form *linear combinations* of vectors. If $\alpha_1, \ldots, \alpha_m \in \mathbb{F}$ and $x_1, \ldots, x_m \in V$, then we can multiply each x_i by the scalar α_i and then add up the resulting vectors to form the linear combination

$$x = \alpha_1 x_1 + \cdots + \alpha_m x_m.$$

We also say that x is a linear combination of the x_is.

If we arrange the vectors in a $1 \times m$ row matrix

$$\begin{bmatrix} x_1 \cdots x_m \end{bmatrix}$$

and the scalars in a column $m \times 1$ matrix, we see that the linear combination can be thought of as a matrix product

$$\sum_{i=1}^{m} \alpha_i x_i = \alpha_1 x_1 + \cdots + \alpha_m x_m = \begin{bmatrix} x_1 \cdots x_m \end{bmatrix} \begin{bmatrix} \alpha_1 \\ \vdots \\ \alpha_m \end{bmatrix}.$$

To be completely rigorous, we should write the linear combination as a 1×1 matrix $[\alpha_1 x_1 + \cdots + \alpha_m x_m]$, but it seems too pedantic to insist on this. Another curiosity here is that matrix multiplication almost forces us to write

$$x_1 \alpha_1 + \cdots + x_m \alpha_m = \begin{bmatrix} x_1 \cdots x_m \end{bmatrix} \begin{bmatrix} \alpha_1 \\ \vdots \\ \alpha_m \end{bmatrix}.$$

This is one reason why we want to be able to multiply by scalars on both the left and right.

Definition 1.5.2. A *finite basis* for V is a finite collection of vectors $x_1, \ldots, x_n \in V$ such that each element $x \in V$ can be written as a linear combination

$$x = \alpha_1 x_1 + \cdots + \alpha_n x_n$$

in precisely one way.

This means that for each $x \in V$, we can find $\alpha_1, \ldots, \alpha_n \in \mathbb{F}$ such that

$$x = \alpha_1 x_1 + \cdots + \alpha_n x_n.$$

Moreover, if we have two linear combinations both yielding x

$$\alpha_1 x_1 + \cdots + \alpha_n x_n = x = \beta_1 x_1 + \cdots + \beta_n x_n,$$

then

$$\alpha_1 = \beta_1, \ldots, \alpha_n = \beta_n.$$

Since each x has a unique linear combination, we also refer to it as the *expansion* of x with respect to the basis. In this way, we get a well-defined correspondence $V \longleftrightarrow \mathbb{F}^n$ by identifying

$$x = \alpha_1 x_1 + \cdots + \alpha_n x_n$$

with the n-tuple $(\alpha_1, \ldots, \alpha_n)$. We note that this correspondence preserves scalar multiplication and vector addition since

$$\alpha x = \alpha \left(\alpha_1 x_1 + \cdots + \alpha_n x_n \right)$$

$$= (\alpha \alpha_1) x_1 + \cdots + (\alpha \alpha_n) x_n,$$

$$x + y = (\alpha_1 x_1 + \cdots + \alpha_n x_n) + (\beta_1 x_1 + \cdots + \beta_n x_n)$$

$$= (\alpha_1 + \beta_1) x_1 + \cdots + (\alpha_n + \beta_n) x_n.$$

This means that the choice of basis makes V equivalent to the more concrete vector space $\mathbb{F}^n$. This idea of making abstract vector spaces more concrete by the use of a basis is developed further in Sects. 1.7 and 1.8.

Note that if $x_1, \ldots, x_n \in V$ form a basis, then any reordering of the basis vectors, such as $x_2, x_1, \ldots, x_n \in V$, also forms a basis. We will think of these two choices as being different bases.

We shall prove in Sect. 1.8 and again in Sect. 1.12 that the number of vectors in such a basis for V is always the same.

Definition 1.5.3. This allows us to define the *dimension* of V over $\mathbb{F}$ to be the number of elements in a basis. Note that the uniqueness condition for the linear combinations guarantees that none of the vectors in a basis can be the zero vector.

Example 1.5.4. The simplest example of a vector space $V = \{0\}$ is a bit special. Its only basis is the empty collection due to the requirement that vectors must have unique expansions with respect to a basis. Since such a choice of basis contains 0 elements, we say that the dimension of the trivial vector space is 0. Note also that in this case, the "choice of basis" is not an ordered collection of vectors. So if one insists on ordered collections of vectors for bases, there will be one logical inconsistency in the theory when one talks about selecting a basis for the trivial vector space.

Another slightly more interesting case that we can cover now is that of one-dimensional spaces.

Lemma 1.5.5. *Let V be a vector space over $\mathbb{F}$. If V has a basis with one element, then any other finite basis also has one element.*

Proof. Let x_1 be a basis for V. If $x \in V$, then $x = \alpha x_1$ for some α. Now, suppose that we have $z_1, \ldots, z_n \in V$, then $z_i = \alpha_i x_1$. If $z_1, \ldots, z_n$ forms a basis, then none of the vectors are zero and consequently $\alpha_i \neq 0$. Thus, for each i, we have $x_1 = \alpha_i^{-1} z_i$. Therefore, if $n > 1$, then we have that x_1 can be written in more than one way as a linear combination of $z_1, \ldots, z_n$. This contradicts the definition of a basis. Whence, $n = 1$ as desired. $\square$

Let us consider some basic examples.

Example 1.5.6. In $\mathbb{F}^n$ define the vectors

$$e_1 = \begin{bmatrix} 1 \\ 0 \\ \vdots \\ 0 \end{bmatrix}, e_2 = \begin{bmatrix} 0 \\ 1 \\ \vdots \\ 0 \end{bmatrix}, \ldots, e_n = \begin{bmatrix} 0 \\ 0 \\ \vdots \\ 1 \end{bmatrix}.$$

Thus, e_i is the vector that is zero in every entry except the ith where it is 1. These vectors evidently form a basis for $\mathbb{F}^n$ since any vector in $\mathbb{F}^n$ has the unique expansion

$$\mathbb{F}^n \ni x = \begin{bmatrix} \alpha_1 \\ \alpha_2 \\ \vdots \\ \alpha_n \end{bmatrix}$$

$$= \alpha_1 \begin{bmatrix} 1 \\ 0 \\ \vdots \\ 0 \end{bmatrix} + \alpha_2 \begin{bmatrix} 0 \\ 1 \\ \vdots \\ 0 \end{bmatrix} + \cdots + \alpha_n \begin{bmatrix} 0 \\ 0 \\ \vdots \\ 1 \end{bmatrix}$$

$$= \alpha_1 e_1 + \alpha_2 e_2 + \cdots + \alpha_n e_n$$

$$= \begin{bmatrix} e_1 & e_2 & \cdots & e_n \end{bmatrix} \begin{bmatrix} \alpha_1 \\ \alpha_2 \\ \vdots \\ \alpha_n \end{bmatrix}.$$

Example 1.5.7. In $\mathbb{F}^2$ consider

$$x_1 = \begin{bmatrix} 1 \\ 0 \end{bmatrix}, x_2 = \begin{bmatrix} 1 \\ 1 \end{bmatrix}.$$

These two vectors also form a basis for $\mathbb{F}^2$ since we can write

$$\begin{bmatrix} \alpha \\ \beta \end{bmatrix} = (\alpha - \beta) \begin{bmatrix} 1 \\ 0 \end{bmatrix} + \beta \begin{bmatrix} 1 \\ 1 \end{bmatrix} = \begin{bmatrix} 1 & 1 \\ 0 & 1 \end{bmatrix} \begin{bmatrix} (\alpha - \beta) \\ \beta \end{bmatrix}.$$

To see that these choices are unique, observe that the coefficient on x_2 must be β and this then uniquely determines the coefficient in front of x_1.

Example 1.5.8. In $\mathbb{F}^2$ consider the slightly more complicated set of vectors

$$x_1 = \begin{bmatrix} 1 \\ -1 \end{bmatrix}, x_2 = \begin{bmatrix} 1 \\ 1 \end{bmatrix}.$$

This time, we see

$$\begin{bmatrix} \alpha \\ \beta \end{bmatrix} = \frac{\alpha - \beta}{2} \begin{bmatrix} 1 \\ -1 \end{bmatrix} + \frac{\alpha + \beta}{2} \begin{bmatrix} 1 \\ 1 \end{bmatrix}$$

$$= \begin{bmatrix} 1 & 1 \\ -1 & 1 \end{bmatrix} \begin{bmatrix} \frac{\alpha - \beta}{2} \\ \frac{\alpha + \beta}{2} \end{bmatrix}.$$

Again, we can see that the coefficients are unique by observing that the system

$$\gamma + \delta = \alpha,$$
$$-\gamma + \delta = \beta$$

has a unique solution. This is because γ, respectively δ, can be found by subtracting, respectively adding, these two equations.

Example 1.5.9. Likewise, the space of matrices $\text{Mat}_{n \times m}(\mathbb{F})$ has a natural basis E_{ij} of nm elements, where E_{ij} is the matrix that is zero in every entry except the (i, j)th where it is 1.

The concept of a basis depends quite a lot on the scalars we use. The field of complex numbers $\mathbb{C}$ is clearly a one-dimensional vector space when we use $\mathbb{C}$ as the scalar field. To be specific, we have that $x_1 = 1$ is a basis for $\mathbb{C}$. If, however, we view $\mathbb{C}$ as a vector space over the real numbers $\mathbb{R}$, then only real numbers in $\mathbb{C}$ are linear combinations of x_1. Therefore, x_1 is no longer a basis when we restrict to real scalars. Evidently, we need to use $x_1 = 1$ and $x_2 = i$ to obtain a basis over $\mathbb{R}$.

It is also possible to have infinite bases. However, some care must be taken in defining this concept as we are not allowed to form infinite linear combinations. We say that a vector space V over $\mathbb{F}$ has a collection $x_i \in V$, where $i \in A$ is some possibly infinite index set, as a basis, if each $x \in V$ is a linear combination of a finite number of the vectors x_i in a unique way. There is, surprisingly, only one important vector space that comes endowed with a natural infinite basis. This is the space $\mathbb{F}[t]$ of polynomials. The collection $x_i = t^i$, $i = 0, 1, 2, \ldots$ evidently gives us a basis. The other spaces $\mathbb{F}[t]$ and Func $(S, \mathbb{F})$, where S is infinite, do not come with any natural bases. There is a rather subtle theorem which asserts that every vector space must have a basis. It is somewhat beyond the scope of this text to prove this theorem as it depends on *Zorn's lemma* or equivalently the *axiom of choice*. It should also be mentioned that it is a mere existence theorem as it does not give a procedure for constructing infinite bases. In order to get around these nasty points, we resort to the trick of saying that a vector space is *infinite-dimensional* if it does not admit a finite basis. Note that in the above lemma, we can also show that if V admits a basis with one element, then it cannot have an infinite basis.

Finally, we need to mention some subtleties in the definition of a basis. In most texts, a distinction is made between an *ordered* basis $x_1, \ldots, x_n$ and a basis as a *subset*

$$\{x_1, \ldots, x_n\} \subset V.$$

There is a fine difference between these two concepts. The collection x_1, x_2 where $x_1 = x_2 = x \in V$ can never be a basis as x can be written as a linear combination of x_1 and x_2 in at least two different ways. As a set, however, we see that $\{x\} = \{x_1, x_2\}$ consists of only one vector, and therefore, this redundancy has disappeared. Throughout this text, we assume that bases are ordered. This is entirely reasonable as most people tend to write down a collection of elements of a set in some, perhaps arbitrary, order. It is also important and convenient to work with ordered bases when time comes to discuss matrix representations. On the few occasions where we shall be working with infinite bases, as with $\mathbb{F}[t]$, they will also be ordered in a natural way using either the natural numbers or the integers.

Exercises

1. Show that $1, t, \ldots, t^n$ form a basis for P_n.
2. Show that if $p_0, \ldots, p_n \in P_n - \{0\}$ satisfy $\deg(p_k) = k$, then they form a basis for P_n.
3. Find a basis $p_1, \ldots, p_4 \in P_3$ such that $\deg(p_i) = 3$ for $i = 1, 2, 3, 4$.
4. For $\alpha \in \mathbb{C}$ consider the subset

$$\mathbb{Q}[\alpha] = \{p(\alpha) : p \in \mathbb{Q}[t]\} \subset \mathbb{C}.$$

Show that:

(a) If $\alpha \in \mathbb{Q}$, then $\mathbb{Q}[\alpha] = \mathbb{Q}$
(b) If α is *algebraic*, i.e., it solves an equation $p(\alpha) = 0$ for some $p \in \mathbb{Q}[t]$, then $\mathbb{Q}[\alpha]$ is a finite-dimensional vector space over $\mathbb{Q}$ with a basis $1, \alpha, \alpha^2, \dots, \alpha^{n-1}$ for some $n \in \mathbb{N}$. Hint: Let n be the smallest number so that α^n is a linear combination of $1, \alpha, \alpha^2, \dots, \alpha^{n-1}$. You must explain why we can find such n.
(c) If α is *algebraic*, then $\mathbb{Q}[\alpha]$ is a field that contains $\mathbb{Q}$. Hint: Show that α must be the root of a polynomial with a nonzero constant term. Use this to find a formula for α^{-1} that depends only on positive powers of α.
(d) Show that α is algebraic if and only if $\mathbb{Q}[\alpha]$ is finite-dimensional over $\mathbb{Q}$.
(e) We say that α is *transcendental* if it is not algebraic. Show that if α is transcendental, then $1, \alpha, \alpha^2, \dots, \alpha^n, \dots$ form an infinite basis for $\mathbb{Q}[\alpha]$. Thus, $\mathbb{Q}[\alpha]$ and $\mathbb{Q}[t]$ represent the same vector space via the substitution $t \longleftrightarrow \alpha$.

5. Show that

$$\begin{bmatrix} 1 \\ 1 \\ 0 \\ 0 \end{bmatrix}, \begin{bmatrix} 1 \\ 0 \\ 1 \\ 0 \end{bmatrix}, \begin{bmatrix} 1 \\ 0 \\ 0 \\ 1 \end{bmatrix}, \begin{bmatrix} 0 \\ 1 \\ 1 \\ 0 \end{bmatrix}, \begin{bmatrix} 0 \\ 1 \\ 0 \\ 1 \end{bmatrix}, \begin{bmatrix} 0 \\ 0 \\ 1 \\ 1 \end{bmatrix}$$

span $\mathbb{C}^4$, i.e., every vector on $\mathbb{C}^4$ can be written as a linear combination of these vectors. Which collections of those six vectors form a basis for $\mathbb{C}^4$?
6. Is it possible to find a basis $x_1, \dots, x_n$ for $\mathbb{F}^n$ so that the ith entry for all of the vectors $x_1, \dots, x_n$ is zero?
7. If $e_1, \dots, e_n$ is the standard basis for $\mathbb{C}^n$, show that both

$$e_1, \dots, e_n, i e_1, \dots, i e_n$$

and

$$e_1, i e_1, \dots, e_n, i e_n$$

form bases for $\mathbb{C}^n$ when viewed as a real vector space.
8. If $x_1, \dots, x_n$ is a basis for the real vector space V, then it is also a basis for the complexification $V_{\mathbb{C}}$ (see Exercise 5 in Sect. 1.4 for the definition of $V_{\mathbb{C}}$).
9. Find a basis for $\mathbb{R}^3$ where all coordinate entries are ± 1.
10. A subspace $M \subset \mathrm{Mat}_{n \times n}(\mathbb{F})$ is called a *two-sided ideal* if for all $X \in \mathrm{Mat}_{n \times n}(\mathbb{F})$ and $A \in M$ also $XA, AX \in M$. Show that if $M \neq \{0\}$, then $M = \mathrm{Mat}_{n \times n}(\mathbb{F})$. Hint: Assume $A \in M$ is such that some entry is nonzero. Make it 1 by multiplying A by an appropriate scalar on the left. Then, show that we can construct the standard basis for $\mathrm{Mat}_{n \times n}(\mathbb{F})$ by multiplying A by the standard basis matrices for $\mathrm{Mat}_{n \times n}(\mathbb{F})$ on the left and right.

11. Let V be a vector space.

   (a) Show that $x, y \in V$ form a basis if and only if $x + y, x - y$ form a basis.
   (b) Show that $x, y, z \in V$ form a basis if and only if $x + y, y + z, z + x$ form a basis.

1.6 Linear Maps

Definition 1.6.1. A map $L : V \to W$ between vector spaces over the same field $\mathbb{F}$ is said to be *linear* if it preserves scalar multiplication and addition in the following way:

$$L(\alpha x) = \alpha L(x),$$
$$L(x + y) = L(x) + L(y),$$

where $\alpha \in \mathbb{F}$ and $x, y \in V$.

It is possible to collect these two properties into one condition as follows:

$$L(\alpha_1 x_1 + \alpha_2 x_2) = \alpha_1 L(x_1) + \alpha_2 L(x_2),$$

where $\alpha_1, \alpha_2 \in \mathbb{F}$ and $x_1, x_2 \in V$. More generally, we have that L preserves linear combinations in the following way:

$$L\left(\begin{bmatrix} x_1 & \cdots & x_m \end{bmatrix} \begin{bmatrix} \alpha_1 \\ \vdots \\ \alpha_m \end{bmatrix} \right) = L(x_1 \alpha_1 + \cdots + x_m \alpha_m)$$
$$= L(x_1)\alpha_1 + \cdots + L(x_m)\alpha_m$$
$$= \begin{bmatrix} L(x_1) & \cdots & L(x_m) \end{bmatrix} \begin{bmatrix} \alpha_1 \\ \vdots \\ \alpha_m \end{bmatrix}.$$

To prove this simple fact, we use induction on m. When $m = 1$, this is simply the fact that L preserves scalar multiplication

$$L(\alpha x) = \alpha L(x).$$

Assuming the induction hypothesis, that the statement holds for $m - 1$, we see that

$$L(x_1 \alpha_1 + \cdots + x_m \alpha_m) = L((x_1 \alpha_1 + \cdots + x_{m-1}\alpha_{m-1}) + x_m \alpha_m)$$
$$= L(x_1 \alpha_1 + \cdots + x_{m-1}\alpha_{m-1}) + L(x_m \alpha_m)$$

$$= (L(x_1)\alpha_1 + \cdots + L(x_{m-1})\alpha_{m-1}) + L(x_m)\alpha_m$$
$$= L(x_1)\alpha_1 + \cdots + L(x_m)\alpha_m.$$

The important feature of linear maps is that they preserve the operations that are allowed on the spaces we work with. Some extra terminology is often used for linear maps.

Definition 1.6.2. If the values are the field itself, i.e., $W = \mathbb{F}$, then we also call $L : V \to \mathbb{F}$ a *linear function* or *linear functional*. If $V = W$, then we call $L : V \to V$ a *linear operator*.

Before giving examples, we introduce some further notation.

Definition 1.6.3. The set of all linear maps $L : V \to W$ is often denoted $\mathrm{Hom}(V, W)$. In case we need to specify the scalars, we add the field as a subscript $\mathrm{Hom}_{\mathbb{F}}(V, W)$.

The abbreviation Hom stands for *homomorphism*. Homomorphisms are in general maps that preserve whatever algebraic structure that is available. Note that

$$\mathrm{Hom}_{\mathbb{F}}(V, W) \subset \mathrm{Map}(V, W)$$

and is a subspace of the latter. Thus, $\mathrm{Hom}_{\mathbb{F}}(V, W)$ is a vector space over $\mathbb{F}$.

It is easy to see that the composition of linear maps always yields a linear map. Thus, if $L_1 : V_1 \to V_2$ and $L_2 : V_2 \to V_3$ are linear maps, then the composition $L_2 \circ L_1 : V_1 \to V_3$ defined by $L_2 \circ L_1(x) = L_2(L_1(x))$ is again a linear map. We often ignore the composition sign $\circ$ and simply write $L_2 L_1$. An important special situation is that one can "multiply" linear operators $L_1, L_2 : V \to V$ via composition. This multiplication is in general not commutative or abelian as it rarely happens that $L_1 L_2$ and $L_2 L_1$ represent the same map. We shall see many examples of this throughout the text.

Finally, still staying with the abstract properties, we note that, if $L : V \to W$ is a linear map and $M \subset V$ is a subspace, then the restriction $L|_M : M \to W$ defined trivially by $L|_M(x) = L(x)$ is also a linear map. We shall often even use the same symbol L for both maps, but beware, many properties for a linear map can quickly change when we restrict it to different subspaces. The restriction leads to another important construction that will become very important in subsequent chapters.

Definition 1.6.4. Let $L : V \to V$ be a linear operator. A subspace $M \subset V$ is said to be *L-invariant* or simply invariant if $L(M) \subset M$.

Thus, the restriction of L to M defines a new linear operator $L|_M : M \to M$.

Example 1.6.5. Define a map $L : \mathbb{F} \to \mathbb{F}$ by scalar multiplication on $\mathbb{F}$ via $L(x) = \lambda x$ for some $\lambda \in \mathbb{F}$. The distributive law says that the map is additive, and the associative law together with the commutative law say that it preserves scalar multiplication. This example can now easily be generalized to scalar multiplication on a vector space V, where we can also define $L : V \to V$ by $L(x) = \lambda x$.

Two special cases are of particular interest. First, the identity transformation 1_V : $V \rightarrow V$ defined by $1_V(x) = x$. This is evidently scalar multiplication by 1. Second, we have the zero transformation $0 = 0_V : V \rightarrow V$ that maps everything to $0 \in V$ and is simply multiplication by 0. The latter map can also be generalized to a zero map $0 : V \rightarrow W$ between different vector spaces. With this in mind, we can always write multiplication by λ as the map $\lambda 1_V$ thus keeping track of what it does, where it does it, and finally keeping track of the fact that we think of the procedure as a map.

Expanding on this theme a bit we can, starting with a linear operator $L : V \rightarrow V$, use powers of L as well as linear combinations to create new operators on V. For instance, $L^2 - 3 \cdot L + 2 \cdot 1_V$ is defined by

$$\left(L^2 - 3 \cdot L + 2 \cdot 1_V\right)(x) = L(L(x)) - 3L(x) + 2x.$$

We shall often do this in quite general situations. The most general construction comes about by selecting a polynomial $p \in \mathbb{F}[t]$ and considering $p(L)$. If $p = \alpha_k t^k + \cdots + \alpha_1 t + \alpha_0$, then

$$p(L) = \alpha_k L^k + \cdots + \alpha_1 L + \alpha_0 1_V.$$

If we think of $t^0 = 1$ as the degree 0 term in the polynomial, then by substituting L, we apparently define $L^0 = 1_V$. So it is still the identity, but the identity in the appropriate set where L lives. Evaluation on $x \in V$ is given by

$$p(L)(x) = \alpha_k L^k(x) + \cdots + \alpha_1 L(x) + \alpha_0 x.$$

Apparently, p simply defines a linear combination of the linear operators $L^k, \ldots, L$, 1_V, and $p(L)(x)$ is a linear combination of the vectors $L^k(x), \ldots, L(x), x$.

Example 1.6.6. Fix $x \in V$. Note that the axioms of scalar multiplication also imply that $L : \mathbb{F} \rightarrow V$ defined by $L(\alpha) = x\alpha$ is linear.

Matrix multiplication is the next level of abstraction. Here we let $V = \mathbb{F}^m$ and $W = \mathbb{F}^n$ and L is represented by an $n \times m$ matrix

$$B = \begin{bmatrix} \beta_{11} & \cdots & \beta_{1m} \\ \vdots & \ddots & \vdots \\ \beta_{n1} & \cdots & \beta_{nm} \end{bmatrix}.$$

The map is defined using matrix multiplication as follows:

$$L(x) = Bx$$

$$= \begin{bmatrix} \beta_{11} & \cdots & \beta_{1m} \\ \vdots & \ddots & \vdots \\ \beta_{n1} & \cdots & \beta_{nm} \end{bmatrix} \begin{bmatrix} \xi_1 \\ \vdots \\ \xi_m \end{bmatrix}$$

$$= \begin{bmatrix} \beta_{11}\xi_1 + \cdots + \beta_{1m}\xi_m \\ \vdots \\ \beta_{n1}\xi_1 + \cdots + \beta_{nm}\xi_m \end{bmatrix}$$

Thus, the ith coordinate of $L(x)$ is given by

$$\sum_{j=1}^{m} \beta_{ij}\xi_j = \beta_{i1}\xi_1 + \cdots + \beta_{im}\xi_m.$$

A similar and very important way of representing this map comes by noting that it creates linear combinations. Write B as a row matrix of its column vectors

$$B = \begin{bmatrix} \beta_{11} & \cdots & \beta_{1m} \\ \vdots & \ddots & \vdots \\ \beta_{n1} & \cdots & \beta_{nm} \end{bmatrix} = \begin{bmatrix} b_1 & \cdots & b_m \end{bmatrix}, \text{ where } b_i = \begin{bmatrix} \beta_{1i} \\ \vdots \\ \beta_{ni} \end{bmatrix}$$

and then observe

$$L(x) = Bx$$

$$= \begin{bmatrix} b_1 & \cdots & b_m \end{bmatrix} \begin{bmatrix} \xi_1 \\ \vdots \\ \xi_m \end{bmatrix}$$

$$= b_1\xi_1 + \cdots + b_m\xi_m.$$

Note that if $m = n$ and the matrix we use is a diagonal matrix with λs down the diagonal and zeros elsewhere, then we obtain the scalar multiplication map $\lambda 1_{\mathbb{F}^n}$. The matrix looks like this

$$\begin{bmatrix} \lambda & 0 & \cdots & 0 \\ 0 & \lambda & & 0 \\ \vdots & & \ddots & \vdots \\ 0 & 0 & \cdots & \lambda \end{bmatrix}$$

A very important observation in connection with linear maps defined by matrix multiplication is that composition of linear maps $L : \mathbb{F}^l \to \mathbb{F}^m$ and $K : \mathbb{F}^m \to \mathbb{F}^n$ is given by the matrix product. The maps are defined by matrix multiplication

$$L(x) = Bx,$$

$$B = \begin{bmatrix} b_1 & \cdots & b_l \end{bmatrix}$$

and

$$K(y) = Cy.$$

The composition can now be computed as follows using that K is linear:

$$(K \circ L)(x) = K(L(x))$$
$$= K(Bx)$$

$$= K\left(\begin{bmatrix} b_1 & \cdots & b_l \end{bmatrix} \begin{bmatrix} \xi_1 \\ \vdots \\ \xi_l \end{bmatrix}\right)$$

$$= \begin{bmatrix} K(b_1) & \cdots & K(b_l) \end{bmatrix} \begin{bmatrix} \xi_1 \\ \vdots \\ \xi_l \end{bmatrix}$$

$$= \left(\begin{bmatrix} Cb_1 & \cdots & Cb_l \end{bmatrix}\right) \begin{bmatrix} \xi_1 \\ \vdots \\ \xi_l \end{bmatrix}$$

$$= \left(C \begin{bmatrix} b_1 & \cdots & b_l \end{bmatrix}\right) \begin{bmatrix} \xi_1 \\ \vdots \\ \xi_l \end{bmatrix}$$

$$= (CB)x.$$

Evidently, this all hinges on the fact that the matrix product CB can be defined by

$$CB = C \begin{bmatrix} b_1 & \cdots & b_l \end{bmatrix}$$
$$= \begin{bmatrix} Cb_1 & \cdots & Cb_l \end{bmatrix},$$

a definition that is completely natural if we think of C as a linear map. It should also be noted that we did not use associativity of matrix multiplication in the form $C(Bx) = (CB)x$. In fact, associativity is a consequence of our calculation.

We can also check things a bit more directly using summation notation. Observe that the ith entry in the composition

$$K\left(L\left(\begin{bmatrix} \alpha_1 \\ \vdots \\ \alpha_l \end{bmatrix}\right)\right) = \begin{bmatrix} \gamma_{11} & \cdots & \gamma_{1m} \\ \vdots & \ddots & \vdots \\ \gamma_{n1} & \cdots & \gamma_{nm} \end{bmatrix} \left(\begin{bmatrix} \beta_{11} & \cdots & \beta_{1l} \\ \vdots & \ddots & \vdots \\ \beta_{m1} & \cdots & \beta_{ml} \end{bmatrix} \begin{bmatrix} \xi_1 \\ \vdots \\ \xi_l \end{bmatrix}\right)$$

satisfies

$$\sum_{j=1}^{m} \gamma_{ij} \left(\sum_{s=1}^{l} \beta_{js}\xi_s\right) = \sum_{j=1}^{m}\sum_{s=1}^{l} \gamma_{ij}\beta_{js}\xi_s$$

$$= \sum_{s=1}^{l} \sum_{j=1}^{m} \gamma_{ij} \beta_{js} \xi_s$$

$$= \sum_{s=1}^{l} \left(\sum_{j=1}^{m} \gamma_{ij} \beta_{js} \right) \xi_s$$

where $\left(\sum_{j=1}^{m} \gamma_{ij} \beta_{js} \right)$ represents the (i, s) entry in the matrix product $[\gamma_{ij}][\beta_{js}]$.

Example 1.6.7. Note that while scalar multiplication on even the simplest vector space $\mathbb{F}$ is the simplest linear map we can have, there are still several levels of complexity depending on the field we use. Let us consider the map $L : \mathbb{C} \to \mathbb{C}$ that is multiplication by i, i.e., $L(x) = ix$. If we write $x = \alpha + i\beta$, we see that $L(x) = -\beta + i\alpha$. Geometrically, what we are doing is rotating x 90°. If we think of $\mathbb{C}$ as the plane $\mathbb{R}^2$, the map is instead given by the matrix

$$\begin{bmatrix} 0 & -1 \\ 1 & 0 \end{bmatrix}$$

which is not at all scalar multiplication if we only think in terms of real scalars. Thus, a supposedly simple operation with complex numbers is somewhat less simple when we forget complex numbers. What we need to keep in mind is that scalar multiplication with real numbers is simply a form of dilation where vectors are made longer or shorter depending on the scalar. Scalar multiplication with complex numbers is from an abstract algebraic viewpoint equally simple to write down, but geometrically, such an operation can involve a rotation from the perspective of a world where only real scalars exist.

Example 1.6.8. The ith coordinate map $\mathbb{F}^n \to \mathbb{F}$ defined by

$$dx_i(x) = dx_i \left(\begin{bmatrix} \xi_1 \\ \vdots \\ \xi_i \\ \vdots \\ \xi_n \end{bmatrix} \right)$$

$$= [0 \cdots 1 \cdots 0] \begin{bmatrix} \xi_1 \\ \vdots \\ \xi_i \\ \vdots \\ \xi_n \end{bmatrix}$$

$$= \xi_i.$$

is a linear map. Here the $1 \times n$ matrix $[0 \cdots 1 \cdots 0]$ is zero everywhere except in the ith entry where it is 1. The notation dx_i is not a mistake, but an incursion from multivariable calculus. While some mystifying words involving infinitesimals are often invoked in connection with such symbols, they have in more advanced and modern treatments of the subject simply been redefined as done here.

A special piece of notation comes in handy here. The Kronecker δ symbol is defined as

$$\delta_{ij} = \begin{cases} 0 & \text{if } i \neq j \\ 1 & \text{if } i = j \end{cases}$$

Thus, the matrix $[0 \cdots 1 \cdots 0]$ can also be written as

$$\begin{bmatrix} 0 \cdots 1 \cdots 0 \end{bmatrix} = \begin{bmatrix} \delta_{i1} & \cdots & \delta_{ii} & \cdots & \delta_{in} \end{bmatrix}$$
$$= \begin{bmatrix} \delta_{i1} & \cdots & \delta_{in} \end{bmatrix}.$$

The matrix representing the identity map $1_{\mathbb{F}^n}$ can then be written as

$$\begin{bmatrix} 1 & \cdots & 0 \\ \vdots & \ddots & \vdots \\ 0 & \cdots & 1 \end{bmatrix} = \begin{bmatrix} \delta_{11} & \cdots & \delta_{1n} \\ \vdots & \ddots & \vdots \\ \delta_{n1} & \cdots & \delta_{nn} \end{bmatrix}.$$

Example 1.6.9. Let us consider the vector space of functions $C^\infty (\mathbb{R}, \mathbb{R})$ that have derivatives of all orders. There are several interesting linear operators $C^\infty (\mathbb{R}, \mathbb{R}) \to C^\infty (\mathbb{R}, \mathbb{R})$

$$D(f)(t) = \frac{df}{dt}(t),$$

$$S(f)(t) = \int_{t_0}^{t} f(s) \, ds,$$

$$T(f)(t) = t \cdot f(t).$$

In a more shorthand fashion, we have the *differentiation operator* $D(f) = f'$, the *integration operator* $S(f) = \int f$, and the *multiplication operator* $T(f) = tf$. Note that the integration operator is not well defined unless we use the definite integral, and even in that case, it depends on the value t_0. Note that the space of polynomials

$$\mathbb{R}[t] \subset C^\infty (\mathbb{R}, \mathbb{R})$$

is an invariant subspace for all three operators. In this case, we usually let $t_0 = 0$ for S. These operators have some interesting relationships. We point out an intriguing one

$$DT - TD = 1.$$

To see this, simply use Leibniz' rule for differentiating a product to obtain

$$D\left(T\left(f\right)\right) = D\left(tf\right)$$
$$= f + tDf$$
$$= f + T\left(D\left(f\right)\right).$$

With some slight changes, the identity $DT - TD = 1$ is the *Heisenberg commutation law*. This law is important in the verification of *Heisenberg's uncertainty principle*.

Definition 1.6.10. The *trace* is a linear map on square matrices that adds the diagonal entries.

$$\text{tr} : \text{Mat}_{n \times n}\left(\mathbb{F}\right) \to \mathbb{F},$$

$$\text{tr}\left(A\right) = \alpha_{11} + \alpha_{22} + \cdots + \alpha_{nn}.$$

The trace satisfies the following important commutation relationship.

Lemma 1.6.11. (Invariance of Trace) *If $A \in \text{Mat}_{m \times n}\left(\mathbb{F}\right)$ and $B \in \text{Mat}_{n \times m}\left(\mathbb{F}\right)$, then $AB \in \text{Mat}_{m \times m}\left(\mathbb{F}\right)$, $BA \in \text{Mat}_{n \times n}\left(\mathbb{F}\right)$, and*

$$\text{tr}\left(AB\right) = \text{tr}\left(BA\right).$$

Proof. We write out the matrices

$$A = \begin{bmatrix} \alpha_{11} & \cdots & \alpha_{1n} \\ \vdots & \ddots & \vdots \\ \alpha_{m1} & \cdots & \alpha_{mn} \end{bmatrix}$$

$$B = \begin{bmatrix} \beta_{11} & \cdots & \beta_{1m} \\ \vdots & \ddots & \vdots \\ \beta_{n1} & \cdots & \beta_{nm} \end{bmatrix}.$$

Thus,

$$AB = \begin{bmatrix} \alpha_{11} & \cdots & \alpha_{1n} \\ \vdots & \ddots & \vdots \\ \alpha_{m1} & \cdots & \alpha_{mn} \end{bmatrix} \begin{bmatrix} \beta_{11} & \cdots & \beta_{1m} \\ \vdots & \ddots & \vdots \\ \beta_{n1} & \cdots & \beta_{nm} \end{bmatrix}$$

$$= \begin{bmatrix} \alpha_{11}\beta_{11} + \cdots + \alpha_{1n}\beta_{n1} & \cdots & \alpha_{11}\beta_{1m} + \cdots + \alpha_{1n}\beta_{nm} \\ \vdots & \ddots & \vdots \\ \alpha_{m1}\beta_{11} + \cdots + \alpha_{mn}\beta_{n1} & \cdots & \alpha_{m1}\beta_{1m} + \cdots + \alpha_{mn}\beta_{nm} \end{bmatrix},$$

$$BA = \begin{bmatrix} \beta_{11} & \cdots & \beta_{1m} \\ \vdots & \ddots & \vdots \\ \beta_{n1} & \cdots & \beta_{nm} \end{bmatrix} \begin{bmatrix} \alpha_{11} & \cdots & \alpha_{1n} \\ \vdots & \ddots & \vdots \\ \alpha_{m1} & \cdots & \alpha_{mn} \end{bmatrix}$$

$$= \begin{bmatrix} \beta_{11}\alpha_{11} + \cdots + \beta_{1m}\alpha_{m1} & \cdots & \beta_{11}\alpha_{1n} + \cdots + \beta_{1m}\alpha_{mn} \\ \vdots & \ddots & \vdots \\ \beta_{n1}\alpha_{11} + \cdots + \beta_{nm}\alpha_{m1} & \cdots & \beta_{n1}\alpha_{1n} + \cdots + \beta_{nm}\alpha_{mn} \end{bmatrix}.$$

This tells us that $AB \in \text{Mat}_{m \times m}(\mathbb{F})$ and $BA \in \text{Mat}_{n \times n}(\mathbb{F})$. To show the identity note that the (i, i) entry in AB is $\sum_{j=1}^{n} \alpha_{ij}\beta_{ji}$, while the (j, j) entry in BA is $\sum_{i=1}^{m} \beta_{ji}\alpha_{ij}$. Thus,

$$\text{tr}(AB) = \sum_{i=1}^{m} \sum_{j=1}^{n} \alpha_{ij}\beta_{ji},$$

$$\text{tr}(BA) = \sum_{j=1}^{n} \sum_{i=1}^{m} \beta_{ji}\alpha_{ij}.$$

By using $\alpha_{ij}\beta_{ji} = \beta_{ji}\alpha_{ij}$ and

$$\sum_{i=1}^{m} \sum_{j=1}^{n} = \sum_{j=1}^{n} \sum_{i=1}^{m},$$

we see that the two traces are equal. $\square$

This allows us to show that *Heisenberg commutation law* cannot be true for matrices.

Corollary 1.6.12. *There are no matrices $A, B \in \text{Mat}_{n \times n}(\mathbb{F})$ such that*

$$AB - BA = 1_{\mathbb{F}^n}.$$

Proof. By the above lemma and linearity, we have that $\text{tr}(AB - BA) = 0$. On the other hand, $\text{tr}(1_{\mathbb{F}^n}) = n$, since the identity matrix has n diagonal entries each of which is 1. $\square$

Remark 1.6.13. Observe that we just used the fact that $n \neq 0$ in $\mathbb{F}$ or, in other words, that $\mathbb{F}$ has characteristic zero. If we allowed ourselves to use the field $\mathbb{F}_2 = \{0, 1\}$ where $1 + 1 = 0$, then we have that $1 = -1$. Thus, we can use the matrices

$$A = \begin{bmatrix} 0 & 1 \\ 0 & 0 \end{bmatrix},$$

$$B = \begin{bmatrix} 0 & 1 \\ 1 & 0 \end{bmatrix},$$

to get the Heisenberg commutation law satisfied:

$$
AB - BA = \begin{bmatrix} 0 & 1 \\ 0 & 0 \end{bmatrix} \begin{bmatrix} 0 & 1 \\ 1 & 0 \end{bmatrix} - \begin{bmatrix} 0 & 1 \\ 1 & 0 \end{bmatrix} \begin{bmatrix} 0 & 1 \\ 0 & 0 \end{bmatrix}
$$

$$
= \begin{bmatrix} 1 & 0 \\ 0 & 0 \end{bmatrix} - \begin{bmatrix} 0 & 0 \\ 0 & 1 \end{bmatrix}
$$

$$
= \begin{bmatrix} 1 & 0 \\ 0 & -1 \end{bmatrix}
$$

$$
= \begin{bmatrix} 1 & 0 \\ 0 & 1 \end{bmatrix}.
$$

We have two further linear maps. Consider $V = \text{Func}(S, \mathbb{F})$ and select $s_0 \in S$; then, the *evaluation map* $\text{ev}_{s_0} : \text{Func}(S, \mathbb{F}) \to \mathbb{F}$ defined by $\text{ev}_{s_0}(f) = f(s_0)$ is linear. More generally, we have the *restriction map* for $T \subset S$ defined as a linear maps $\text{Func}(S, \mathbb{F}) \to \text{Func}(T, \mathbb{F})$, by mapping f to $f|_T$. The notation $f|_T$ means that we only consider f as mapping from T into $\mathbb{F}$. In other words, we have forgotten that f maps all of S into $\mathbb{F}$ and only remembered what it did on T.

Linear maps play a big role in multivariable calculus and are used in a number of ways to clarify and understand certain constructions. The fact that linear algebra is the basis for multivariable calculus should not be surprising as linear algebra is merely a generalization of vector algebra.

Let $F : \Omega \to \mathbb{R}^n$ be a differentiable function defined on some open domain $\Omega \subset \mathbb{R}^m$, i.e., for each $x_0 \in \Omega$, we can find a linear map $L : \mathbb{R}^m \to \mathbb{R}^n$ satisfying

$$
\lim_{|h| \to 0} \frac{|F(x_0 + h) - F(x_0) - L(h)|}{|h|} = 0.
$$

It is easy to see that such a linear map must be unique. It is also called the differential of F at $x_0 \in \Omega$ and denoted by $L = DF_{x_0} : \mathbb{R}^m \to \mathbb{R}^n$. The differential DF_{x_0} is also represented by the $n \times m$ matrix of partial derivatives

$$
DF_{x_0}(h) = DF_{x_0}\left(\begin{bmatrix} h_1 \\ \vdots \\ h_m \end{bmatrix} \right)
$$

$$
= \begin{bmatrix} \frac{\partial F_1}{\partial x_1} & \cdots & \frac{\partial F_1}{\partial x_m} \\ \vdots & \ddots & \vdots \\ \frac{\partial F_n}{\partial x_1} & \cdots & \frac{\partial F_n}{\partial x_m} \end{bmatrix} \begin{bmatrix} h_1 \\ \vdots \\ h_m \end{bmatrix}
$$

$$= \begin{bmatrix} \frac{\partial F_1}{\partial x_1} h_1 + \cdots + \frac{\partial F_1}{\partial x_m} h_m \\ \vdots \\ \frac{\partial F_n}{\partial x_1} h_1 + \cdots + \frac{\partial F_n}{\partial x_m} h_m \end{bmatrix}$$

One of the main ideas in differential calculus (of several variables) is that linear maps are simpler to work with and that they give good local approximations to differentiable maps. This can be made more precise by observing that we have the *first-order approximation*

$$F(x_0 + h) = F(x_0) + DF_{x_0}(h) + o(h),$$

$$\lim_{|h| \to 0} \frac{|o(h)|}{|h|} = 0$$

One of the goals of differential calculus is to exploit knowledge of the linear map DF_{x_0} and then use this first-order approximation to get a better understanding of the map F itself.

In case $f : \Omega \to \mathbb{R}$ is a function, one often sees the differential of f defined as the expression

$$\mathrm{d}f = \frac{\partial f}{\partial x_1} \mathrm{d}x_1 + \cdots + \frac{\partial f}{\partial x_m} \mathrm{d}x_m.$$

Having now interpreted $\mathrm{d}x_i$ as a linear function, we then observe that $\mathrm{d}f$ itself is a linear function whose matrix description is given by

$$\mathrm{d}f(h) = \frac{\partial f}{\partial x_1} \mathrm{d}x_1(h) + \cdots + \frac{\partial f}{\partial x_m} \mathrm{d}x_m(h)$$

$$= \frac{\partial f}{\partial x_1} h_1 + \cdots + \frac{\partial f}{\partial x_m} h_m$$

$$= \begin{bmatrix} \frac{\partial f}{\partial x_1} & \cdots & \frac{\partial f}{\partial x_m} \end{bmatrix} \begin{bmatrix} h_1 \\ \vdots \\ h_m \end{bmatrix}.$$

More generally, if we write

$$F = \begin{bmatrix} F_1 \\ \vdots \\ F_n \end{bmatrix},$$

then

$$DF_{x_0} = \begin{bmatrix} \mathrm{d}F_1 \\ \vdots \\ \mathrm{d}F_n \end{bmatrix}$$

with the understanding that

$$DF_{x_0}(h) = \begin{bmatrix} dF_1(h) \\ \vdots \\ dF_n(h) \end{bmatrix}.$$

Note how this conforms nicely with the above matrix representation of the differential.

Exercises

1. Let V, W be vector spaces over $\mathbb{Q}$. Show that any *additive* map $L : V \to W$, i.e.,

$$L(x_1 + x_2) = L(x_1) + L(x_2),$$

is linear.

2. Let $D : \mathbb{F}[t] \to \mathbb{F}[t]$ be defined by

$$D(\alpha_0 + \alpha_1 t + \cdots + \alpha_n t^n) = \alpha_1 + 2\alpha_2 t + \cdots + n\alpha_n t^{n-1}.$$

   (a) Show that this defines a linear operator.
   (b) Show directly, i.e., without using differential calculus, that this operator satisfies Leibniz' rule

$$D(pq) = pD(q) + (D(p))q.$$

   (c) Show that the subspace $P_n \subset \mathbb{F}[t]$ of polynomials of degree $\leq n$ is invariant.

3. Let $L : V \to V$ be a linear operator and V a vector space over $\mathbb{F}$. Show that the map $K : \mathbb{F}[t] \to \text{Hom}_{\mathbb{F}}(V, V)$ defined by $K(p) = p(L)$ is a linear map.

4. Let $L : V \to V$ be a linear operator and V a vector space over $\mathbb{F}$. Show that if $M \subset V$ is L-invariant and $p \in \mathbb{F}[t]$, then M is also invariant under $p(L)$.

5. Let $T : V \to W$ be a linear map, and $\tilde{V}$ is a vector space, all over the same field. Show that right composition

$$R_T : \text{Hom}(W, \tilde{V}) \to \text{Hom}(V, \tilde{V})$$

defined by $R_T(K) = K \circ T$ and left composition

$$L_T : \text{Hom}(\tilde{V}, V) \to \text{Hom}(\tilde{V}, W)$$

defined by $L_T(K) = T \circ K$ are linear maps.

6. Assume that $A \in \mathrm{Mat}_{n \times n}(\mathbb{F})$ has a block decomposition

$$A = \begin{bmatrix} A_{11} & A_{12} \\ A_{21} & A_{22} \end{bmatrix},$$

where $A_{11} \in \mathrm{Mat}_{k \times k}(\mathbb{F})$.

(a) Show that the subspace $\mathbb{F}^k = \{(\alpha_1, \ldots, \alpha_k, 0, \ldots, 0)\} \subset \mathbb{F}^n$ is invariant if and only if $A_{21} = 0$.
(b) Show that the subspace $M^{n-k} = \{(0, \ldots, 0, \alpha_{k+1}, \ldots, \alpha_n)\} \subset \mathbb{F}^n$ is invariant if and only if $A_{12} = 0$.

7. Let $A \in \mathrm{Mat}_{n \times n}(\mathbb{F})$ be upper triangular, i.e., $\alpha_{ij} = 0$ for $i > j$ or

$$A = \begin{bmatrix} \alpha_{11} & \alpha_{12} & \cdots & \alpha_{1n} \\ 0 & \alpha_{22} & \cdots & \alpha_{2n} \\ \vdots & \vdots & \ddots & \vdots \\ 0 & 0 & \cdots & \alpha_{nn} \end{bmatrix},$$

and $p \in \mathbb{F}[t]$. Show that $p(A)$ is also upper triangular and the diagonal entries are $p(\alpha_{ii})$, i.e.,

$$p(A) = \begin{bmatrix} p(\alpha_{11}) & * & \cdots & * \\ 0 & p(\alpha_{22}) & \cdots & * \\ \vdots & \vdots & \ddots & \vdots \\ 0 & 0 & \cdots & p(\alpha_{nn}) \end{bmatrix}.$$

8. Let $t_1, \ldots, t_n \in \mathbb{R}$ and define

$$L : C^\infty(\mathbb{R}, \mathbb{R}) \to \mathbb{R}^n$$
$$L(f) = (f(t_1), \ldots, f(t_n)).$$

Show that L is linear.

9. Let $t_0 \in \mathbb{R}$ and define

$$L : C^\infty(\mathbb{R}, \mathbb{R}) \to \mathbb{R}^n$$
$$L(f) = (f(t_0), (Df)(t_0), \ldots, (D^{n-1}f)(t_0)).$$

Show that L is linear.

10. Let $A \in \mathrm{Mat}_{n \times n}(\mathbb{R})$ be symmetric, i.e., the (i, j) entry is the same as the (j, i) entry. Show that $A = 0$ if and only if $\mathrm{tr}(A^2) = 0$.

11. For each $n \geq 2$, find $A \in \mathrm{Mat}_{n \times n}(\mathbb{F})$ such that $A \neq 0$, but $\mathrm{tr}(A^k) = 0$ for all $k = 1, 2, \ldots$.

12. Find $A \in \mathrm{Mat}_{2 \times 2}(\mathbb{R})$ such that $\mathrm{tr}(A^2) < 0$.

1.7   Linear Maps as Matrices

We saw above that quite a lot of linear maps can be defined using matrices. In this section, we shall generalize this construction and show that all abstractly defined linear maps between finite-dimensional vector spaces come from some basic matrix constructions.

To warm up, we start with the simplest situation.

Lemma 1.7.1. *Assume V is one-dimensional over $\mathbb{F}$, then any $L : V \to V$ is of the form $L = \lambda 1_V$.*

Proof. Assume x_1 is a basis. Then, $L(x_1) = \lambda x_1$ for some $\lambda \in \mathbb{F}$. Now, any $x = \alpha x_1$ so $L(x) = L(\alpha x_1) = \alpha L(x_1) = \alpha \lambda x_1 = \lambda x$ as desired.   □

This gives us a very simple canonical form for linear maps in this elementary situation. The rest of the section tries to explain how one can generalize this to vector spaces with finite bases.

Possibly, the most important abstractly defined linear map comes from considering linear combinations. We fix a vector space V over $\mathbb{F}$ and select $x_1, \ldots, x_m \in V$. Then, we have a linear map

$$L : \mathbb{F}^m \to V$$

$$L\left(\begin{bmatrix} \alpha_1 \\ \vdots \\ \alpha_m \end{bmatrix}\right) = \begin{bmatrix} x_1 & \cdots & x_m \end{bmatrix} \begin{bmatrix} \alpha_1 \\ \vdots \\ \alpha_m \end{bmatrix} = x_1 \alpha_1 + \cdots + x_m \alpha_m.$$

The fact that it is linear follows from knowing that $L : \mathbb{F} \to V$ defined by $L(\alpha) = \alpha x$ is linear together with the fact that sums of linear maps are linear. We shall denote this map by its row matrix

$$L = \begin{bmatrix} x_1 & \cdots & x_m \end{bmatrix},$$

where the entries are vectors. Using the standard basis $e_1, \ldots, e_m$ for $\mathbb{F}^m$ we observe that the entries x_i (think of them as column vectors) satisfy

$$L(e_i) = \begin{bmatrix} x_1 & \cdots & x_m \end{bmatrix} e_i = x_i.$$

Thus, the vectors that form the columns for the matrix for L are the images of the basis vectors for $\mathbb{F}^m$. With this in mind, we can show

Lemma 1.7.2. *Any linear map $L : \mathbb{F}^m \to V$ is of the form*

$$L = \begin{bmatrix} x_1 & \cdots & x_m \end{bmatrix},$$

where $x_i = L(e_i)$.

Proof. Define $L(e_i) = x_i$ and use linearity of L to see that

$$
L\left(\begin{bmatrix} \alpha_1 \\ \vdots \\ \alpha_m \end{bmatrix}\right) = L\left(\begin{bmatrix} e_1 \cdots e_m \end{bmatrix} \begin{bmatrix} \alpha_1 \\ \vdots \\ \alpha_m \end{bmatrix}\right)
$$

$$
= L(e_1\alpha_1 + \cdots + e_m\alpha_m)
$$

$$
= L(e_1)\alpha_1 + \cdots + L(e_m)\alpha_m
$$

$$
= \begin{bmatrix} L(e_1) \cdots L(e_m) \end{bmatrix} \begin{bmatrix} \alpha_1 \\ \vdots \\ \alpha_m \end{bmatrix}
$$

$$
= \begin{bmatrix} x_1 \cdots x_m \end{bmatrix} \begin{bmatrix} \alpha_1 \\ \vdots \\ \alpha_m \end{bmatrix}.
$$

$\square$

If we specialize to the situation where $V = \mathbb{F}^n$, then vectors $x_1, \ldots, x_m$ really are $n \times 1$ column matrices. More explicitly,

$$
x_i = \begin{bmatrix} \beta_{1i} \\ \vdots \\ \beta_{ni} \end{bmatrix},
$$

and

$$
\begin{bmatrix} x_1 \cdots x_m \end{bmatrix} \begin{bmatrix} \alpha_1 \\ \vdots \\ \alpha_m \end{bmatrix} = x_1\alpha_1 + \cdots + x_m\alpha_m
$$

$$
= \begin{bmatrix} \beta_{11} \\ \vdots \\ \beta_{n1} \end{bmatrix} \alpha_1 + \cdots + \begin{bmatrix} \beta_{1m} \\ \vdots \\ \beta_{nm} \end{bmatrix} \alpha_m
$$

$$
= \begin{bmatrix} \beta_{11}\alpha_1 \\ \vdots \\ \beta_{n1}\alpha_1 \end{bmatrix} + \cdots + \begin{bmatrix} \beta_{1m}\alpha_m \\ \vdots \\ \beta_{nm}\alpha_m \end{bmatrix}
$$

$$
= \begin{bmatrix} \beta_{11}\alpha_1 + \cdots + \beta_{1m}\alpha_m \\ \vdots \\ \beta_{n1}\alpha_1 + \cdots + \beta_{nm}\alpha_m \end{bmatrix}
$$

$$
= \begin{bmatrix} \beta_{11} & \cdots & \beta_{1m} \\ \vdots & \ddots & \vdots \\ \beta_{n1} & \cdots & \beta_{nm} \end{bmatrix} \begin{bmatrix} \alpha_1 \\ \vdots \\ \alpha_m \end{bmatrix}.
$$

Hence, any linear map $\mathbb{F}^m \to \mathbb{F}^n$ is given by matrix multiplication, and the columns of the matrix are the images of the basis vectors of $\mathbb{F}^m$.

We can also use this to study maps $V \to W$ as long as we have bases $e_1, \ldots, e_m$ for V and $f_1, \ldots, f_n$ for W. Each $x \in V$ has a unique expansion

$$
x = \begin{bmatrix} e_1 \cdots e_m \end{bmatrix} \begin{bmatrix} \alpha_1 \\ \vdots \\ \alpha_m \end{bmatrix}.
$$

So if $L : V \to W$ is linear, then

$$
L(x) = L\left(\begin{bmatrix} e_1 \cdots e_m \end{bmatrix} \begin{bmatrix} \alpha_1 \\ \vdots \\ \alpha_m \end{bmatrix} \right)
$$

$$
= \begin{bmatrix} L(e_1) \cdots L(e_m) \end{bmatrix} \begin{bmatrix} \alpha_1 \\ \vdots \\ \alpha_m \end{bmatrix}
$$

$$
= \begin{bmatrix} x_1 \cdots x_m \end{bmatrix} \begin{bmatrix} \alpha_1 \\ \vdots \\ \alpha_m \end{bmatrix},
$$

where $x_i = L(e_i)$. In effect, we have proven that

$$
L \circ \begin{bmatrix} e_1 \cdots e_m \end{bmatrix} = \begin{bmatrix} L(e_1) \cdots L(e_m) \end{bmatrix}
$$

if we interpret

$$
\begin{bmatrix} e_1 \cdots e_m \end{bmatrix} : \mathbb{F}^m \to V,
$$

$$
\begin{bmatrix} L(e_1) \cdots L(e_m) \end{bmatrix} : \mathbb{F}^m \to W
$$

as linear maps.

Expanding $L(e_i) = x_i$ with respect to the basis for W gives us

$$x_i = \begin{bmatrix} f_1 & \cdots & f_n \end{bmatrix} \begin{bmatrix} \beta_{1i} \\ \vdots \\ \beta_{ni} \end{bmatrix}$$

and

$$\begin{bmatrix} x_1 & \cdots & x_m \end{bmatrix} = \begin{bmatrix} f_1 & \cdots & f_n \end{bmatrix} \begin{bmatrix} \beta_{11} & \cdots & \beta_{1m} \\ \vdots & \ddots & \vdots \\ \beta_{n1} & \cdots & \beta_{nm} \end{bmatrix}.$$

This gives us the *matrix representation* for a linear map $V \to W$ with respect to the specified bases.

$$L(x) = \begin{bmatrix} x_1 & \cdots & x_m \end{bmatrix} \begin{bmatrix} \alpha_1 \\ \vdots \\ \alpha_m \end{bmatrix}$$

$$= \begin{bmatrix} f_1 & \cdots & f_n \end{bmatrix} \begin{bmatrix} \beta_{11} & \cdots & \beta_{1m} \\ \vdots & \ddots & \vdots \\ \beta_{n1} & \cdots & \beta_{nm} \end{bmatrix} \begin{bmatrix} \alpha_1 \\ \vdots \\ \alpha_m \end{bmatrix}.$$

We will often use the notation

$$[L] = \begin{bmatrix} \beta_{11} & \cdots & \beta_{1m} \\ \vdots & \ddots & \vdots \\ \beta_{n1} & \cdots & \beta_{nm} \end{bmatrix}$$

for the matrix representing L. The way to remember the formula for $[L]$ is to use

$$L \circ \begin{bmatrix} e_1 & \cdots & e_m \end{bmatrix} = \begin{bmatrix} L(e_1) & \cdots & L(e_m) \end{bmatrix}$$
$$= \begin{bmatrix} f_1 & \cdots & f_n \end{bmatrix} [L].$$

In the special case where $L : V \to V$ is a linear operator, one usually only selects one basis $e_1, \ldots, e_n$. In this case, we get the relationship

$$L \circ \begin{bmatrix} e_1 & \cdots & e_n \end{bmatrix} = \begin{bmatrix} L(e_1) & \cdots & L(e_n) \end{bmatrix}$$
$$= \begin{bmatrix} e_1 & \cdots & e_n \end{bmatrix} [L]$$

for the matrix representation.

Example 1.7.3. Let

$$P_n = \{\alpha_0 + \alpha_1 t + \cdots + \alpha_n t^n : \alpha_0, \alpha_1, \ldots, \alpha_n \in \mathbb{F}\}$$

be the space of polynomials of degree $\leq n$ and $D : P_n \to P_n$ the differentiation operator

$$D(\alpha_0 + \alpha_1 t + \cdots + \alpha_n t^n) = \alpha_1 + \cdots + n\alpha_n t^{n-1}.$$

If we use the basis $1, t, \ldots, t^n$ for P_n, then

$$D(t^k) = kt^{k-1},$$

and thus, the $(n+1) \times (n+1)$ matrix representation is computed via

$$\left[D(1) \; D(t) \; D(t^2) \; \cdots \; D(t^n) \right]$$
$$= \left[0 \; 1 \; 2t \; \cdots \; nt^{n-1} \right]$$
$$= \left[1 \; t \; t^2 \; \cdots \; t^n \right] \begin{bmatrix} 0 & 1 & 0 & \cdots & 0 \\ 0 & 0 & 2 & \cdots & 0 \\ 0 & 0 & 0 & \ddots & 0 \\ \vdots & \vdots & \vdots & \ddots & n \\ 0 & 0 & 0 & \cdots & 0 \end{bmatrix}.$$

Next, consider the maps $T, S : P_n \to P_{n+1}$ defined by

$$T(\alpha_0 + \alpha_1 t + \cdots + \alpha_n t^n) = \alpha_0 t + \alpha_1 t^2 + \cdots + \alpha_n t^{n+1},$$
$$S(\alpha_0 + \alpha_1 t + \cdots + \alpha_n t^n) = \alpha_0 t + \frac{\alpha_1}{2} t^2 + \cdots + \frac{\alpha_n}{n+1} t^{n+1}.$$

This time, the image space and domain are not the same but the choices for basis are at least similar. We get the $(n+2) \times (n+1)$ matrix representations

$$\left[T(1) \; T(t) \; T(t^2) \; \cdots \; T(t^n) \right]$$
$$= \left[t \; t^2 \; t^3 \; \cdots \; t^{n+1} \right]$$
$$= \left[1 \; t \; t^2 \; t^3 \; \cdots \; t^{n+1} \right] \begin{bmatrix} 0 & 0 & 0 & \cdots & 0 \\ 1 & 0 & 0 & \cdots & 0 \\ 0 & 1 & 0 & \cdots & 0 \\ 0 & 0 & 1 & \ddots & \vdots \\ \vdots & \vdots & \vdots & \ddots & 0 \\ 0 & 0 & 0 & \cdots & 1 \end{bmatrix}$$

$$\left[S\left(1\right)\, S\left(t\right)\, S\left(t^2\right)\, \cdots\, S\left(t^n\right) \right]$$

$$= \left[t \; \tfrac{1}{2}t^2 \; \tfrac{1}{3}t^3 \; \cdots \; \tfrac{1}{n+1}t^{n+1} \right]$$

$$= \left[1\; t\; t^2\; t^3 \cdots t^{n+1} \right]
\begin{bmatrix}
0 & 0 & 0 & \cdots & 0 \\
1 & 0 & 0 & \cdots & 0 \\
0 & \tfrac{1}{2} & 0 & \cdots & 0 \\
0 & 0 & \tfrac{1}{3} & \ddots & \vdots \\
\vdots & \vdots & \vdots & \ddots & 0 \\
0 & 0 & 0 & \cdots & \tfrac{1}{n}
\end{bmatrix}.$$

Doing a matrix representation of a linear map that is already given as a matrix can get a little confusing, but the procedure is obviously the same.

Example 1.7.4. Let

$$L = \begin{bmatrix} 1 & 1 \\ 0 & 2 \end{bmatrix} : \mathbb{F}^2 \to \mathbb{F}^2$$

and consider the basis

$$x_1 = \begin{bmatrix} 1 \\ 0 \end{bmatrix}, x_2 = \begin{bmatrix} 1 \\ 1 \end{bmatrix}.$$

Then,

$$L\left(x_1\right) = x_1,$$

$$L\left(x_2\right) = \begin{bmatrix} 2 \\ 2 \end{bmatrix} = 2x_2.$$

So

$$\left[L\left(x_1\right)\, L\left(x_2\right) \right] = \left[x_1 \; x_2 \right] \begin{bmatrix} 1 & 0 \\ 0 & 2 \end{bmatrix}.$$

Example 1.7.5. Again, let

$$L = \begin{bmatrix} 1 & 1 \\ 0 & 2 \end{bmatrix} : \mathbb{F}^2 \to \mathbb{F}^2$$

but consider instead the basis

$$y_1 = \begin{bmatrix} 1 \\ -1 \end{bmatrix}, y_2 = \begin{bmatrix} 1 \\ 1 \end{bmatrix}.$$

Then,

$$L(y_1) = \begin{bmatrix} 0 \\ -2 \end{bmatrix} = y_1 - y_2,$$

$$L(y_2) = \begin{bmatrix} 2 \\ 2 \end{bmatrix} = 2y_2.$$

So

$$[L(y_1) \ L(y_2)] = [y_1 \ y_2] \begin{bmatrix} 1 & 0 \\ -1 & 2 \end{bmatrix}.$$

Example 1.7.6. Let

$$A = \begin{bmatrix} a & c \\ b & d \end{bmatrix} \in \mathrm{Mat}_{2\times2}(\mathbb{F})$$

and consider

$$L_A : \mathrm{Mat}_{2\times2}(\mathbb{F}) \to \mathrm{Mat}_{2\times2}(\mathbb{F})$$

$$L_A(X) = AX.$$

We use the basis E_{ij} for $\mathrm{Mat}_{n\times n}(\mathbb{F})$ where the ij entry in E_{ij} is 1 and all other entries are zero. Next, order the basis $E_{11}, E_{21}, E_{12}, E_{22}$. This means that we think of $\mathrm{Mat}_{2\times2}(\mathbb{F}) \approx \mathbb{F}^4$ where the columns are stacked on top of each other with the first column being the top most. With this choice of basis, we note that

$$[L_A(E_{11}) \ L_A(E_{21}) \ L_A(E_{12}) \ L_A(E_{22})]$$

$$= [AE_{11} \ AE_{21} \ AE_{12} \ AE_{22}]$$

$$= \left[\begin{bmatrix} a & 0 \\ b & 0 \end{bmatrix} \begin{bmatrix} c & 0 \\ d & 0 \end{bmatrix} \begin{bmatrix} 0 & a \\ 0 & b \end{bmatrix} \begin{bmatrix} 0 & c \\ 0 & d \end{bmatrix} \right]$$

$$= [E_{11} \ E_{21} \ E_{12} \ E_{22}] \begin{bmatrix} a & c & 0 & 0 \\ b & d & 0 & 0 \\ 0 & 0 & a & c \\ 0 & 0 & b & d \end{bmatrix}$$

Thus, L_A has the block diagonal form

$$\begin{bmatrix} A & 0 \\ 0 & A \end{bmatrix}$$

This problem easily generalizes to the case of $n \times n$ matrices, where L_A will have a block diagonal form that looks like

$$\begin{bmatrix} A & 0 & \cdots & 0 \\ 0 & A & & 0 \\ \vdots & & \ddots & \vdots \\ 0 & 0 & \cdots & A \end{bmatrix}$$

Example 1.7.7. Let $L : \mathbb{F}^n \to \mathbb{F}^n$ be a linear map that maps basis vectors to basis vectors. Thus, $L(e_j) = e_{\sigma(j)}$, where

$$\sigma : \{1,\ldots,n\} \to \{1,\ldots,n\}.$$

If σ is one-to-one and onto, then it is called a *permutation*. Apparently, it permutes the elements of $\{1,\ldots,n\}$. The corresponding linear map is denoted L_σ. The matrix representation of L_σ can be computed from the simple relationship $L_\sigma(e_j) = e_{\sigma(j)}$. Thus, the jth column has zeros everywhere except for a 1 in the $\sigma(j)$ entry. This means that $[L_\sigma] = [\delta_{i,\sigma(j)}]$. The matrix $[L_\sigma]$ is also known as a *permutation matrix*.

Example 1.7.8. Let $L : V \to V$ be a linear map whose matrix representation with respect to the basis x_1, x_2 is given by

$$\begin{bmatrix} 1 & 2 \\ 0 & 1 \end{bmatrix}.$$

We wish to compute the matrix representation of $K = 2L^2 + 3L - 1_V$. We know that

$$\begin{bmatrix} L(x_1) & L(x_2) \end{bmatrix} = \begin{bmatrix} x_1 & x_2 \end{bmatrix} \begin{bmatrix} 1 & 2 \\ 0 & 1 \end{bmatrix}$$

or equivalently

$$L(x_1) = x_1,$$
$$L(x_2) = 2x_1 + x_2.$$

Thus,

$$\begin{aligned} K(x_1) &= 2L(L(x_1)) + 3L(x_1) - 1_V(x_1) \\ &= 2L(x_1) + 3x_1 - x_1 \\ &= 2x_1 + 3x_1 - x_1 \\ &= 4x_1, \\ K(x_2) &= 2L(L(x_2)) + 3L(x_2) - 1_V(x_2) \\ &= 2L(2x_1 + x_2) + 3(2x_1 + x_2) - x_2 \\ &= 2(2x_1 + (2x_1 + x_2)) + 3(2x_1 + x_2) - x_2 \\ &= 14x_1 + 4x_2, \end{aligned}$$

and

$$\begin{bmatrix} K\,(x_1) & K\,(x_2) \end{bmatrix} = \begin{bmatrix} x_1 & x_2 \end{bmatrix} \begin{bmatrix} 4 & 14 \\ 0 & 4 \end{bmatrix}.$$

Exercises

1. (a) Show that, $t^3, t^3 + t^2, t^3 + t^2 + t, t^3 + t^2 + t + 1$ form a basis for P_3.
   (b) Compute the image of $(1, 2, 3, 4)$ under the coordinate map

$$\begin{bmatrix} t^3 & t^3 + t^2 & t^3 + t^2 + t & t^3 + t^2 + t + 1 \end{bmatrix} : \mathbb{F}^4 \to P_3$$

   (c) Find the vector in $\mathbb{F}^4$ whose image is $4t^3 + 3t^2 + 2t + 1$.
2. Find the matrix representation for $D : P_3 \to P_3$ with respect to the basis t^3, $t^3 + t^2, t^3 + t^2 + t, t^3 + t^2 + t + 1$.
3. Find the matrix representation for

$$D^2 + 2D + 1_{P_3} : P_3 \to P_3$$

   with respect to the standard basis $1, t, t^2, t^3$.
4. Show that, if $L : V \to V$ is a linear operator on a finite-dimensional vector space and $p\,(t) \in \mathbb{F}\,[t]$, then the matrix representations for L and $p\,(L)$ with respect to some fixed basis are related by $[p\,(L)] = p\,([L])$.
5. Consider the two linear maps $K, L : P_n \to \mathbb{C}^{n+1}$ defined by

$$K\,(f) = (f\,(t_0)\,, (Df)\,(t_0)\,, \ldots, (D^n f)\,(t_0))\,,$$
$$L\,(f) = (f\,(t_0)\,, \ldots, f\,(t_n))\,.$$

   (a) Find a basis $p_0, \ldots, p_n$ for P_n such that $K\,(p_i) = e_{i+1}$, where $e_1, \ldots, e_{n+1}$ is the standard (aka *canonical*) basis for $\mathbb{C}^{n+1}$.
   (b) Provided $t_0, \ldots, t_n$ are distinct, find a basis $q_0, \ldots, q_n$ for P_n such that $L\,(q_i) = e_{i+1}$.
6. Let

$$A = \begin{bmatrix} a & c \\ b & d \end{bmatrix}$$

   and consider the linear map $R_A : \mathrm{Mat}_{2\times 2}\,(\mathbb{F}) \to \mathrm{Mat}_{2\times 2}\,(\mathbb{F})$ defined by $R_A\,(X) = XA$. Compute the matrix representation of this linear maps with respect to the basis

$$E_{11} = \begin{bmatrix} 1 & 0 \\ 0 & 0 \end{bmatrix},$$

$$E_{21} = \begin{bmatrix} 0 & 0 \\ 1 & 0 \end{bmatrix},$$

$$E_{12} = \begin{bmatrix} 0 & 1 \\ 0 & 0 \end{bmatrix},$$

$$E_{22} = \begin{bmatrix} 0 & 0 \\ 0 & 1 \end{bmatrix}.$$

7. Compute a matrix representation for $L : \text{Mat}_{2\times 2}(\mathbb{F}) \to \text{Mat}_{1\times 2}(\mathbb{F})$ defined by $L(X) = \begin{bmatrix} 1 & -1 \end{bmatrix} X$ using the standard bases.

8. Let $A \in \text{Mat}_{n\times m}(\mathbb{F})$ and E_{ij} the matrix that has 1 in the ij entry and is zero elsewhere.

   (a) If $E_{ij} \in \text{Mat}_{k\times n}(\mathbb{F})$, then $E_{ij} A \in \text{Mat}_{k\times m}(\mathbb{F})$ is the matrix whose ith row is the jth row of A and all other entries are zero.
   (b) If $E_{ij} \in \text{Mat}_{n\times k}(\mathbb{F})$, then $A E_{ij} \in \text{Mat}_{n\times k}(\mathbb{F})$ is the matrix whose jth column is the ith column of A and all other entries are zero.

9. Let e_1, e_2 be the standard basis for $\mathbb{C}^2$ and consider the two real bases e_1, e_2, ie_1, ie_2 and e_1, ie_1, e_2, ie_2. If $\lambda = \alpha + i\beta$ is a complex number, then compute the real matrix representations for $\lambda 1_{\mathbb{C}^2}$ with respect to both bases.

10. Show that if $L : V \to V$ has a lower triangular representation with respect to the basis $x_1, \ldots, x_n$, then it has an upper triangular representation with respect to $x_n, \ldots, x_1$.

11. Let V and W be vector spaces with bases $e_1, \ldots, e_m$ and $f_1, \ldots, f_n$, respectively. Define $E_{ij} \in \text{Hom}(V, W)$ as the linear map that sends e_j to f_i and all other e_ks go to zero, i.e., $E_{ij}(e_k) = \delta_{jk} f_i$.

   (a) Show that the matrix representation for E_{ij} is 1 in the ij entry and 0 otherwise.
   (b) Show that E_{ij} form a basis for $\text{Hom}(V, W)$.
   (c) Let $L \in \text{Hom}(V, W)$ and expand $L = \sum_{i,j} \alpha_{ij} E_{ij}$. Show that $[L] = [\alpha_{ij}]$ with respect to these bases.

1.8   Dimension and Isomorphism

We are now almost ready to prove that the number of elements in a basis for a fixed vector space is always the same.

Definition 1.8.1. We say that a linear map $L : V \to W$ is an *isomorphism* if we can find $K : W \to V$ such that $LK = 1_W$ and $KL = 1_V$.

One can also describe the equations $LK = 1_W$ and $KL = 1_V$ in an interesting little diagram of maps

$$
\begin{array}{ccc}
V & \xrightarrow{\;L\;} & W \\
1_V \uparrow & & \uparrow 1_W \\
V & \xleftarrow{\;K\;} & W,
\end{array}
$$

where the vertical arrows are the identity maps.

Definition 1.8.2. Two vector spaces V and W over $\mathbb{F}$ are said to be *isomorphic* if we can find an isomorphism $L : V \to W$.

Note that if V_1 and V_2 are isomorphic and V_2 and V_3 are isomorphic, then V_1 and V_3 are also isomorphic. The isomorphism is the composition of the given isomorphisms.

Recall that a map $f : S \to T$ between sets is *one-to-one* or *injective* if $f(x_1) = f(x_2)$ implies that $x_1 = x_2$. A better name for this concept is two-to-two as pointed out by Arens, since injective maps evidently take two distinct points to two distinct points. We say that $f : S \to T$ is *onto* or *surjective* if every $y \in T$ is of the form $y = f(x)$ for some $x \in S$. In others words, $f(S) = T$. A map that is both one-to-one and onto is said to be bijective. Such a map always has an inverse f^{-1} defined via $f^{-1}(y) = x$ if $f(x) = y$. Note that for each $y \in T$, such an x exists since f is onto and that this x is unique since f is one-to-one. The relationship between f and f^{-1} is $f \circ f^{-1}(y) = y$ and $f^{-1} \circ f(x) = x$. Observe that $f^{-1} : T \to S$ is also a bijection and has inverse $\left(f^{-1} \right)^{-1} = f$.

Lemma 1.8.3. *V and W are* isomorphic *if and only if there is a bijective linear map $L : V \to W$.*

The "if and only if" part asserts that the two statements:

- V and W are *isomorphic.*
- There is a bijective linear map $L : V \to W$.

are equivalent. In other words, if one statement is true, then so is the other. To establish the proposition, it is therefore necessary to prove two things, namely, that the first statement implies the second and that the second implies the first.

Proof. If V and W are isomorphic, then we can find linear maps $L : V \to W$ and $K : W \to V$ so that $LK = 1_W$ and $KL = 1_V$. Then, for any $y \in W$

$$
y = 1_W(y) = L(K(y)).
$$

Thus, $y = L(x)$ if $x = K(y)$. This means L is onto. If $L(x_1) = L(x_2)$, then

$$
x_1 = 1_V(x_1) = KL(x_1) = KL(x_2) = 1_V(x_2) = x_2.
$$

Showing that L is one-to-one.

Conversely, assume $L : V \to W$ is linear and a bijection. Then, we have an inverse map L^{-1} that satisfies $L \circ L^{-1} = 1_W$ and $L^{-1} \circ L = 1_V$. In order for this inverse to be allowable as K, we need to check that it is linear. Thus, select $\alpha_1, \alpha_2 \in \mathbb{F}$ and $y_1, y_2 \in W$. Let $x_i = L^{-1}(y_i)$ so that $L(x_i) = y_i$. Then we have

$$
\begin{aligned}
L^{-1}(\alpha_1 y_1 + \alpha_2 y_2) &= L^{-1}(\alpha_1 L(x_1) + \alpha_2 L(x_2)) \\
&= L^{-1}(L(\alpha_1 x_1 + \alpha_2 x_2)) \\
&= 1_V(\alpha_1 x_1 + \alpha_2 x_2) \\
&= \alpha_1 x_1 + \alpha_2 x_2 \\
&= \alpha_1 L^{-1}(y_1) + \alpha_2 L^{-1}(y_2).
\end{aligned}
$$

$\square$

Recall that a finite basis for V over $\mathbb{F}$ consists of a collection of vectors $x_1, \ldots, x_n \in V$ so that each $x \in V$ has a unique expansion $x = x_1 \alpha_1 + \cdots + x_n \alpha_n$, $\alpha_1, \ldots, \alpha_n \in \mathbb{F}$. This means that the linear map

$$
\begin{bmatrix} x_1 & \cdots & x_n \end{bmatrix} : \mathbb{F}^n \to V
$$

is a bijection and hence by the above lemma an isomorphism. We saw in Lemma 1.7.2 that any linear map $\mathbb{F}^m \to V$ must be of this form. In particular, any isomorphism $\mathbb{F}^m \to V$ gives rise to a basis for V. Since $\mathbb{F}^n$ is our prototype for an n-dimensional vector space over $\mathbb{F}$, it is natural to say that a vector space has dimension n if it is isomorphic to $\mathbb{F}^n$. As we have just seen, this is equivalent to saying that V has a basis consisting of n vectors. The only problem is that we do not know if two spaces $\mathbb{F}^m$ and $\mathbb{F}^n$ can be isomorphic when $m \neq n$. This is taken care of next.

Theorem 1.8.4. (Uniqueness of Dimension) *If $\mathbb{F}^m$ and $\mathbb{F}^n$ are isomorphic over $\mathbb{F}$, then $n = m$.*

Proof. Suppose we have $L : \mathbb{F}^m \to \mathbb{F}^n$ and $K : \mathbb{F}^n \to \mathbb{F}^m$ such that $LK = 1_{\mathbb{F}^n}$ and $KL = 1_{\mathbb{F}^m}$. In Sect. 1.7, we showed that the linear maps L and K are represented by matrices, i.e., $[L] \in \operatorname{Mat}_{n \times m}(\mathbb{F})$ and $[K] \in \operatorname{Mat}_{m \times n}(\mathbb{F})$. Using invariance of trace (Lemma 1.6.11), we then see that

$$
\begin{aligned}
n &= \operatorname{tr}([1_{\mathbb{F}^n}]) \\
&= \operatorname{tr}([L][K]) \\
&= \operatorname{tr}([K][L]) \\
&= \operatorname{tr}([1_{\mathbb{F}^m}]) \\
&= m.
\end{aligned}
$$

$\square$

This proof has the defect of only working when the field has characteristic 0. The result still holds in the more general situation where the characteristic is nonzero. Other more standard proofs that work in these more general situations can be found in Sects. 1.12 and 1.13.

Definition 1.8.5. We can now unequivocally denote and define the *dimension* of a vector space V over $\mathbb{F}$ as $\dim_{\mathbb{F}} V = n$ if V is isomorphic to $\mathbb{F}^n$. In case V is not isomorphic to any $\mathbb{F}^n$, we say that V is *infinite-dimensional* and write $\dim_{\mathbb{F}} V = \infty$.

Note that for some vector spaces, it is possible to change the choice of scalars. Such a change can have a rather drastic effect on what the dimension is. For example, $\dim_{\mathbb{C}} \mathbb{C} = 1$, while $\dim_{\mathbb{R}} \mathbb{C} = 2$. If we consider $\mathbb{R}$ as a vector space over $\mathbb{Q}$, something even worse happens: $\dim_{\mathbb{Q}} \mathbb{R} = \infty$. This is because $\mathbb{R}$ is not countably infinite, while all of the vector spaces $\mathbb{Q}^n$ are countably infinite. More precisely, it is possible to find a bijective map $f : \mathbb{N} \to \mathbb{Q}^n$, but, as first observed by Cantor using his famous diagonal argument, there is no bijective map $f : \mathbb{N} \to \mathbb{R}$. Thus, $\dim_{\mathbb{Q}} \mathbb{R} = \infty$ for set-theoretic reasons related to the (non)existence of bijective maps between sets.

Corollary 1.8.6. *If V and W are finite-dimensional vector spaces over $\mathbb{F}$, then* $\text{Hom}_{\mathbb{F}}(V, W)$ *is also finite-dimensional and*

$$\dim_{\mathbb{F}} \text{Hom}_{\mathbb{F}}(V, W) = (\dim_{\mathbb{F}} W) \cdot (\dim_{\mathbb{F}} V)$$

Proof. By choosing bases for V and W, we showed in Sect. 1.7 that there is a natural map:

$$\text{Hom}_{\mathbb{F}}(V, W) \to \text{Mat}_{(\dim_{\mathbb{F}} W) \times (\dim_{\mathbb{F}} V)}(\mathbb{F}) \simeq \mathbb{F}^{(\dim_{\mathbb{F}} W) \cdot (\dim_{\mathbb{F}} V)}.$$

This map is both one-to-one and onto as the matrix representation is uniquely determined by the linear map and every matrix yields a linear map. Finally, one easily checks that the map is linear. □

In the special case where $V = W$ and we have a basis for the n-dimensional space V, the linear isomorphism

$$\text{Hom}_{\mathbb{F}}(V, V) \longleftrightarrow \text{Mat}_{n \times n}(\mathbb{F})$$

also preserves composition and products. Thus, for $L, K : V \to V$, we have

$$[LK] = [L][K].$$

The composition in $\text{Hom}_{\mathbb{F}}(V, V)$ and matrix product in $\text{Mat}_{n \times n}(\mathbb{F})$ give an extra product structure on the vector spaces that make them into so-called *algebras*. Algebras are vector spaces that also have a product structure. This product structure must satisfy the associative law, the distributive law, and also commute with scalar multiplication. Unlike a field, it is not required that all nonzero elements have inverses. The above isomorphism is what we call an algebra isomorphism.

Exercises

1. Let $L, K : V \to V$ be linear maps between finite-dimensional vector spaces that satisfy $L \circ K = 0$. Is it true that $K \circ L = 0$?
2. Let $L : V \to W$ be a linear map between finite-dimensional vector spaces. Show that L is an isomorphism if and only if it maps a basis for V to a basis for W.
3. If V is finite-dimensional, show that V and $\mathrm{Hom}_{\mathbb{F}}(V, \mathbb{F})$ have the same dimension.
4. Show that a linear map $L : V \to W$ is one-to-one if and only if $L(x) = 0$ implies that $x = 0$.
5. Let V be a vector space over $\mathbb{F}$. Consider the map

$$K : V \to \mathrm{Hom}_{\mathbb{F}}(\mathrm{Hom}_{\mathbb{F}}(V, \mathbb{F}), \mathbb{F})$$

    defined by the condition that

$$K(x) \in \mathrm{Hom}_{\mathbb{F}}(\mathrm{Hom}_{\mathbb{F}}(V, \mathbb{F}), \mathbb{F})$$

    is the linear functional on $\mathrm{Hom}_{\mathbb{F}}(V, \mathbb{F})$ such that

$$K(x)(L) = L(x), \text{ for } L \in \mathrm{Hom}_{\mathbb{F}}(V, \mathbb{F}).$$

    Show that this map is one-to-one. Show that it is also onto when V is finite-dimensional.
6. Let $V \neq \{0\}$ be finite-dimensional and assume that

$$L_1, \ldots, L_n : V \to V$$

    are linear operators. Show that if $L_1 \circ \cdots \circ L_n = 0$, then at least one of the maps L_i is not one-to-one.
7. Let $t_0, \ldots, t_n \in \mathbb{R}$ be distinct and consider $P_n \subset \mathbb{C}[t]$. Define $L : P_n \to \mathbb{C}^{n+1}$ by $L(p) = (p(t_0), \ldots, p(t_n))$. Show that L is an isomorphism. (This problem will be easier to solve later in the text.)
8. Let $t_0 \in \mathbb{F}$ and consider $P_n \subset \mathbb{F}[t]$. Show that $L : P_n \to \mathbb{F}^{n+1}$ defined by

$$L(p) = (p(t_0), (Dp)(t_0), \ldots, (D^n p)(t_0))$$

    is an isomorphism. Hint: Think of a Taylor expansion at t_0.
9. (a) Let V be finite-dimensional. Show that if $L_1, L_2 : \mathbb{F}^n \to V$ are isomorphisms, then for any linear operator $L : V \to V$

$$\mathrm{tr}\left(L_1^{-1} \circ L \circ L_1\right) = \mathrm{tr}\left(L_2^{-1} \circ L \circ L_2\right).$$

    This means we can define $\mathrm{tr}(L)$. Hint: Try not to use explicit matrix representations.

(b) Let V and W be finite-dimensional and $L_1 : V \to W$ and $L_2 : W \to V$
linear maps. Show that

$$\text{tr}\,(L_1 \circ L_2) = \text{tr}\,(L_2 \circ L_1)$$

10. Construct an isomorphism $V \to \text{Hom}_{\mathbb{F}}\,(\mathbb{F}, V)$ with selecting bases for the spaces.

11. Let V be a complex vector space. Is the identity map $V \to V^*$ an isomorphism? (see Exercise 6 in Sect. 1.4 for the definition of V^*.)

12. Assume that V and W are finite-dimensional. Define

$$\text{Hom}_{\mathbb{F}}\,(V, W) \to \text{Hom}_{\mathbb{F}}\,(\text{Hom}_{\mathbb{F}}\,(W, V)\,, \mathbb{F})\,,$$

$$L \to [A \to \text{tr}\,(A \circ L)]\,.$$

Thus, the linear map $L : V \to W$ is mapped to a linear map $\text{Hom}_{\mathbb{F}}\,(W, V) \to \mathbb{F}$ that simply takes $A \in \text{Hom}_{\mathbb{F}}\,(W, V)$ to $\text{tr}\,(A \circ L)$. Show that this map is an isomorphism.

13. Consider the map

$$\Psi : \mathbb{C} \to \text{Mat}_{2 \times 2}\,(\mathbb{R})$$

defined by

$$\Psi\,(\alpha + i\beta) = \begin{bmatrix} \alpha & -\beta \\ \beta & \alpha \end{bmatrix}.$$

(a) Show that this is $\mathbb{R}$-linear and one-to-one but not onto. Find an example of a matrix in $\text{Mat}_{2 \times 2}\,(\mathbb{R})$ that does not come from $\mathbb{C}$.

(b) Extend this map to a linear map

$$\Psi : \text{Mat}_{n \times n}\,(\mathbb{C}) \to \text{Mat}_{2n \times 2n}\,(\mathbb{R})$$

and show that this map is also $\mathbb{R}$-linear and one-to-one but not onto. Conclude that there must be matrices in $\text{Mat}_{2n \times 2n}\,(\mathbb{R})$ that do not come from complex matrices in $\text{Mat}_{n \times n}\,(\mathbb{C})$.

(c) Show that $\dim_{\mathbb{R}} \text{Mat}_{n \times n}\,(\mathbb{C}) = 2n^2$, while $\dim_{\mathbb{R}} \text{Mat}_{2n \times 2n}\,(\mathbb{R}) = 4n^2$.

14. For $A = [\alpha_{ij}] \in \text{Mat}_{n \times m}\,(\mathbb{F})$, define the *transpose* $A^t = [\beta_{ij}] \in \text{Mat}_{m \times n}\,(\mathbb{F})$ by $\beta_{ij} = \alpha_{ji}$. Thus, A^t is gotten from A by reflecting in the diagonal entries.

(a) Show that $A \to A^t$ is a linear map which is also an isomorphism whose inverse is given by $B \to B^t$.

(b) If $A \in \text{Mat}_{n \times m}\,(\mathbb{F})$ and $B \in \text{Mat}_{m \times n}\,(\mathbb{F})$, show that $(AB)^t = B^t A^t$.

(c) Show that if $A \in \text{Mat}_{n \times n}\,(\mathbb{F})$ is invertible, i.e., there exists $A^{-1} \in \text{Mat}_{n \times n}\,(\mathbb{F})$ such that

$$AA^{-1} = A^{-1}A = 1_{\mathbb{F}^n}\,,$$

then A^t is also invertible and $(A^t)^{-1} = (A^{-1})^t$.

1.9   Matrix Representations Revisited

While the number of elements in a basis is always the same, there is unfortunately not a clear choice of a basis for many abstract vector spaces. This necessitates a discussion on the relationship between expansions of vectors in different bases.

Using the idea of isomorphism in connection with a choice of basis, we can streamline the procedure for expanding vectors and constructing the matrix representation of a linear map.

Fix a linear map $L : V \to W$ and bases $e_1, \ldots, e_m$ for V and $f_1, \ldots, f_n$ for W. One can then encode all of the necessary information in a diagram of maps

$$
\begin{array}{ccc}
V & \xrightarrow{\; L \;} & W \\
\uparrow & & \uparrow \\
\mathbb{F}^m & \xrightarrow{\; [L] \;} & \mathbb{F}^n
\end{array}
$$

In this diagram, the top horizontal arrow represents L and the bottom horizontal arrow represents the matrix for L interpreted as a linear map $[L] : \mathbb{F}^m \to \mathbb{F}^n$. The two vertical arrows are the basis isomorphisms defined by the choices of bases for V and W, i.e.,

$$
\left[e_1 \cdots e_m \right] : \mathbb{F}^m \to V,
$$
$$
\left[f_1 \cdots f_n \right] : \mathbb{F}^n \to W.
$$

Thus, we have the formulae relating L and $[L]$

$$
L = \left[f_1 \cdots f_n \right] \circ [L] \circ \left[e_1 \cdots e_m \right]^{-1},
$$
$$
[L] = \left[f_1 \cdots f_n \right]^{-1} \circ L \circ \left[e_1 \cdots e_m \right].
$$

Note that a basis isomorphism

$$
\left[x_1 \cdots x_m \right] : \mathbb{F}^m \to \mathbb{F}^m
$$

is a matrix

$$
\left[x_1 \cdots x_m \right] \in \mathrm{Mat}_{m \times m} (\mathbb{F})
$$

provided we write the vectors $x_1, \ldots, x_m$ as column vectors. As such, the map can be inverted using the standard matrix inverse. That said, it is not an easy problem to invert matrices or linear maps in general.

It is important to be aware of the fact that different bases will yield different matrix representations. To see what happens abstractly let us assume that we have two bases $x_1, \ldots, x_n$ and $y_1, \ldots, y_n$ for a vector space V. If we think of $x_1, \ldots, x_n$ as a basis for the domain and $y_1, \ldots, y_n$ as a basis for the image, then the identity map $1_V : V \to V$ has a matrix representation that is computed via

$$\begin{bmatrix} x_1 & \cdots & x_n \end{bmatrix} = \begin{bmatrix} y_1 & \cdots & y_n \end{bmatrix} \begin{bmatrix} \beta_{11} & \cdots & \beta_{1n} \\ \vdots & \ddots & \vdots \\ \beta_{n1} & \cdots & \beta_{nn} \end{bmatrix}$$

$$= \begin{bmatrix} y_1 & \cdots & y_n \end{bmatrix} B.$$

The matrix B, being the matrix representation for an isomorphism, is itself invertible, and we see that by multiplying by B^{-1} on the right, we obtain

$$\begin{bmatrix} y_1 & \cdots & y_n \end{bmatrix} = \begin{bmatrix} x_1 & \cdots & x_n \end{bmatrix} B^{-1}.$$

This is the matrix representation for $1_V^{-1} = 1_V$ when we switch the bases around. Differently stated, we have

$$B = \begin{bmatrix} y_1 & \cdots & y_n \end{bmatrix}^{-1} \begin{bmatrix} x_1 & \cdots & x_n \end{bmatrix},$$

$$B^{-1} = \begin{bmatrix} x_1 & \cdots & x_n \end{bmatrix}^{-1} \begin{bmatrix} y_1 & \cdots & y_n \end{bmatrix}.$$

We next check what happens to a vector $x \in V$

$$x = \begin{bmatrix} x_1 & \cdots & x_n \end{bmatrix} \begin{bmatrix} \alpha_1 \\ \vdots \\ \alpha_n \end{bmatrix}$$

$$= \begin{bmatrix} y_1 & \cdots & y_n \end{bmatrix} \begin{bmatrix} \beta_{11} & \cdots & \beta_{1n} \\ \vdots & \ddots & \vdots \\ \beta_{n1} & \cdots & \beta_{nn} \end{bmatrix} \begin{bmatrix} \alpha_1 \\ \vdots \\ \alpha_n \end{bmatrix}.$$

Thus, if we know the coordinates for x with respect to $x_1, \ldots, x_n$, then we immediately obtain the coordinates for x with respect to $y_1, \ldots, y_n$ by changing

$$\begin{bmatrix} \alpha_1 \\ \vdots \\ \alpha_n \end{bmatrix}$$

to

$$\begin{bmatrix} \beta_{11} & \cdots & \beta_{1n} \\ \vdots & \ddots & \vdots \\ \beta_{n1} & \cdots & \beta_{nn} \end{bmatrix} \begin{bmatrix} \alpha_1 \\ \vdots \\ \alpha_n \end{bmatrix}.$$

We can evidently also go backwards using the inverse B^{-1} rather than B.

Example 1.9.1. In $\mathbb{F}^2$, let e_1, e_2 be the standard basis and

$$x_1 = \begin{bmatrix} 1 \\ 0 \end{bmatrix}, \ x_2 = \begin{bmatrix} 1 \\ 1 \end{bmatrix}.$$

Then B_1^{-1} is easily found using

$$\begin{aligned} \begin{bmatrix} 1 & 1 \\ 0 & 1 \end{bmatrix} &= \begin{bmatrix} x_1 & x_2 \end{bmatrix} \\ &= \begin{bmatrix} e_1 & e_2 \end{bmatrix} B_1^{-1} \\ &= \begin{bmatrix} 1 & 0 \\ 0 & 1 \end{bmatrix} B_1^{-1} \\ &= B_1^{-1} \end{aligned}$$

B_1 itself requires solving

$$\begin{bmatrix} e_1 & e_2 \end{bmatrix} = \begin{bmatrix} x_1 & x_2 \end{bmatrix} B_1 \text{ or}$$

$$\begin{bmatrix} 1 & 0 \\ 0 & 1 \end{bmatrix} = \begin{bmatrix} 1 & 1 \\ 0 & 1 \end{bmatrix} B_1.$$

Thus,

$$\begin{aligned} B_1 &= \begin{bmatrix} x_1 & x_2 \end{bmatrix}^{-1} \\ &= \begin{bmatrix} 1 & 1 \\ 0 & 1 \end{bmatrix}^{-1} \\ &= \begin{bmatrix} 1 & -1 \\ 0 & 1 \end{bmatrix} \end{aligned}$$

Example 1.9.2. In $\mathbb{F}^2$, let

$$y_1 = \begin{bmatrix} 1 \\ -1 \end{bmatrix}, \ y_2 = \begin{bmatrix} 1 \\ 1 \end{bmatrix}$$

and

$$x_1 = \begin{bmatrix} 1 \\ 0 \end{bmatrix}, \ x_2 = \begin{bmatrix} 1 \\ 1 \end{bmatrix}.$$

Then, B_2 is found by

$$B_2 = \begin{bmatrix} x_1 & x_2 \end{bmatrix}^{-1} \begin{bmatrix} y_1 & y_2 \end{bmatrix}$$

$$= \begin{bmatrix} 1 & -1 \\ 0 & 1 \end{bmatrix} \begin{bmatrix} 1 & 1 \\ -1 & 1 \end{bmatrix}$$

$$= \begin{bmatrix} 2 & 0 \\ -1 & 1 \end{bmatrix}$$

and

$$B_2^{-1} = \begin{bmatrix} \frac{1}{2} & 0 \\ \frac{1}{2} & 1 \end{bmatrix}.$$

Recall that we know

$$\begin{bmatrix} \alpha \\ \beta \end{bmatrix} = \alpha e_1 + \beta e_2$$

$$= \frac{\alpha - \beta}{2} y_1 + \frac{\alpha + \beta}{2} y_2$$

$$= (\alpha - \beta) x_1 + \beta x_2.$$

Thus, it should be true that

$$\begin{bmatrix} (\alpha - \beta) \\ \beta \end{bmatrix} = \begin{bmatrix} 2 & 0 \\ -1 & 1 \end{bmatrix} \begin{bmatrix} \frac{\alpha-\beta}{2} \\ \frac{\alpha+\beta}{2} \end{bmatrix},$$

which indeed is the case.

Now, suppose that we have a linear operator $L : V \to V$. It will have matrix representations with respect to both bases. First, let us do this in a diagram of maps

$$\mathbb{F}^n \xrightarrow{A_1} \mathbb{F}^n$$
$$\downarrow \qquad \downarrow$$
$$V \xrightarrow{L} V$$
$$\uparrow \qquad \uparrow$$
$$\mathbb{F}^n \xrightarrow{A_2} \mathbb{F}^n$$

Here the downward arrows come from the isomorphism

$$\begin{bmatrix} x_1 & \cdots & x_n \end{bmatrix} : \mathbb{F}^n \to V,$$

and the upward arrows are

$$\begin{bmatrix} y_1 & \cdots & y_n \end{bmatrix} : \mathbb{F}^n \to V.$$

Thus,

$$L = \begin{bmatrix} x_1 & \cdots & x_n \end{bmatrix} A_1 \begin{bmatrix} x_1 & \cdots & x_n \end{bmatrix}^{-1}$$

$$L = \begin{bmatrix} y_1 & \cdots & y_n \end{bmatrix} A_2 \begin{bmatrix} y_1 & \cdots & y_n \end{bmatrix}^{-1}$$

We wish to discover what the relationship between A_1 and A_2 is. To figure this out, we simply note that

$$\begin{bmatrix} x_1 & \cdots & x_n \end{bmatrix} A_1 \begin{bmatrix} x_1 & \cdots & x_n \end{bmatrix}^{-1}$$

$$= L$$

$$= \begin{bmatrix} y_1 & \cdots & y_n \end{bmatrix} A_2 \begin{bmatrix} y_1 & \cdots & y_n \end{bmatrix}^{-1}.$$

Hence,

$$A_1 = \begin{bmatrix} x_1 & \cdots & x_n \end{bmatrix}^{-1} \begin{bmatrix} y_1 & \cdots & y_n \end{bmatrix} A_2 \begin{bmatrix} y_1 & \cdots & y_n \end{bmatrix}^{-1} \begin{bmatrix} x_1 & \cdots & x_n \end{bmatrix}$$

$$= B^{-1} A_2 B.$$

To memorize this formula, keep in mind that B transforms from the $x_1, \ldots, x_n$ basis to the $y_1, \ldots, y_n$ basis while B^{-1} reverses this process. The matrix product $B^{-1} A_2 B$ then indicates that starting from the right, we have gone from $x_1, \ldots, x_n$ to $y_1, \ldots, y_n$ then used A_2 on the $y_1, \ldots, y_n$ basis and then transformed back from the $y_1, \ldots, y_n$ basis to the $x_1, \ldots, x_n$ basis in order to find what A_1 does with respect to the $x_1, \ldots, x_n$ basis.

Definition 1.9.3. Two matrices $A_1, A_2 \in \mathrm{Mat}_{n \times n}(\mathbb{F})$ are said to be *similar* if there is an invertible matrix $B \in \mathrm{Mat}_{n \times n}(\mathbb{F})$ such that

$$A_1 = B^{-1} A_2 B.$$

We have evidently shown that any two matrix representations of the same linear operator are always similar.

Example 1.9.4. We have the representations for

$$L = \begin{bmatrix} 1 & 1 \\ 0 & 2 \end{bmatrix}$$

with respect to the three bases we just studied earlier in Sect. 1.7

$$\begin{bmatrix} L(e_1) & L(e_2) \end{bmatrix} = \begin{bmatrix} e_1 & e_2 \end{bmatrix} \begin{bmatrix} 1 & 1 \\ 0 & 2 \end{bmatrix},$$

$$\left[\, L\left(x_1\right) \; L\left(x_2\right)\,\right] = \left[\, x_1 \; x_2 \,\right] \begin{bmatrix} 1 & 0 \\ 0 & 2 \end{bmatrix},$$

$$\left[\, L\left(y_1\right) \; L\left(y_2\right)\,\right] = \left[\, y_1 \; y_2 \,\right] \begin{bmatrix} 1 & 0 \\ -1 & 2 \end{bmatrix}.$$

Using the changes of basis calculated above, we can check the following relationships:

$$\begin{bmatrix} 1 & 0 \\ 0 & 2 \end{bmatrix} = B_1 \begin{bmatrix} 1 & 1 \\ 0 & 2 \end{bmatrix} B_1^{-1}$$

$$= \begin{bmatrix} 1 & -1 \\ 0 & 1 \end{bmatrix} \begin{bmatrix} 1 & 1 \\ 0 & 2 \end{bmatrix} \begin{bmatrix} 1 & 1 \\ 0 & 1 \end{bmatrix}$$

$$\begin{bmatrix} 1 & 0 \\ 0 & 2 \end{bmatrix} = B_2 \begin{bmatrix} 1 & 0 \\ -1 & 2 \end{bmatrix} B_2^{-1}$$

$$= \begin{bmatrix} 2 & 0 \\ -1 & 1 \end{bmatrix} \begin{bmatrix} 1 & 0 \\ -1 & 2 \end{bmatrix} \begin{bmatrix} \frac{1}{2} & 0 \\ \frac{1}{2} & 1 \end{bmatrix}.$$

One can more generally consider $L : V \to W$ and see what happens if we change bases in both V and W. The analysis is similar as long as we keep in mind that there are four bases in play. The key diagram looks like

$$\begin{array}{ccc} \mathbb{F}^m & \xrightarrow{A_1} & \mathbb{F}^n \\ \downarrow & & \downarrow \\ V & \xrightarrow{L} & W \\ \uparrow & & \uparrow \\ \mathbb{F}^m & \xrightarrow{A_2} & \mathbb{F}^n \end{array}$$

One of the goals in the study of linear operators or just square matrices is to find a suitable basis that makes the matrix representation as simple as possible. This is a rather complicated theory which the rest of the book will try to uncover.

Exercises

1. Let $V = \{ f \in \text{Func}\,(\mathbb{R}, \mathbb{C}) : f\,(t) = \alpha \cos\,(t) + \beta \sin\,(t) \,, \alpha, \beta \in \mathbb{C} \}$.

   (a) Show that $\cos\,(t)\,, \sin\,(t)$ and $\exp\,(it)\,, \exp\,(-it)$ both form a basis for V.
   (b) Find the change of basis matrix.

(c) Find the matrix representation of $D : V \rightarrow V$ with respect to both bases and check that the change of basis matrix gives the correct relationship between these two matrices.

2. Let

$$A = \begin{bmatrix} 0 & -1 \\ 1 & 0 \end{bmatrix} : \mathbb{R}^2 \rightarrow \mathbb{R}^2$$

and consider the basis

$$x_1 = \begin{bmatrix} 1 \\ -1 \end{bmatrix}, x_2 = \begin{bmatrix} 1 \\ 1 \end{bmatrix}.$$

(a) Compute the matrix representation of A with respect to x_1, x_2.
(b) Compute the matrix representation of A with respect to $\frac{1}{\sqrt{2}}x_1, \frac{1}{\sqrt{2}}x_2$.
(c) Compute the matrix representation of A with respect to $x_1, x_1 + x_2$.

3. Let e_1, e_2 be the standard basis for $\mathbb{C}^2$ and consider the two real bases e_1, e_2, ie_1, ie_2 and e_1, ie_1, e_2, ie_2. If $\lambda = \alpha + i\beta$ is a complex number, compute the real matrix representations for $\lambda 1_{\mathbb{C}^2}$ with respect to both bases. Show that the two matrices are related via the change of basis formula.

4. If $x_1, \ldots, x_n$ is a basis for V, then what is the change of basis matrix from $x_1, \ldots, x_n$ to $x_n, \ldots, x_1$? How does the matrix representation of an operator on V change with this change of basis?

5. Let $L : V \rightarrow V$ be a linear operator, $p(t) \in \mathbb{F}[t]$, a polynomial and $K : V \rightarrow W$ an isomorphism. Show that

$$p\left(K \circ L \circ K^{-1}\right) = K \circ p(L) \circ K^{-1}.$$

6. Let A be a permutation matrix (see Example 1.7.7 for the definition.) Will the matrix representation for A still be a permutation matrix if we select a different basis?

7. What happens to the matrix representation of a linear map if the change of basis matrix is a permutation matrix (see Example 1.7.7 for the definition)?

1.10 Subspaces

We are now ready for a more in-depth study of subspaces. Recall that a nonempty subset $M \subset V$ of a vector space V is said to be a *subspace* if it is closed under addition and scalar multiplication:

$$x, y \in M \Rightarrow x + y \in M,$$

$$\alpha \in \mathbb{F} \text{ and } x \in M \Rightarrow \alpha x \in M$$

The two axioms for a subspace can be combined into one as follows:

$$\alpha_1, \alpha_2 \in \mathbb{F} \text{ and } x_1, x_2 \in M \Rightarrow \alpha_1 x_1 + \alpha_2 x_2 \in M$$

Any vector space always has two *trivial subspaces*, namely, V and $\{0\}$. Some more interesting examples come below.

Example 1.10.1. Let M_i be the ith coordinate axis in $\mathbb{F}^n$, i.e., the set consisting of the vectors where all but the ith coordinate are zero. Thus,

$$M_i = \{(0, \ldots, 0, \alpha_i, 0, \ldots, 0) : \alpha_i \in \mathbb{F}\}.$$

Example 1.10.2. Polynomials in $\mathbb{F}[t]$ of degree $\leq n$ form a subspace denoted P_n.

Example 1.10.3. The set of continuous functions $C^0([a, b], \mathbb{R})$ on an interval $[a, b] \subset \mathbb{R}$ is evidently a subspace of Func $([a, b], \mathbb{R})$. Likewise, the space of functions that have derivatives of all orders is a subspace

$$C^\infty([a, b], \mathbb{R}) \subset C^0([a, b], \mathbb{R}).$$

If we regard polynomials as functions on $[a, b]$, then it too becomes a subspace

$$\mathbb{R}[t] \subset C^\infty([a, b], \mathbb{R}).$$

Example 1.10.4. Solutions to simple types of equations often form subspaces:

$$\{3\alpha_1 - 2\alpha_2 + \alpha_3 = 0 : (\alpha_1, \alpha_2, \alpha_3) \in \mathbb{F}^3\}.$$

However, something like

$$\{3\alpha_1 - 2\alpha_2 + \alpha_3 = 1 : (\alpha_1, \alpha_2, \alpha_3) \in \mathbb{F}^3\}$$

does not yield a subspace as it does not contain the origin.

Example 1.10.5. There are other interesting examples of subspaces of $C^\infty(\mathbb{R}, \mathbb{C})$. If $\omega > 0$ is some fixed number, then we consider

$$C_\omega^\infty(\mathbb{R}, \mathbb{C}) = \{f \in C^\infty(\mathbb{R}, \mathbb{C}) : f(t) = f(t + \omega) \text{ for all } t \in \mathbb{R}\}.$$

These are the *periodic functions* with period ω. Note that

$$f(t) = \exp(i 2\pi t / \omega)$$
$$= \cos(2\pi t / \omega) + i \sin(2\pi t / \omega)$$

is an example of a periodic function.

Subspaces allow for a generalized type of calculus. That is, we can "add" and "multiply" them to form other subspaces. However, it is not possible to find inverses for either operation.

Definition 1.10.6. If $M, N \subset V$ are subspaces, then we can form two new subspaces, the *sum* and the *intersection*:

$$M + N = \{x + y : x \in M \text{ and } y \in N\},$$
$$M \cap N = \{x : x \in M \text{ and } x \in N\}.$$

It is certainly true that both of these sets contain the origin. The intersection is most easily seen to be a subspace so let us check the sum. If $\alpha \in \mathbb{F}$ and $x \in M$, $y \in N$, then we have $\alpha x \in M$, $\alpha y \in N$ so

$$\alpha x + \alpha y = \alpha (x + y) \in M + N.$$

In this way, we see that $M + N$ is closed under scalar multiplication. To check that it is closed under addition is equally simple.

We can think of $M + N$ as addition of subspaces and $M \cap N$ as a kind of multiplication. The element that acts as zero for addition is the trivial subspace $\{0\}$ as $M + \{0\} = M$, while $M \cap V = M$ implies that V is the identity for intersection. Beyond this, it is probably not that useful to think of these subspace operations as arithmetic operations e.g., the distributive law does not hold.

Definition 1.10.7. If $S \subset V$ is a subset of a vector space, then the *span* of S is defined as

$$\text{span} (S) = \bigcap_{S \subset M \subset V} M,$$

where $M \subset V$ is always a subspace of V. Thus, the span is the intersection of all subspaces that contain S. This is a subspace of V and must in fact be the smallest subspace containing S.

We immediately get the following elementary properties.

Proposition 1.10.8. *Let V be a vector space and $S, T \subset V$ subsets.*

(1) If $S \subset T$, then $\text{span} (S) \subset \text{span} (T)$.
(2) If $M \subset V$ is a subspace, then $\text{span} (M) = M$.
(3) $\text{span} (\text{span} (S)) = \text{span} (S)$.
(4) $\text{span} (S) = \text{span} (T)$ *if and only if* $S \subset \text{span} (T)$ *and* $T \subset \text{span} (S)$.

Proof. The first property is obvious from the definition of span.

To prove the second property, we first note that we always have that $S \subset \text{span} (S)$. In particular, $M \subset \text{span} (M)$. On the other hand, as M is a subspace that contains M, it must also follow that $\text{span} (M) \subset M$.

The third property follows from the second as $\text{span} (S)$ is a subspace.

To prove the final property, we first observe that if $\text{span} (S) \subset \text{span} (T)$, then $S \subset \text{span} (T)$. Thus, it is clear that if $\text{span} (S) = \text{span} (T)$, then $S \subset \text{span} (T)$

and $T \subset \text{span}(S)$. Conversely, we have from the first and third properties that if $S \subset \text{span}(T)$, then $\text{span}(S) \subset \text{span}(\text{span}(T)) = \text{span}(T)$. This shows that if $S \subset \text{span}(T)$ and $T \subset \text{span}(S)$, then $\text{span}(S) = \text{span}(T)$. $\qquad\square$

The following lemma gives an alternate and very convenient description of the span.

Lemma 1.10.9. (Characterization of $\text{span}(M)$) *Let $S \subset V$ be a nonempty subset of M. Then,* $\text{span}(S)$ *consists of all linear combinations of vectors in S.*

Proof. Let C be the set of all linear combinations of vectors in S. Since $\text{span}(S)$ is a subspace, it must be true that $C \subset \text{span}(S)$. Conversely, if $x, y \in C$, then we note that also $\alpha x + \beta y$ is a linear combination of vectors from S. Thus, $\alpha x + \beta y \in C$ and hence C is a subspace. This means that $\text{span}(S) \subset C$. $\qquad\square$

Definition 1.10.10. We say that two subspaces $M, N \subset V$ have *trivial intersection* provided $M \cap N = \{0\}$, i.e., their intersection is the trivial subspace. We say that M and N are *transversal* provided $M + N = V$.

Both concepts are important in different ways. Transversality also plays a very important role in the more advanced subject of differentiable topology. Differentiable topology is the study of smooth maps and manifolds.

Definition 1.10.11. If we combine the two concepts of transversality and trivial intersection, we arrive at another important idea. Two subspaces are said to be *complementary* if they are transversal and have trivial intersection.

Lemma 1.10.12. *Two subspaces $M, N \subset V$ are complementary if and only if each vector $z \in V$ can be written as $z = x + y$, where $x \in M$ and $y \in N$ in one and only one way.*

Before embarking on the proof, let us explain the use of "one and only one." The idea is first that z can be written like that in (at least) one way the second part is that this is the only way in which to do it. In other words, having found x and y so that $z = x + y$, there cannot be any other ways in which to decompose z into a sum of elements from M and N.

Proof. First assume that M and N are complementary. Since $V = M + N$, we know that $z = x + y$ for some $x \in M$ and $y \in N$. If we have

$$x_1 + y_1 = z = x_2 + y_2,$$

where $x_1, x_2 \in M$ and $y_1, y_2 \in N$, then by moving each of x_2 and y_1 to the other side, we get

$$M \ni x_1 - x_2 = y_2 - y_1 \in N.$$

This means that

$$x_1 - x_2 = y_2 - y_1 \in M \cap N = \{0\}$$

and hence that

$$x_1 - x_2 = y_2 - y_1 = 0.$$

Thus, $x_1 = x_2$ and $y_1 = y_2$ and we have established that z has the desired unique decomposition.

Conversely, assume that any $z = x + y$, for unique $x \in M$ and $y \in N$. First, we see that this means $V = M + N$. To see that $M \cap N = \{0\}$, we simply select $z \in M \cap N$. Then, $z = z + 0 = 0 + z$ where $z \in M, 0 \in N$ and $0 \in M, z \in N$. Since such decompositions are assumed to be unique, we must have that $z = 0$ and hence $M \cap N = \{0\}$.   □

Definition 1.10.13. When we have two complementary subspaces $M, N \subset V$, we also say that V is a *direct sum* of M and N and we write this symbolically as $V = M \oplus N$. The special sum symbol indicates that indeed, $V = M + N$ and also that the two subspaces have trivial intersection. Using what we have learned so far about subspaces, we get a result that is often quite useful.

Corollary 1.10.14. *Let $M, N \subset V$ be subspaces. If $M \cap N = \{0\}$, then*

$$M + N = M \oplus N,$$

and if both are finite-dimensional, then

$$\dim (M + N) = \dim (M) + \dim (N).$$

Proof. The first statement follows immediately from the definition. The second statement is proven by selecting bases $e_1, \ldots, e_k$ for M and $f_1, \ldots, f_l$ for N and then showing that the concatenation $e_1, \ldots, e_k, f_1, \ldots, f_l$ is a basis for $M + N$.
                                                                                    □

We also have direct sum decompositions for more than two subspaces. If $M_1, \ldots, M_k \subset V$ are subspaces, we say that V is a direct sum of $M_1, \ldots, M_k$ and write

$$V = M_1 \oplus \cdots \oplus M_k$$

provided any vector $z \in V$ can be decomposed as

$$z = x_1 + \cdots + x_k,$$

$$x_1 \in M_1, \ldots, x_k \in M_k$$

in one and only one way.

Here are some examples of direct sums.

Example 1.10.15. The prototypical example of a direct sum comes from the plane, where $V = \mathbb{R}^2$ and

$$M = \{(x, 0) : x \in \mathbb{R}\}$$

is the first coordinate axis and

$$N = \{(0, y) : y \in \mathbb{R}\}$$

the second coordinate axis.

Example 1.10.16. Direct sum decompositions are by no means unique, as can be seen using $V = \mathbb{R}^2$ and

$$M = \{(x, 0) : x \in \mathbb{R}\}$$

and

$$N = \{(y, y) : y \in \mathbb{R}\}$$

the diagonal. We can easily visualize and prove that the intersection is trivial. As for transversality, just observe that

$$(x, y) = (x - y, 0) + (y, y).$$

Example 1.10.17. We also have the direct sum decomposition

$$\mathbb{F}^n = M_1 \oplus \cdots \oplus M_n,$$

where

$$M_i = \{(0, \ldots, 0, \alpha_i, 0, \ldots, 0) : \alpha_i \in \mathbb{F}\}.$$

Example 1.10.18. Here is a more abstract example that imitates the first. Partition the set

$$\{1, 2, \ldots, n\} = \{i_1, \ldots, i_k\} \cup \{j_1, \ldots, j_{n-k}\}$$

into two complementary sets. Let

$$V = \mathbb{F}^n,$$
$$M = \{(\alpha_1, \ldots, \alpha_n) \in \mathbb{F}^n : \alpha_{j_1} = \cdots = \alpha_{j_{n-k}} = 0\},$$
$$N = \{(\alpha_1, \ldots, \alpha_n) : \alpha_{i_1} = \cdots = \alpha_{i_k} = 0\}.$$

Thus,

$$M = M_{i_1} \oplus \cdots \oplus M_{i_k},$$
$$N = M_{j_1} \oplus \cdots \oplus M_{j_{n-k}},$$

and $\mathbb{F}^n = M \oplus N$. Note that M is isomorphic to $\mathbb{F}^k$ and N to $\mathbb{F}^{n-k}$ but with different indices for the axes. Thus, we have the more or less obvious decomposition: $\mathbb{F}^n = \mathbb{F}^k \times \mathbb{F}^{n-k}$. Note, however, that when we use $\mathbb{F}^k$ rather than M, we do not think of $\mathbb{F}^k$ as a subspace of $\mathbb{F}^n$, as vectors in $\mathbb{F}^k$ are k-tuples of the form $(\alpha_{i_1}, \ldots, \alpha_{i_k})$. Thus, there is a subtle difference between writing $\mathbb{F}^n$ as a product or direct sum.

Example 1.10.19. Another very interesting decomposition is that of separating functions into odd and even parts. Recall that a function $f : \mathbb{R} \to \mathbb{R}$ is said to be odd, respectively even, if $f(-t) = -f(t)$, respectively, $f(-t) = f(t)$. Note that constant functions are even, while functions whose graphs are lines through the origin are odd. We denote the subsets of odd and even functions by $\text{Func}^{\text{odd}}(\mathbb{R}, \mathbb{R})$ and $\text{Func}^{\text{ev}}(\mathbb{R}, \mathbb{R})$. It is easily seen that these subsets are subspaces. Also, $\text{Func}^{\text{odd}}(\mathbb{R}, \mathbb{R}) \cap \text{Func}^{\text{ev}}(\mathbb{R}, \mathbb{R}) = \{0\}$ since only the zero function can be both odd and even. Finally, any $f \in \text{Func}(\mathbb{R}, \mathbb{R})$ can be decomposed as follows:

$$f(t) = f_{\text{ev}}(t) + f_{\text{odd}}(t),$$

$$f_{\text{ev}}(t) = \frac{f(t) + f(-t)}{2},$$

$$f_{\text{odd}}(t) = \frac{f(t) - f(-t)}{2}.$$

A specific example of such a decomposition is

$$e^t = \cosh(t) + \sinh(t),$$

$$\cosh(t) = \frac{e^t + e^{-t}}{2},$$

$$\sinh(t) = \frac{e^t - e^{-t}}{2}.$$

If we consider complex-valued functions $\text{Func}(\mathbb{R}, \mathbb{C})$, we still have the same concepts of even and odd and also the desired direct sum decomposition. Here, another similar and very interesting decomposition is Euler's formula

$$e^{it} = \cos(t) + i\sin(t)$$

$$\cos(t) = \frac{e^{it} + e^{-it}}{2},$$

$$\sin(t) = \frac{e^{it} - e^{-it}}{2i}.$$

Some interesting questions come to mind with the definitions encountered here. What is the relationship between $\dim_{\mathbb{F}} M$ and $\dim_{\mathbb{F}} V$ for a subspace $M \subset V$? Do all subspaces have a complement? How are subspaces and linear maps related?

At this point, we can show that subspaces of finite-dimensional vector spaces do have complements. This result is central to almost all of the subsequent developments in this chapter. As such, it is worth noting that the result is so basic that it does not even depend on the concept of dimension. It is also noteworthy that it gives us a stronger conclusion than stated here (see Corollary 1.12.6).

Theorem 1.10.20. (Existence of Complements) *Let $M \subset V$ be a subspace and assume that $V = \text{span}\{x_1, \ldots, x_n\}$. If $M \neq V$, then it is possible to choose $x_{i_1}, \ldots, x_{i_k} \in \{x_1, \ldots, x_n\}$ such that*

$$V = M \oplus \text{span}\{x_{i_1}, \ldots, x_{i_k}\}$$

Proof. Successively choose $x_{i_1}, \ldots, x_{i_k}$ such that

$$x_{i_1} \notin M,$$
$$x_{i_2} \notin M + \text{span}\{x_{i_1}\},$$
$$\vdots \quad \vdots$$
$$x_{i_k} \notin M + \text{span}\{x_{i_1}, \ldots, x_{i_{k-1}}\}.$$

This process can be continued until

$$V = M + \text{span}\{x_{i_1}, \ldots, x_{i_k}\},$$

and since

$$\text{span}\{x_1, \ldots, x_n\} = V,$$

we know that this will happen for some $k \leq n$. It now only remains to be seen that

$$\{0\} = M \cap \text{span}\{x_{i_1}, \ldots, x_{i_k}\}.$$

To check, this suppose that

$$x \in M \cap \text{span}\{x_{i_1}, \ldots, x_{i_k}\}$$

and write

$$x = \alpha_{i_1} x_{i_1} + \cdots + \alpha_{i_k} x_{i_k} \in M.$$

If $\alpha_{i_1} = \cdots = \alpha_{i_k} = 0$, there is nothing to worry about. Otherwise, we can find the largest l so that $\alpha_{i_l} \neq 0$. Then,

$$\frac{1}{\alpha_{i_l}} x = \frac{\alpha_{i_1}}{\alpha_{i_l}} x_{i_1} + \cdots + \frac{\alpha_{i_{l-1}}}{\alpha_{i_l}} x_{i_{l-1}} + x_{i_l} \in M$$

which implies the contradictory statement that

$$x_{i_l} \in M + \text{span}\{x_{i_1}, \ldots, x_{i_{l-1}}\}.$$

$\square$

If we use Corollary 1.10.14, then we see that this theorem shows that dim $(M) \leq$ dim (V) as long as we know that both M and V are finite-dimensional. Thus, the important point lies in showing that M is finite-dimensional. We will establish this in the next section.

Exercises

1. Show that the subset of linear maps $L : \mathbb{R}^3 \to \mathbb{R}^2$ defined by

$$S = \left\{ L : \mathbb{R}^3 \to \mathbb{R}^2 : L(1,2,3) = 0, \ (2,3) = L(x) \text{ for some } x \in \mathbb{R}^2 \right\}$$

   is not a subspace of Hom $(\mathbb{R}^3, \mathbb{R}^2)$.
2. Find a one-dimensional complex subspace $M \subset \mathbb{C}^2$ such that $\mathbb{R}^2 \cap M = \{0\}$.
3. Let $L : V \to W$ be a linear map and $N \subset W$ a subspace. Show that

$$L^{-1}(N) = \{x \in V : L(x) \in N\}$$

   is a subspace of V.
4. Is it true that subspaces satisfy the distributive law

$$M \cap (N_1 + N_2) = M \cap N_1 + M \cap N_2?$$

   If not, give a counter example.
5. Show that if V is finite-dimensional, then Hom (V, V) is a direct sum of the two subspaces $M = \text{span}\{1_V\}$ and $N = \{L : \text{tr}L = 0\}$.
6. Show that $\text{Mat}_{n \times n}(\mathbb{R})$ is the direct sum of the following three subspaces (you also have to show that they are subspaces):

$$I = \text{span}\{1_{\mathbb{R}^n}\},$$
$$S_0 = \{A : \text{tr}A = 0 \text{ and } A^t = A\},$$
$$A = \{A : A^t = -A\}.$$

   (A^t is defined in Exercise 14 in Sect. 1.8.)
7. Let V be a vector space over a field $\mathbb{F}$ of characteristic zero. Let $M_1, \ldots, M_k \subsetneq V$ be proper subspaces of a finite-dimensional vector space and $N \subset V$ a subspace. Show that if $N \subset M_1 \cup \cdots \cup M_k$, then $N \subset M_i$ for some i. Conclude that, if N is not contained in any of the M_is, then we can find $x \in N$ such that $x \notin M_1, \ldots, x \notin M_k$. Hint: Do the case where $k = 2$ first.
8. An *affine subspace* $A \subset V$ of a vector space is a subset such that *affine linear combinations* of vectors in A lie in A, i.e., if $\alpha_1 + \cdots + \alpha_n = 1$ and $x_1, \ldots, x_n \in A$, then $\alpha_1 x_1 + \cdots + \alpha_n x_n \in A$.

(a) Show that A is an affine subspace if and only if there is a point $x_0 \in V$ and a subspace $M \subset V$ such that

$$A = x_0 + M = \{x_0 + x : x \in M\}.$$

(b) Show that A is an affine subspace if and only if there is a subspace $M \subset V$ with the properties: (1) if $x, y \in A$, then $x - y \in M$ and (2) if $x \in A$ and $z \in M$, then $x + z \in A$.

(c) Show that the subspaces constructed in parts (a) and (b) are equal.

(d) Show that the set of monic polynomials of degree n in P_n, i.e., the coefficient in front of t^n is 1, is an affine subspace with $M = P_{n-1}$.

9. Show that the two spaces below are subspaces of $C_{2\pi}^\infty (\mathbb{R}, \mathbb{R})$ that are not equal to each other:

$$V_1 = \{b_1 \sin (t) + b_2 \sin (2t) + b_3 \sin (3t) : b_1, b_2, b_3 \in \mathbb{R}\},$$
$$V_2 = \{b_1 \sin (t) + b_2 \sin^2 (t) + b_3 \sin^3 (t) : b_1, b_2, b_3 \in \mathbb{R}\}.$$

What is their intersection?

10. Show that if $M \subset V$ and $N \subset W$ are subspaces, then $M \times N \subset V \times W$ is also a subspace.

11. If $A \in \mathrm{Mat}_{n \times n} (\mathbb{F})$ has $\mathrm{tr}\,(A) = 0$, show that

$$A = A_1 B_1 - B_1 A_1 + \cdots + A_m B_m - B_m A_m$$

for suitable $A_i, B_i \in \mathrm{Mat}_{n \times n} (\mathbb{F})$. Hint: Show that

$$M = \mathrm{span}\,\{XY - YX : X, Y \in \mathrm{Mat}_{n \times n} (\mathbb{F})\}$$

has dimension $n^2 - 1$ by exhibiting a suitable basis.

12. Let $L : V \to W$ be a linear map and consider the graph

$$G_L = \{(x, L(x)) : x \in V\} \subset V \times W.$$

(a) Show that G_L is a subspace.

(b) Show that the map $V \to G_L$ that sends x to $(x, L(x))$ is an isomorphism.

(c) Show that L is one-to-one if and only if the projection $P_W : V \times W \to W$ is one-to-one when restricted to G_L.

(d) Show that L is onto if and only if the projection $P_W : V \times W \to W$ is onto when restricted to G_L.

(e) Show that a subspace $N \subset V \times W$ is the graph of a linear map $K : V \to W$ if and only if the projection $P_V : V \times W \to V$ is an isomorphism when restricted to N.

(f) Show that a subspace $N \subset V \times W$ is the graph of a linear map $K : V \to W$ if and only if $V \times W = N \oplus (\{0\} \times W)$.

1.11   Linear Maps and Subspaces

Linear maps generate a lot of interesting subspaces and can also be used to understand certain important aspects of subspaces. Conversely, the subspaces associated to a linear map give us crucial information as to whether the map is one-to-one or onto.

Definition 1.11.1. Let $L : V \to W$ be a linear map between vector spaces. The *kernel* or *nullspace* of L is

$$\ker (L) = \mathrm{N}\,(L) = L^{-1}\,(0) = \{x \in V : L\,(x) = 0\}.$$

The *image* or *range* of L is

$$\mathrm{im}\,(L) = \mathrm{R}\,(L) = L\,(V) = \{y \in W : y = L\,(x) \text{ for some } x \in V\}.$$

Both of these spaces are subspaces.

Lemma 1.11.2. $\ker (L)$ *is a subspace of* V *and* $\mathrm{im}\,(L)$ *is a subspace of* W.

Proof. Assume that $\alpha_1, \alpha_2 \in \mathbb{F}$ and that $x_1, x_2 \in \ker (L)$, then

$$L\,(\alpha_1 x_1 + \alpha_2 x_2) = \alpha_1 L\,(x_1) + \alpha_2 L\,(x_2) = 0.$$

More generally, if we only assume $x_1, x_2 \in V$, then we have

$$\alpha_1 L\,(x_1) + \alpha_2 L\,(x_2) = L\,(\alpha_1 x_1 + \alpha_2 x_2) \in \mathrm{im}\,(L).$$

This proves the claim. $\square$

The same proof shows that $L\,(M) = \{L\,(x) : x \in M\}$ is a subspace of W when M is a subspace of V.

Lemma 1.11.3. L *is one-to-one if and only if* $\ker (L) = \{0\}$.

Proof. We know that $L\,(0 \cdot 0) = 0 \cdot L\,(0) = 0$, so if L is one-to-one, we have that $L\,(x) = 0 = L\,(0)$ implies that $x = 0$. Hence, $\ker (L) = \{0\}$.

Conversely, assume that $\ker (L) = \{0\}$. If $L\,(x_1) = L\,(x_2)$, then linearity of L tells us that $L\,(x_1 - x_2) = 0$. Then, $\ker (L) = \{0\}$ implies $x_1 - x_2 = 0$, which shows that $x_1 = x_2$. $\square$

If we have a direct sum decomposition $V = M \oplus N$, then we can construct what is called the *projection* of V onto M along N.

Definition 1.11.4. The map $E : V \to V$ is defined as follows. For $z \in V$, we write $z = x + y$ for unique $x \in M$, $y \in N$ and define

$$E\,(z) = x.$$

Thus, $\mathrm{im}\,(E) = M$ and $\ker (E) = N$.

Note that

$$(1_V - E)(z) = z - x = y.$$

This means that $1_V - E$ is the projection of V onto N along M. So the decomposition $V = M \oplus N$ gives us similar resolution of 1_V using these two projections: $1_V = E + (1_V - E)$.

Using all of the examples of direct sum decompositions, we get several examples of projections. Note that each projection E onto M leads in a natural way to a linear map $P : V \to M$. This map has the same definition $P(z) = P(x + y) = x$, but it is not E as it is not defined as an operator $V \to V$. It is perhaps pedantic to insist on having different names but note that as it stands we are not allowed to compose P with itself as it does not map into V.

We are now ready to establish several extremely important results relating linear maps, subspaces, and dimensions.

Recall that complements to a fixed subspace are usually not unique; however, they do have the same dimension as the next result shows.

Lemma 1.11.5. (Uniqueness of Complements) *If* $V = M_1 \oplus N = M_2 \oplus N$, *then* M_1 *and* M_2 *are isomorphic.*

Proof. Let $P : V \to M_2$ be the projection whose kernel is N. We contend that the map $P|_{M_1} : M_1 \to M_2$ is an isomorphism. The kernel can be computed as

$$\ker(P|_{M_1}) = \{x \in M_1 : P(x) = 0\}$$
$$= \{x \in V : P(x) = 0\} \cap M_1$$
$$= N \cap M_1$$
$$= \{0\}.$$

Thus, $P|_{M_1}$ is one-to-one by Lemma 1.11.3. To check that the map is onto select $x_2 \in M_2$. Next, write $x_2 = x_1 + y_1$, where $x_1 \in M_1$ and $y_1 \in N$. Then,

$$x_2 = P(x_2)$$
$$= P(x_1 + y_1)$$
$$= P(x_1) + P(y_1)$$
$$= P(x_1)$$
$$= P|_{M_1}(x_1).$$

This establishes the claim. □

Theorem 1.11.6. (The Subspace Theorem) *Assume that* V *is finite-dimensional and that* $M \subset V$ *is a subspace. Then,* M *is finite-dimensional and*

$$\dim_{\mathbb{F}} M \leq \dim_{\mathbb{F}} V.$$

Moreover, if $V = M \oplus N$, then

$$\dim_{\mathbb{F}} V = \dim_{\mathbb{F}} M + \dim_{\mathbb{F}} N.$$

Proof. If $M = V$, we are finished. Otherwise, select a basis $x_1, \ldots, x_m$ for V and use Theorem 1.10.20 to extract a complement to M in V

$$V = M \oplus \mathrm{span}\,\{x_{i_1}, \ldots, x_{i_k}\}.$$

On the other hand, we also know that

$$V = \mathrm{span}\,\{x_{j_1}, \ldots, x_{j_l}\} \oplus \mathrm{span}\,\{x_{i_1}, \ldots, x_{i_k}\},$$

where $k + l = m$ and

$$\{1, \ldots, m\} = \{j_1, \ldots, j_l\} \cup \{i_1, \ldots, i_k\}\,.v$$

Lemma 1.11.5 then shows that M and $\mathrm{span}\,\{x_{j_1}, \ldots, x_{j_l}\}$ are isomorphic. Thus,

$$\dim_{\mathbb{F}} M = l < m.$$

In addition, we see that if $V = M \oplus N$, then Lemma 1.11.5 also shows that

$$\dim_{\mathbb{F}} N = k.$$

This proves the theorem.   □

Theorem 1.11.7. (The Dimension Formula) *Let V be finite-dimensional and $L : V \to W$ a linear map, then $\mathrm{im}\,(L)$ is finite-dimensional and*

$$\dim_{\mathbb{F}} V = \dim_{\mathbb{F}} \ker (L) + \dim_{\mathbb{F}} \mathrm{im}\,(L)\,.$$

Proof. We know that $\dim_{\mathbb{F}} \ker (L) \le \dim_{\mathbb{F}} V$ and that it has a complement $N \subset V$ of dimension $k = \dim_{\mathbb{F}} V - \dim_{\mathbb{F}} \ker (L)$. Since $N \cap \ker (L) = \{0\}$, the linear map L must be one-to-one when restricted to N. Thus, $L|_N : N \to \mathrm{im}\,(L)$ is an isomorphism. This proves the theorem.   □

Definition 1.11.8. The number nullity $(L) = \dim_{\mathbb{F}} \ker (L)$ is called the *nullity* of L, and rank $(L) = \dim_{\mathbb{F}} \mathrm{im}\,(L)$ is known as the *rank* of L.

Corollary 1.11.9. *If M is a subspace of V and $\dim_{\mathbb{F}} M = \dim_{\mathbb{F}} V = n < \infty$, then $M = V$.*

Proof. If $M \neq V$, there must be a complement of dimension > 0. This gives us a contradiction with the subspace theorem.   □

Corollary 1.11.10. *Assume that $L : V \to W$ and $\dim_{\mathbb{F}} V = \dim_{\mathbb{F}} W < \infty$. Then, L is an isomorphism if either nullity $(L) = 0$ or rank $(L) = \dim W$.*

Proof. The dimension theorem shows that if either nullity$(L) = 0$ or rank $(L) =$ dim W, then also rank $(L) =$ dim V or nullity$(L) = 0$, thus showing that L is an isomorphism. □

Knowing that the vector spaces are abstractly isomorphic can therefore help us in checking when a given linear map might be an isomorphism.

Many of these results are not true in infinite-dimensional spaces. The differentiation operator $D : C^\infty (\mathbb{R}, \mathbb{R}) \to C^\infty (\mathbb{R}, \mathbb{R})$ is onto and has a kernel consisting of all constant functions. The multiplication operator $T : C^\infty (\mathbb{R}, \mathbb{R}) \to C^\infty (\mathbb{R}, \mathbb{R})$ on the other hand is one-to-one but is not onto as $T (f) (0) = 0$ for all $f \in C^\infty (\mathbb{R}, \mathbb{R})$.

Corollary 1.11.11. *If $L : V \to W$ is a linear map between finite-dimensional spaces, then we can find bases $e_1, \dots, e_m$ for V and $f_1, \dots, f_n$ for W so that*

$$L (e_1) = f_1,$$

$$\vdots \quad \vdots$$

$$L (e_k) = f_k,$$
$$L (e_{k+1}) = 0,$$

$$\vdots \quad \vdots$$

$$L (e_m) = 0,$$

where $k =$ rank (L).

Proof. Simply decompose $V = \ker (L) \oplus M$. Then, choose a basis $e_1, \dots, e_k$ for M and a basis $e_{k+1}, \dots, e_m$ for $\ker (L)$. Combining these two bases gives us a basis for V. Then, define $f_1 = L (e_1), \dots, f_k = L (e_k)$. Since $L|_M : M \to$ im (L) is an isomorphism, this implies that $f_1, \dots, f_k$ form a basis for im (L). We then get the desired basis for W by letting $f_{k+1}, \dots, f_n$ be a basis for a complement to im (L) in W. □

While this certainly gives the nicest possible matrix representation for L, it is not very useful. The complete freedom one has in the choice of both bases somehow also means that aside from the rank, no other information is encoded in the matrix. The real goal will be to find the best matrix for a linear operator $L : V \to V$ with respect to one basis. In the general situation $L : V \to W$, we will have something more to say in case V and W are inner product spaces and the bases are orthonormal (see Sects. 4.8, 4.9, and 4.10).

Finally, it is worth mentioning that projections as a class of linear operators on V can be characterized in a surprisingly simple manner.

Theorem 1.11.12. (Characterization of Projections) *Projections all satisfy the functional relationship $E^2 = E$. Conversely, any $E : V \to V$ that satisfies $E^2 = E$ is a projection.*

Proof. First assume that E is the projection onto M along N coming from $V = M \oplus N$. If $z = x + y \in M \oplus N$, then

$$E^2 (z) = E (E (z))$$
$$= E (x)$$
$$= x$$
$$= E (z).$$

Conversely, assume that $E^2 = E$, then $E (x) = x$ provided $x \in \text{im} (E)$. Thus, we have

$$\text{im} (E) \cap \text{ker} (E) = \{0\}, \text{ and}$$
$$\text{im} (E) + \text{ker} (E) = \text{im} (E) \oplus \text{ker} (E)$$

From the dimension theorem, we also have that

$$\dim (\text{im} (E)) + \dim (\text{ker} (E)) = \dim (V).$$

This shows that $\text{im} (E) + \text{ker} (E)$ is a subspace of dimension $\dim (V)$ and hence all of V. Finally, if we write $z = x + y$, $x \in \text{im} (E)$ and $y \in \text{ker} (E)$, then $E (x + y) = E (x) = x$, so E is the projection onto $\text{im} (E)$ along $\text{ker} (E)$. □

In this way, we have shown that there is a natural identification between direct sum decompositions and projections, i.e., maps satisfying $E^2 = E$.

Exercises

1. Let $L, K : V \to V$ be linear maps that satisfy $L \circ K = 1_V$. Show that

   (a) If V is finite-dimensional, then $K \circ L = 1_V$.
   (b) If V is infinite-dimensional give an example where $K \circ L \neq 1_V$.

2. Let $M \subset V$ be a k-dimensional subspace of an n-dimensional vector space. Show that any isomorphism $L : M \to \mathbb{F}^k$ can be extended to an isomorphism $\hat{L} : V \to \mathbb{F}^n$, such that $\hat{L}|_M = L$. Here we have identified $\mathbb{F}^k$ with the subspace in $\mathbb{F}^n$ where the last $n - k$ coordinates are zero.

3. Let $L : V \to V$ be a linear operator on a vector space over $\mathbb{F}$. Show that

   (a) If $K : V \to V$ commutes with L, i.e., $K \circ L = L \circ K$, then $\text{ker} K \subset V$ is an L-invariant subspace.
   (b) If $p \in \mathbb{F} [t]$, then $\text{ker} p (L) \subset V$ is an L-invariant subspace.

4. Let $L : V \to W$ be a linear map.

   (a) If L has rank k, show that it can be factored through $\mathbb{F}^k$, i.e., we can find
       $K_1 : V \to \mathbb{F}^k$ and $K_2 : \mathbb{F}^k \to W$ such that $L = K_2 K_1$.
   (b) Show that any matrix $A \in \mathrm{Mat}_{n \times m}(\mathbb{F})$ of rank k can be factored $A = BC$,
       where $B \in \mathrm{Mat}_{n \times k}(\mathbb{F})$ and $C \in \mathrm{Mat}_{k \times m}(\mathbb{F})$.
   (c) Conclude that any rank 1 matrix $A \in \mathrm{Mat}_{n \times m}(\mathbb{F})$ looks like

   $$A = \begin{bmatrix} \alpha_1 \\ \vdots \\ \alpha_n \end{bmatrix} \begin{bmatrix} \beta_1 & \cdots & \beta_m \end{bmatrix}.$$

5. Assume $L_1 : V_1 \to V_2$ and $L_2 : V_2 \to V_3$ are linear maps between finite-dimensional vector spaces. Show:

   (a) $\mathrm{im}\,(L_2 \circ L_1) \subset \mathrm{im}\,(L_2)$. In particular, if $L_2 \circ L_1$ is onto, then so is L_2.
   (b) $\ker(L_1) \subset \ker(L_2 \circ L_1)$. In particular, if $L_2 \circ L_1$ is one-to-one, then so is L_1.
   (c) Give an example where $L_2 \circ L_1$ is an isomorphism but L_1 and L_2 are not.
   (d) What happens in (c) if we assume that the vector spaces all have the same dimension?
   (e) (Sylvester's rank inequality) Show that

   $$\mathrm{rank}\,(L_1) + \mathrm{rank}\,(L_2) - \dim\,(V_2) \le \mathrm{rank}\,(L_2 \circ L_1)$$
   $$\le \min\,\{\mathrm{rank}\,(L_1), \mathrm{rank}\,(L_2)\}.$$

   (e) Show that

   $$\dim\,(\ker L_2 \circ L_1) \le \dim\,(\ker L_1) + \dim\,(\ker L_2).$$

6. Let $L : V \to V$ be a linear operator on a finite-dimensional vector space.

   (a) Show that $L = \lambda 1_V$ if and only if $L(x) \in \mathrm{span}\,\{x\}$ for all $x \in V$.
   (b) Show that $L = \lambda 1_V$ if and only if $L \circ K = K \circ L$ for all $K \in \mathrm{Hom}\,(V, V)$.
   (c) Show that $L = \lambda 1_V$ if and only if $L \circ K = K \circ L$ for all isomorphisms $K : V \to V$.

7. Show that two 2-dimensional subspaces of a 3-dimensional vector space must have a nontrivial intersection.

8. (Dimension formula for subspaces) Let $M_1, M_2 \subset V$ be subspaces of a finite-dimensional vector space. Show that

   $$\dim\,(M_1 \cap M_2) + \dim\,(M_1 + M_2) = \dim\,(M_1) + \dim\,(M_2).$$

Conclude that if M_1 and M_2 are transverse, then $M_1 \cap M_2$ has the "expected" dimension $(\dim(M_1) + \dim(M_2)) - \dim V$. Hint: Use the dimension formula on the linear map $L : M_1 \times M_2 \to V$ defined by $L(x_1, x_2) = x_1 - x_2$. Alternatively, select a suitable basis for $M_1 + M_2$ by starting with a basis for $M_1 \cap M_2$.

9. Let $M \subset V$ be a subspace and V, W finite-dimensional vector spaces. Show that the subset of $\operatorname{Hom}_{\mathbb{F}}(V, W)$ consisting of maps that vanish on M, i.e., $L|_M = 0$, is a subspace of dimension $\dim_{\mathbb{F}} W \cdot (\dim_{\mathbb{F}} V - \dim_{\mathbb{F}} M)$.

10. We say that a linear map $L : V \to V$ is *reduced* by a direct sum decomposition $V = M \oplus N$ if both M and N are invariant under L and neither subspace is a trivial subspace. We also say that $L : V \to V$ is *decomposable* if we can find a nontrivial decomposition that reduces $L : V \to V$.

   (a) Show that for $L = \begin{bmatrix} 0 & 1 \\ 0 & 0 \end{bmatrix}$ with $M = \ker(L) = \operatorname{im}(L)$, it is not possible to find N such that $V = M \oplus N$ reduces L.
   (b) Show more generally that one cannot find a nontrivial decomposition that reduces L.

11. Let $L : V \to V$ be a linear transformation and $M \subset V$ a subspace. Show:

   (a) If E is a projection onto M and $ELE = LE$, then M is invariant under L.
   (b) If M is invariant under L, then $ELE = LE$ for all projections onto M.
   (c) If $V = M \oplus N$ and E is the projection onto M along N, then $M \oplus N$ reduces (see previous exercise) L if and only if $EL = LE$.

12. Assume $V = M \oplus N$.

   (a) Show that any linear map $L : V \to V$ has a 2×2 matrix type decomposition

   $$\begin{bmatrix} A & B \\ C & D \end{bmatrix},$$

   where $A : M \to M$, $B : N \to M$, $C : M \to N$, $D : N \to N$.
   (b) Show that the projection onto M along N looks like

   $$E = 1_M \oplus 0_N = \begin{bmatrix} 1_M & 0 \\ 0 & 0_N \end{bmatrix}$$

   (c) Show that if $L(M) \subset M$, then $C = 0$.
   (d) Show that if $L(M) \subset M$ and $L(N) \subset N$, then $B = 0$ and $C = 0$. In this case, L is reduced by $M \oplus N$, and we write

   $$L = A \oplus D$$

   $$= L|_M \oplus L|_N.$$

13. Let $M_1, M_2 \subset V$ be subspaces of a finite-dimensional vector space. Show that

    (a) If $M_1 \cap M_2 = \{0\}$ and $\dim(M_1) + \dim(M_2) \geq \dim V$, then $V = M_1 \oplus M_2$.
    (b) If $M_1 + M_2 = V$ and $\dim(M_1) + \dim(M_2) \leq \dim V$, then $V = M_1 \oplus M_2$.

14. Let $A \in \text{Mat}_{n \times l}(\mathbb{F})$ and consider $L_A : \text{Mat}_{l \times m}(\mathbb{F}) \to \text{Mat}_{n \times m}(\mathbb{F})$ defined by $L_A(X) = AX$. Find the kernel and image of this map.

15. Let

$$0 \xrightarrow{L_0} V_1 \xrightarrow{L_1} V_2 \xrightarrow{L_2} \cdots \xrightarrow{L_{n-1}} V_n \xrightarrow{L_n} 0$$

be a sequence of linear maps. Note that L_0 and L_n are both the trivial linear maps with image $\{0\}$. Show that

$$\sum_{i=1}^{n} (-1)^i \dim V_i = \sum_{i=1}^{n} (-1)^i \left(\dim(\ker(L_i)) - \dim(\text{im}(L_{i-1})) \right).$$

Hint: First, try the case where $n = 2$.

16. Show that the matrix

$$\begin{bmatrix} 0 & 1 \\ 0 & 0 \end{bmatrix}$$

as a linear map satisfies $\ker(L) = \text{im}(L)$.

17. Show that

$$\begin{bmatrix} 0 & 0 \\ \alpha & 1 \end{bmatrix}$$

defines a projection for all $\alpha \in \mathbb{F}$. Compute the kernel and image.

18. For any integer $n > 1$, give examples of linear maps $L : \mathbb{C}^n \to \mathbb{C}^n$ such that

    (a) $\mathbb{C}^n = \ker(L) \oplus \text{im}(L)$ is a nontrivial direct sum decomposition.
    (b) $\{0\} \neq \ker(L) \subset \text{im}(L)$.

19. For $P_n \subset \mathbb{R}[t]$ and $2(n+1)$ points $a_0 < b_0 < a_1 < b_1 < \cdots < a_n < b_n$, consider the map $L : P_n \to \mathbb{R}^{n+1}$ defined by

$$L(p) = \begin{bmatrix} \frac{1}{b_0 - a_0} \int_{a_0}^{b_0} p(t)\, dt \\ \vdots \\ \frac{1}{b_n - a_n} \int_{a_n}^{b_n} p(t)\, dt \end{bmatrix}.$$

Show that L is a linear isomorphism.

1.12   Linear Independence

In this section, we finally come around to studying the concepts of linear dependence and independence as well as how they tie in with kernels and images of linear maps.

Definition 1.12.1. Let $x_1, \ldots, x_m$ be vectors in a vector space V. We say that $x_1, \ldots, x_m$ are *linearly independent* if

$$x_1 \alpha_1 + \cdots + x_m \alpha_m = 0$$

implies that

$$\alpha_1 = \cdots = \alpha_m = 0.$$

In other words, if $L : \mathbb{F}^m \to V$ is the linear map defined by $L = \begin{bmatrix} x_1 \cdots x_m \end{bmatrix}$, then $x_1, \ldots, x_m$ are linearly independent if and only if $\ker(L) = \{0\}$.

The image of the map L can be identified with $\operatorname{span}\{x_1, \ldots, x_m\}$ and is described as

$$\{x_1 \alpha_1 + \cdots + x_m \alpha_m : \alpha_1, \ldots, \alpha_m \in \mathbb{F}\}.$$

Note that $x_1, \ldots, x_m$ is a basis precisely when $\ker(L) = \{0\}$ and $V = \operatorname{im}L$. The notions of kernel and image therefore enter our investigations of dimension in a very natural way.

Definition 1.12.2. Conversely, we say that $x_1, \ldots, x_m$ are *linearly dependent* if they are not linearly independent, i.e., we can find $\alpha_1, \ldots, \alpha_m \in \mathbb{F}$ not all zero so that $x_1 \alpha_1 + \cdots + x_m \alpha_m = 0$.

In the next section, we shall see how Gauss elimination helps us decide when a selection of vectors in $\mathbb{F}^n$ is linearly dependent or independent.

   We give here a characterization of linear dependence that is quite useful in both concrete and abstract situations.

Lemma 1.12.3. (Characterization of Linear Dependence) *Let $x_1, \ldots, x_n \in V$. Then, $x_1, \ldots, x_n$ are linearly dependent if and only if either $x_1 = 0$ or we can find a smallest $k \geq 2$ such that x_k is a linear combination of $x_1, \ldots, x_{k-1}$.*

Proof. First observe that if $x_1 = 0$, then $1 x_1 = 0$ is a nontrivial linear combination. Next, if

$$x_k = \alpha_1 x_1 + \cdots + \alpha_{k-1} x_{k-1},$$

then we also have a nontrivial linear combination

$$\alpha_1 x_1 + \cdots + \alpha_{k-1} x_{k-1} + (-1) x_k = 0.$$

   Conversely, assume that $x_1, \ldots, x_n$ are linearly dependent. Select a nontrivial linear combination such that

$$\alpha_1 x_1 + \cdots + \alpha_n x_n = 0.$$

Then, we can pick k so that $\alpha_k \neq 0$ and $\alpha_{k+1} = \cdots = \alpha_n = 0$. If $k = 1$, then we must have $x_1 = 0$ and we are finished. Otherwise,

$$x_k = -\frac{\alpha_1}{\alpha_k}x_1 - \cdots - \frac{\alpha_{k-1}}{\alpha_k}x_{k-1}.$$

Thus, the set of ks with the property that x_k is a linear combination of $x_1, \ldots, x_{k-1}$ is a nonempty set that contains some integer ≥ 2. Now, simply select the smallest integer in this set to get the desired choice for k. $\square$

This immediately leads us to the following criterion for linear independence.

Corollary 1.12.4. (Characterization of Linear Independence) *Let $x_1, \ldots, x_n \in V$. Then, $x_1, \ldots, x_n$ are linearly independent if and only if $x_1 \neq 0$ and for each $k \geq 2$*

$$x_k \notin \text{span}\{x_1, \ldots, x_{k-1}\}.$$

Example 1.12.5. Let $A \in \text{Mat}_{n \times n}(\mathbb{F})$ be an upper triangular matrix with k nonzero entries on the diagonal. We claim that the rank of A is $\geq k$. Select the k column vectors $x_1, \ldots, x_k$ that correspond to the nonzero diagonal entries from left to right. Thus, $x_1 \neq 0$ and

$$x_l \notin \text{span}\{x_1, \ldots, x_{l-1}\}$$

since x_l has a nonzero entry that lies below all of the nonzero entries for $x_1, \ldots, x_{l-1}$. Using the dimension formula (Theorem 1.11.7), we see that $\dim(\ker(A)) \leq n - k$.

It is possible for A to have rank $> k$. Consider, e.g.,

$$A = \begin{bmatrix} 1 & 0 & 0 \\ 0 & 0 & 1 \\ 0 & 0 & 0 \end{bmatrix}.$$

This matrix has rank 2 but only one nonzero entry on the diagonal.

Recall from Theorem 1.10.20 that we can choose complements to a subspace by selecting appropriate vectors from a set that spans the vector space. The proof of that result actually supplies us with a bit more information.

Corollary 1.12.6. *Let $M \subset V$ be a subspace and assume that $V = \text{span}\{x_1, \ldots, x_n\}$. If $M \neq V$, then it is possible to select linearly independent $x_{i_1}, \ldots, x_{i_k} \in \{x_1, \ldots, x_n\}$ such that*

$$V = M \oplus \text{span}\{x_{i_1}, \ldots, x_{i_k}\}$$

Proof. Recall that $x_{i_1}, \ldots, x_{i_k}$ were selected so that

$$x_{i_1} \notin M,$$
$$x_{i_2} \notin M + \mathrm{span}\{x_{i_1}\},$$
$$\vdots$$
$$x_{i_k} \notin M + \mathrm{span}\{x_{i_1}, \ldots, x_{i_{k-1}}\},$$
$$V = M + \mathrm{span}\{x_{i_1}, \ldots, x_{i_k}\}.$$

In particular, $x_{i_1} \neq 0$ and $x_{i_l} \notin \mathrm{span}\{x_{i_1}, \ldots, x_{i_{l-1}}\}$ for $l = 2, \ldots, k$ so Corollary 1.12.4 proves the claim. □

A more traditional method for establishing that all bases for a vector space have the same number of elements is based on the following classical result, often referred to as the *replacement theorem*.

Theorem 1.12.7. (Steinitz Replacement) *Let $y_1, \ldots, y_m \in V$ be linearly independent and $V = \mathrm{span}\{x_1, \ldots, x_n\}$. Then, $m \leq n$ and V has a basis of the form $y_1, \ldots, y_m, x_{i_1}, \ldots, x_{i_l}$ where $l \leq n - m$.*

Proof. Corollary 1.12.6 immediately gives us linearly independent $x_{i_1}, \ldots, x_{i_l}$ such that $\mathrm{span}\{x_{i_1}, \ldots, x_{i_l}\}$ is a complement to $M = \mathrm{span}\{y_1, \ldots, y_m\}$. Thus, $y_1, \ldots, y_m, x_{i_1}, \ldots, x_{i_l}$ must form a basis for V.

The subspace theorem (Theorem 1.11.6) tells us that $m + l = \dim(V)$. The fact that $n \geq \dim(V)$ is a direct application of Corollary 1.12.6 with $M = \{0\}$.

It is, however, possible to give a more direct argument that does not refer to the concept of dimension. Instead, we use a simple algorithm that shows directly that $l \leq n - m$.

Observe that $y_1, x_1, \ldots, x_n$ are linearly dependent since y_1 is a linear combination of $x_1, \ldots, x_n$. As $y_1 \neq 0$, this shows that some x_i is a linear combination of the previous vectors. Thus, also

$$\mathrm{span}\{y_1, x_1, \ldots, \hat{x}_i, \ldots, x_n\} = V,$$

where $\hat{x}_i$ refers to having deleted x_i. Now, repeat the argument with y_2 in place of y_1 and $y_1, x_1, \ldots, \hat{x}_i, \ldots, x_n$ in place of $x_1, \ldots, x_n$. Thus,

$$y_2, y_1, x_1, \ldots, \hat{x}_i, \ldots, x_n$$

is linearly dependent, and since y_2, y_1 are linearly independent, some x_j is a linear combination of the previous vectors. Continuing in this fashion, we get a set of n vectors

$$y_m, \ldots, y_1, x_{j_1}, \ldots x_{j_{n-m}}$$

that spans V. Finally, we can use Corollary 1.12.6 to eliminate vectors to obtain a basis. Note that the proof (Corollary 1.12.6) shows that the basis will be of the form

$$y_m, \ldots, y_1, x_{i_1}, \ldots x_{i_l}$$

as $y_m, \ldots, y_1$ are linearly independent. This shows that $l \le n - m$.   □

Remark 1.12.8. This theorem leads us to a new proof of the fact that any two bases must contain the same number of elements. It also shows that a linearly independent collection of vectors contains no more vectors than a basis, while a spanning set contains no fewer elements than a basis.

Next, we prove a remarkable theorem for matrices, that we shall revisit many more times in this text.

Definition 1.12.9. For $A = [\alpha_{ij}] \in \text{Mat}_{n \times m}(\mathbb{F})$, define the *transpose* $A^t = [\beta_{ij}] \in \text{Mat}_{m \times n}(\mathbb{F})$ by $\beta_{ij} = \alpha_{ji}$. Thus, the columns of A^t are the rows of A (see also Exercise 14 in Sect. 1.8).

Definition 1.12.10. The *column rank* of a matrix is the dimension of the column space, i.e., the space spanned by the column vectors. In other words, it is the maximal number of linearly independent column vectors. This is also the dimension of the image of the matrix viewed as a linear map. Similarly, the *row rank* is the dimension of the row space, i.e., the space spanned by the row vectors. This is the dimension of the image of the transposed matrix.

Theorem 1.12.11. (The Rank Theorem) *Any $n \times m$ matrix has the property that the row rank is equal to the column rank.*

Proof. Let $A \in \text{Mat}_{n \times m}(\mathbb{F})$ and $x_1, \ldots, x_r \in \mathbb{F}^n$ be a basis for the column space of A. Next, write the columns of A as linear combinations of this basis

$$A = \begin{bmatrix} x_1 & \cdots & x_r \end{bmatrix} \begin{bmatrix} \beta_{11} & \beta_{1m} \\ \beta_{r1} & \beta_{rm} \end{bmatrix}$$

$$= \begin{bmatrix} x_1 & \cdots & x_r \end{bmatrix} B$$

By taking transposes, we obtain

$$A^t = B^t \begin{bmatrix} x_1 & \cdots & x_r \end{bmatrix}^t.$$

But this shows that the columns of A^t, i.e., the rows of A, are linear combinations of the r vectors that form the columns of B^t

$$\begin{bmatrix} \beta_{11} \\ \vdots \\ \beta_{1m} \end{bmatrix}, \ldots, \begin{bmatrix} \beta_{r1} \\ \vdots \\ \beta_{rm} \end{bmatrix}$$

Thus, the row space is spanned by r vectors. This shows that there cannot be more than r linearly independent rows.

A similar argument starting with a basis for the row space of A shows that the reverse inequality also holds.   □

There is a very interesting example associated to the rank theorem.

Example 1.12.12. Let $t_1, \ldots, t_n \in \mathbb{F}$ be distinct. We claim that the vectors

$$\begin{bmatrix} 1 \\ t_1 \\ \vdots \\ t_1^{n-1} \end{bmatrix}, \ldots, \begin{bmatrix} 1 \\ t_n \\ \vdots \\ t_n^{n-1} \end{bmatrix}$$

are a basis for $\mathbb{F}^n$. To show this, we have to show that the rank of the corresponding matrix

$$\begin{bmatrix} 1 & 1 & \cdots & 1 \\ t_1 & t_2 & & t_n \\ \vdots & & & \vdots \\ t_1^{n-1} & t_2^{n-1} & \cdots & t_n^{n-1} \end{bmatrix}$$

is n. The simplest way to do this is by considering the row rank. If the rows are linearly dependent, then we can find $\alpha_0, \ldots, \alpha_{n-1} \in \mathbb{F}$ so that

$$\alpha_0 \begin{bmatrix} 1 \\ 1 \\ \vdots \\ 1 \end{bmatrix} + \alpha_1 \begin{bmatrix} t_1 \\ t_2 \\ \vdots \\ t_n \end{bmatrix} + \cdots + \alpha_{n-1} \begin{bmatrix} t_1^{n-1} \\ t_2^{n-1} \\ \vdots \\ t_n^{n-1} \end{bmatrix} = 0.$$

Thus, the polynomial

$$p(t) = \alpha_0 + \alpha_1 t + \cdots + \alpha_{n-1} t^{n-1}$$

has $t_1, \ldots, t_n$ as roots. In other words we have a polynomial of degree $\leq n - 1$ with n roots. This is not possible unless $\alpha_1 = \cdots = \alpha_{n-1} = 0$ (see also Sect. 2.1).

The criteria for linear dependence lead to an important result about the powers of a linear operator. Before going into that, we observe that there is a connection between polynomials and linear combinations of powers of a linear operator. Let $L : V \to V$ be a linear operator on an n-dimensional vector space. If

$$p(t) = \alpha_k t^k + \cdots + \alpha_1 t + \alpha_0 \in \mathbb{F}[t],$$

then

$$p(L) = \alpha_k L^k + \cdots + \alpha_1 L + \alpha_0 1_V$$

is a linear combination of

$$L^k, \ldots, L, 1_V.$$

Conversely, any linear combination of $L^k, \ldots, L, 1_V$ must look like this.

Since $\text{Hom}\,(V, V)$ has dimension n^2, it follows that $1_V, L, L^2, \ldots, L^{n^2}$ are linearly dependent. This means that we can find a smallest positive integer $k \leq n^2$ such that $1_V, L, L^2, \ldots, L^k$ are linearly dependent. Thus, $1_V, L, L^2, \ldots, L^l$ are linearly independent for $l < k$ and

$$L^k \in \text{span}\left\{1_V, L, L^2, \ldots, L^{k-1}\right\}.$$

In Sect. 2.7, we shall show that $k \leq n$. The fact that

$$L^k \in \text{span}\left\{1_V, L, L^2, \ldots, L^{k-1}\right\}$$

means that we have a polynomial

$$\mu_L(t) = t^k + \alpha_{k-1}t^{k-1} + \cdots + \alpha_1 t + \alpha_0$$

such that

$$\mu_L(L) = 0.$$

This is the so-called *minimal polynomial* for L. Apparently, there is no polynomial of smaller degree that has L as a root. For a more in-depth analysis of the minimal polynomial, see Sect. 2.4.

Recall that we characterized projections as linear operators that satisfy $L^2 = L$ (see Theorem 1.11.12). Thus, nontrivial projections are precisely the operators whose minimal polynomial is $\mu_L(t) = t^2 - t$. Note that the two trivial projections 1_V and 0_V have minimal polynomials $\mu_{1_V} = t - 1$ and $\mu_{0_V} = t$.

Example 1.12.13. Let

$$A = \begin{bmatrix} \lambda & 1 \\ 0 & \lambda \end{bmatrix}$$

$$B = \begin{bmatrix} \lambda & 0 & 0 \\ 0 & \lambda & 1 \\ 0 & 0 & \lambda \end{bmatrix}$$

$$C = \begin{bmatrix} 0 & -1 & 0 \\ 1 & 0 & 0 \\ 0 & 0 & i \end{bmatrix}.$$

We note that A is not proportional to 1_V, so μ_A cannot have degree 1. But

$$A^2 = \begin{bmatrix} \lambda & 1 \\ 0 & \lambda \end{bmatrix}^2$$

$$= \begin{bmatrix} \lambda^2 & 2\lambda \\ 0 & \lambda^2 \end{bmatrix}$$

$$= 2\lambda \begin{bmatrix} \lambda & 1 \\ 0 & \lambda \end{bmatrix} - \lambda^2 \begin{bmatrix} 1 & 0 \\ 0 & 1 \end{bmatrix}.$$

Thus,

$$\mu_A(t) = t^2 - 2\lambda t + \lambda^2 = (t - \lambda)^2.$$

The calculation for B is similar and evidently yields the same minimal polynomial

$$\mu_B(t) = t^2 - 2\lambda t + \lambda^2 = (t - \lambda)^2.$$

Finally, for C, we note that

$$C^2 = \begin{bmatrix} -1 & 0 & 0 \\ 0 & -1 & 0 \\ 0 & 0 & -1 \end{bmatrix}.$$

Thus,

$$\mu_C(t) = t^2 + 1.$$

In the theory of differential equations, it is also important to understand when functions are linearly independent. We start with vector-valued functions $x_1(t), \ldots, x_k(t) : I \to \mathbb{F}^n$, where I is any set but usually an interval. These k functions are linearly independent provided they are linearly independent at just one point $t_0 \in I$. In other words, if the k vectors $x_1(t_0), \ldots, x_k(t_0) \in \mathbb{F}^n$ are linearly independent, then the functions are also linearly independent. The converse statement is, not true in general. To see why this is we give a specific example.

Example 1.12.14. It is an important fact from analysis that there are functions $\phi(t) \in C^\infty(\mathbb{R}, \mathbb{R})$ such that

$$\phi(t) = \begin{cases} 0 & t \le 0, \\ 1 & t \ge 1. \end{cases}$$

These can easily be pictured, but it takes some work to construct them. Given this function, we consider $x_1, x_2 : \mathbb{R} \to \mathbb{R}^2$ defined by

$$x_1(t) = \begin{bmatrix} \phi(t) \\ 0 \end{bmatrix},$$

$$x_2(t) = \begin{bmatrix} 0 \\ \phi(-t) \end{bmatrix}.$$

When $t \leq 0$, we have that $x_1 = 0$ so the two functions are linearly dependent on $(-\infty, 0]$. When $t \geq 0$, we have that $x_2(t) = 0$ so the functions are also linearly dependent on $[0, \infty)$. Now, assume that we can find $\lambda_1, \lambda_2 \in \mathbb{R}$ such that

$$\lambda_1 x_1(t) + \lambda_2 x_2(t) = 0 \text{ for all } t \in \mathbb{R}.$$

If $t \geq 1$, this implies that

$$0 = \lambda_1 x_1(t) + \lambda_2 x_2(t)$$
$$= \lambda_1 \begin{bmatrix} 1 \\ 0 \end{bmatrix} + \lambda_2 \begin{bmatrix} 0 \\ 0 \end{bmatrix}$$
$$= \lambda_1 \begin{bmatrix} 1 \\ 0 \end{bmatrix}.$$

Thus, $\lambda_1 = 0$. Similarly, we have for $t \leq -1$

$$0 = \lambda_1 x_1(t) + \lambda_2 x_2(t)$$
$$= \lambda_1 \begin{bmatrix} 0 \\ 0 \end{bmatrix} + \lambda_2 \begin{bmatrix} 1 \\ 0 \end{bmatrix}$$
$$= \lambda_2 \begin{bmatrix} 1 \\ 0 \end{bmatrix}.$$

So $\lambda_2 = 0$. This shows that the two functions x_1 and x_2 are linearly independent as functions on $\mathbb{R}$ even though the vectors $x_1(t), x_2(t)$ are linearly dependent for each $t \in \mathbb{R}$.

Next, we want to study what happens in the special case where $n = 1$, i.e., we have functions $x_1(t), \ldots, x_k(t) : I \to \mathbb{F}$. In this case, the above strategy for determining linear independence at a point completely fails as the values lie in a one-dimensional vector space. We can, however, construct auxiliary vector-valued functions by taking derivatives. In order to be able to take derivatives, we have to assume either that $I = \mathbb{F}$ and $x_i \in \mathbb{F}[t]$ are polynomials with the formal derivatives defined as in Exercise 2 in Sect. 1.6 or that $I \subset \mathbb{R}$ is an interval, $\mathbb{F} = \mathbb{C}$, and $x_i \in C^\infty(I, \mathbb{C})$. In either case, we can then construct new vector-valued functions $z_1, \ldots, z_k : I \to \mathbb{F}^k$ by listing x_i and its first $k-1$ derivatives in column form

$$z_i(t) = \begin{bmatrix} x_i(t) \\ (Dx_i)(t) \\ \\ (D^{k-1}x_i)(t) \end{bmatrix}.$$

First, we claim that $x_1, \ldots, x_k$ are linearly dependent if and only if $z_1, \ldots, z_k$ are linearly dependent. This is quite simple and depends on the fact that D^n is linear. We only need to observe that

$$
\alpha_1 z_1 + \cdots + \alpha_k z_k = \alpha_1 \begin{bmatrix} x_1 \\ Dx_1 \\ \vdots \\ D^{k-1}x_1 \end{bmatrix} + \cdots + \alpha_k \begin{bmatrix} x_k \\ Dx_k \\ \vdots \\ D^{k-1}x_k \end{bmatrix}
$$

$$
= \begin{bmatrix} \alpha_1 x_1 \\ \alpha_1 Dx_1 \\ \vdots \\ \alpha_1 D^{k-1}x_1 \end{bmatrix} + \cdots + \begin{bmatrix} \alpha_k x_k \\ \alpha_k Dx_k \\ \vdots \\ \alpha_k D^{k-1}x_k \end{bmatrix}
$$

$$
= \begin{bmatrix} \alpha_1 x_1 + \cdots + \alpha_k x_k \\ \alpha_1 Dx_1 + \cdots + \alpha_k Dx_k \\ \vdots \\ \alpha_1 D^{k-1}x_1 + \cdots + \alpha_k D^{k-1}x_k \end{bmatrix}
$$

$$
= \begin{bmatrix} \alpha_1 x_1 + \cdots + \alpha_k x_k \\ D\left(\alpha_1 x_1 + \cdots + \alpha_k x_k\right) \\ \vdots \\ D^{k-1}\left(\alpha_1 x_1 + \cdots + \alpha_k x_k\right) \end{bmatrix}.
$$

Thus, $\alpha_1 z_1 + \cdots + \alpha_k z_k = 0$ if and only if $\alpha_1 x_1 + \cdots + \alpha_k x_k = 0$. This shows the claim. Let us now see how this works in action.

Example 1.12.15. Let $x_i(t) = \exp(\lambda_i t)$, where $\lambda_i \in \mathbb{C}$ are distinct. Then,

$$
z_i(t) = \begin{bmatrix} \exp(\lambda_i t) \\ \lambda_i \exp(\lambda_i t) \\ \vdots \\ \lambda_i^{k-1} \exp(\lambda_i t) \end{bmatrix} = \begin{bmatrix} 1 \\ \lambda_i \\ \vdots \\ \lambda_i^{k-1} \end{bmatrix} \exp(\lambda_i t).
$$

Thus, $\exp(\lambda_1 t), \ldots, \exp(\lambda_k t)$ are linearly independent as we saw in Example 1.12.12 that the vectors

$$
\begin{bmatrix} 1 \\ \lambda_1 \\ \vdots \\ \lambda_1^{k-1} \end{bmatrix}, \ldots, \begin{bmatrix} 1 \\ \lambda_k \\ \vdots \\ \lambda_k^{k-1} \end{bmatrix}
$$

are linearly independent.

Example 1.12.16. Let $x_k(t) = \cos(kt)$, $k = 0, 1, 2, \ldots, n$. In this case, direct check will involve a matrix that has both cosines and sines in alternating rows. Instead, we can use Euler's formula that

$$x_k(t) = \cos(kt) = \frac{1}{2}e^{ikt} - \frac{1}{2}e^{-ikt}.$$

We know from the previous exercise that the $2n + 1$ functions $\exp(ikt)$, $k = 0, \pm 1, \ldots, \pm n$ are linearly independent. Thus, the original $n + 1$ cosine functions are also linearly independent.

Note that if we added the n sine functions $y_k(t) = \sin(kt)$, $k = 1, \ldots, n$, we have $2n + 1$ cosine and sine functions that also become linearly independent.

Exercises

1. *(Characterization of Linear Independence)* Show that, $x_1, \ldots, x_n \in V - \{0\}$ are linearly independent if and only if

$$\mathrm{span}\{x_1, \ldots, \hat{x}_i, \ldots, x_n\} \neq \mathrm{span}\{x_1, \ldots, x_n\}$$

   for all $i = 1, \ldots, n$. Here the "hat" $\hat{x}_i$ over a vector means that it has been deleted from the set.
2. *(Characterization of Linear Independence)* Show that $x_1, \ldots, x_n \in V - \{0\}$ are linearly independent if and only if

$$\mathrm{span}\{x_1, \ldots, x_n\} = \mathrm{span}\{x_1\} \oplus \cdots \oplus \mathrm{span}\{x_n\}.$$

3. Assume that we have nonzero vectors $x_1, \ldots, x_k \in V$ and a direct sum of subspaces
$$M_1 + \cdots + M_k = M_1 \oplus \cdots \oplus M_k.$$
   Show that if $x_i \in M_i$, then $x_1, \ldots, x_k$ are linearly independent.
4. Show that $t^3 + t^2 + 1, t^3 + t^2 + t, t^3 + t + 2$ are linearly independent in P_3. Which of the standard basis vectors $1, t, t^2, t^3$ can be added to this collection to create a basis for P_3?
5. Show that, if $p_0(t), \ldots, p_n(t) \in \mathbb{F}[t]$ all have degree $\leq n$ and all vanish at t_0, then they are linearly dependent.
6. Assume that we have two fields $\mathbb{F} \subset \mathbb{L}$, such as $\mathbb{R} \subset \mathbb{C}$. Show that

   (a) If $x_1, \ldots, x_m$ form a basis for $\mathbb{F}^m$, then they also form a basis for $\mathbb{L}^m$.
   (b) If $x_1, \ldots, x_k$ are linearly independent in $\mathbb{F}^m$, then they are also linearly independent in $\mathbb{L}^m$.
   (c) If $x_1, \ldots, x_k$ are linearly dependent in $\mathbb{F}^m$, then they are also linearly dependent in $\mathbb{L}^m$.

(d) If $x_1, \ldots, x_k \in \mathbb{F}^m$, then

$$\dim_{\mathbb{F}} \mathrm{span}_{\mathbb{F}} \{x_1, \ldots, x_k\} = \dim_{\mathbb{L}} \mathrm{span}_{\mathbb{L}} \{x_1, \ldots, x_k\}.$$

(e) If $M \subset \mathbb{F}^m$ is a subspace, then

$$M = \mathrm{span}_{\mathbb{L}} (M) \cap \mathbb{F}^m.$$

(f) Let $A \in \mathrm{Mat}_{n \times m} (\mathbb{F})$. Then, $A : \mathbb{F}^m \to \mathbb{F}^n$ is one-to-one (resp. onto) if and only if $A : \mathbb{L}^m \to \mathbb{L}^n$ is one-to-one (resp. onto).

7. Show that $\dim_{\mathbb{F}} V \leq n$ if and only if every collection of $n + 1$ vectors is linearly dependent.

8. Let $L : V \to W$ be a linear map.

   (a) Show that if $x_1, \ldots, x_k$ span V and L is not one-to-one, then $L(x_1), \ldots, L(x_k)$ are linearly dependent.
   (b) Show that if $x_1, \ldots, x_k$ are linearly dependent, then $L(x_1), \ldots, L(x_k)$ are linearly dependent.
   (c) Show that if $L(x_1), \ldots, L(x_k)$ are linearly independent, then $x_1, \ldots, x_k$ are linearly independent.

9. Let $A \in \mathrm{Mat}_{n \times m} (\mathbb{F})$ and assume that $y_1, \ldots, y_m \in V$

$$\left[y_1 \cdots y_m \right] = \left[x_1 \cdots x_n \right] A,$$

where $x_1, \ldots, x_n$ form a basis for V.

   (a) Show that $y_1, \ldots, y_m$ span V if and only if A has rank n. Conclude that $m \geq n$.
   (b) Show that $y_1, \ldots, y_m$ are linearly independent if and only if $\ker (A) = \{0\}$. Conclude that $m \leq n$.
   (c) Show that $y_1, \ldots, y_m$ form a basis for V if and only if A is invertible. Conclude that $m = n$.

1.13 Row Reduction

In this section, we give a brief and rigorous outline of the standard procedures involved in solving systems of linear equations. The goal in the context of what we have already learned is to find a way of computing the image and kernel of a linear map that is represented by a matrix. Along the way, we shall reprove that the dimension is well defined as well as the dimension formula for linear maps.

The usual way of writing n equations with m variables is

$$a_{11}x_1 + \cdots + a_{1m}x_m = b_1$$
$$\vdots \qquad\qquad \vdots \quad \vdots$$
$$a_{n1}x_1 + \cdots + a_{nm}x_m = b_n,$$

where the variables are $x_1, \ldots, x_m$. The goal is to understand for which choices of constants a_{ij} and b_i such systems can be solved and then list all the solutions. To conform to our already specified notation, we change the system so that it looks like

$$\alpha_{11}\xi_1 + \cdots + \alpha_{1m}\xi_m = \beta_1$$
$$\vdots \qquad\qquad \vdots \quad \vdots$$
$$\alpha_{n1}\xi_1 + \cdots + \alpha_{nm}\xi_m = \beta_n.$$

In matrix form, this becomes

$$\begin{bmatrix} \alpha_{11} & \cdots & \alpha_{1m} \\ \vdots & \ddots & \vdots \\ \alpha_{n1} & \cdots & \alpha_{nm} \end{bmatrix} \begin{bmatrix} \xi_1 \\ \vdots \\ \xi_m \end{bmatrix} = \begin{bmatrix} \beta_1 \\ \vdots \\ \beta_n \end{bmatrix}$$

and can be abbreviated to

$$Ax = b.$$

As such, we can easily use the more abstract language of linear algebra to address some general points.

Proposition 1.13.1. *Let $L : V \to W$ be a linear map.*

(1) $L(x) = b$ can be solved if and only if $b \in \mathrm{im}\,(L)$.
(2) If $L(x_0) = b$ and $x \in \ker(L)$, then $L(x + x_0) = b$.
(3) If $L(x_0) = b$ and $L(x_1) = b$, then $x_0 - x_1 \in \ker(L)$.

Therefore, we can find all solutions to $L(x) = b$ provided we can find the kernel $\ker(L)$ and just one solution x_0. Note that the kernel consists of the solutions to what we call the *homogeneous system*: $L(x) = 0$.

Definition 1.13.2. With this behind us, we are now ready to address the issue of how to make the necessary calculations that allow us to find a solution to

$$\begin{bmatrix} \alpha_{11} & \cdots & \alpha_{1m} \\ \vdots & \ddots & \vdots \\ \alpha_{n1} & \cdots & \alpha_{nm} \end{bmatrix} \begin{bmatrix} \xi_1 \\ \vdots \\ \xi_m \end{bmatrix} = \begin{bmatrix} \beta_1 \\ \vdots \\ \beta_n \end{bmatrix}.$$

The usual method is through *elementary row operations*. To keep things more conceptual, think of the actual linear equations

$$\alpha_{11}\xi_1 + \cdots + \alpha_{1m}\xi_m = \beta_1$$
$$\vdots \qquad\qquad \vdots \quad \vdots$$
$$\alpha_{n1}\xi_1 + \cdots + \alpha_{nm}\xi_m = \beta_n$$

and observe that we can perform the following three operations without changing the solutions to the equations:

(1) Interchanging equations (or rows).
(2) Adding a multiple of an equation (or row) to a different equation (or row).
(3) Multiplying an equation (or row) by a nonzero number.

Using these operations, one can put the system in *row echelon form*. This is most easily done by considering the *augmented matrix*, where the variables have disappeared

$$\begin{bmatrix} \alpha_{11} & \cdots & \alpha_{1m} & \beta_1 \\ \vdots & \ddots & \vdots & \vdots \\ \alpha_{n1} & \cdots & \alpha_{nm} & \beta_n \end{bmatrix}$$

and then performing the above operations, now on rows, until it takes the special form where

1. The first nonzero entry in each row is normalized to be 1. This is also called the *leading* 1 for the row.
2. The leading 1s appear in echelon form, i.e., as we move down along the rows the leading 1s will appear farther to the right.

The method by which we put a matrix into row echelon form is called *Gauss elimination*. Having put the system into this simple form, one can then solve it by starting from the last row or equation.

When doing the process on A itself, we denote the resulting row echelon matrix by A_{ref}. There are many ways of doing row reductions so as to come up with a row echelon form for A, and it is quite likely that one ends up with different echelon forms. To see why, consider

$$A = \begin{bmatrix} 1 & 1 & 0 \\ 0 & 1 & 1 \\ 0 & 0 & 1 \end{bmatrix}.$$

This matrix is clearly in row echelon form. However, we can subtract the second row from the first row to obtain a new matrix which is still in row echelon form:

$$\begin{bmatrix} 1 & 0 & -1 \\ 0 & 1 & 1 \\ 0 & 0 & 1 \end{bmatrix}.$$

It is now possible to use two more elementary row operations to arrive at

$$
\begin{bmatrix}
1 & 0 & 0 \\
0 & 1 & 0 \\
0 & 0 & 1
\end{bmatrix}.
$$

The important information about A_{ref} is the placement of the leading 1 in each row, and this placement will always be the same for any row echelon form. To get a unique row echelon form, we need to reduce the matrix using *Gauss-Jordan elimination*. This process is what we just performed on the above matrix A to get it into final form. The idea is to first arrive at some row echelon form A_{ref} and then, starting with the second row, eliminate all entries above the leading 1; this is then repeated with row three and so on. In this way, we end up with a matrix that is still in row echelon form, but also has the property that all entries below and above the leading 1 in each row are zero. We say that such a matrix is in *reduced row echelon form*. If we start with a matrix A, then the resulting reduced row echelon form is denoted A_{rref}. For example, if we have

$$
A_{\text{ref}} =
\begin{bmatrix}
0 & 1 & 4 & 1 & 0 & 3 & -1 \\
0 & 0 & 0 & 1 & -2 & 5 & -4 \\
0 & 0 & 0 & 0 & 0 & 0 & 1 \\
0 & 0 & 0 & 0 & 0 & 0 & 0
\end{bmatrix},
$$

then we can reduce further to get a new reduced row echelon form

$$
A_{\text{rref}} =
\begin{bmatrix}
0 & 1 & 4 & 0 & 2 & -2 & 0 \\
0 & 0 & 0 & 1 & -2 & 5 & 0 \\
0 & 0 & 0 & 0 & 0 & 0 & 1 \\
0 & 0 & 0 & 0 & 0 & 0 & 0
\end{bmatrix}.
$$

The row echelon form and reduced row echelon form of a matrix can more abstractly be characterized as follows. Suppose that we have an $n \times m$ matrix $A = \begin{bmatrix} x_1 & \cdots & x_m \end{bmatrix}$, where $x_1, \ldots, x_m \in \mathbb{F}^n$ correspond to the columns of A. Let $e_1, \ldots, e_n \in \mathbb{F}^n$ be the canonical basis. The matrix is in row echelon form if we can find $1 \leq j_1 < \cdots < j_k \leq m$, where $k \leq n$, such that

$$
x_{j_s} = e_s + \sum_{i < s} \alpha_{i j_s} e_i
$$

for $s = 1, \ldots, k$. For all other indices j, we have

$$
x_j = 0, \text{ if } j < j_1,
$$

$$
x_j \in \text{span} \{e_1, \ldots, e_s\}, \text{ if } j_s < j < j_{s+1},
$$

$$
x_j \in \text{span} \{e_1, \ldots, e_k\}, \text{ if } j_k < j.
$$

Moreover, the matrix is in reduced row echelon form if in addition we assume that

$$x_{j_s} = e_s.$$

Below, we shall prove that the reduced row echelon form of a matrix is unique, but before doing so, it is convenient to reinterpret the row operations as matrix multiplication.

Let $A \in \text{Mat}_{n \times m}(\mathbb{F})$ be the matrix we wish to row reduce. The row operations we have described can be accomplished by multiplying A by certain invertible $n \times n$ matrices on the left. These matrices are called *elementary matrices*.

Definition 1.13.3. To define these matrices, we use the standard basis matrices E_{kl} where the kl entry is 1 while all other entries are 0. The matrix product $E_{kl}A$ is a matrix whose kth row is the lth row of A and all other rows vanish.

1. Interchanging rows k and l: This can be accomplished by the matrix multiplication $I_{kl}A$, where

$$I_{kl} = E_{kl} + E_{lk} + \sum_{i \neq k,l} E_{ii}$$
$$= E_{kl} + E_{lk} + 1_{\mathbb{F}^n} - E_{kk} - E_{ll},$$

or in other words, the ij entries α_{ij} in I_{kl} satisfy $\alpha_{kl} = \alpha_{lk} = 1$, $\alpha_{ii} = 1$ if $i \neq k, l$, and $\alpha_{ij} = 0$ otherwise. Note that $I_{kl} = I_{lk}$ and $I_{kl}I_{lk} = 1_{\mathbb{F}^n}$. Thus I_{kl} is invertible.

2. Multiplying row l by $\alpha \in \mathbb{F}$ and adding it to row $k \neq l$. This can be accomplished via $R_{kl}(\alpha) A$, where

$$R_{kl}(\alpha) = 1_{\mathbb{F}^n} + \alpha E_{kl},$$

or in other words, the ij entries α_{ij} in $R_{kl}(\alpha)$ look like $\alpha_{ii} = 1$, $\alpha_{kl} = \alpha$, and $\alpha_{ij} = 0$ otherwise. This time, we note that $R_{kl}(\alpha) R_{kl}(-\alpha) = 1_{\mathbb{F}^n}$.

3. Multiplying row k by $\alpha \in \mathbb{F} - \{0\}$. This can be accomplished by $M_k(\alpha) A$, where

$$M_k(\alpha) = \alpha E_{kk} + \sum_{i \neq k} E_{ii}$$
$$= 1_{\mathbb{F}^n} + (\alpha - 1) E_{kk},$$

or in other words, the ij entries α_{ij} of $M_k(\alpha)$ are $\alpha_{kk} = \alpha$, $\alpha_{ii} = 1$ if $i \neq k$, and $\alpha_{ij} = 0$ otherwise. Clearly, $M_k(\alpha) M_k(\alpha^{-1}) = 1_{\mathbb{F}^n}$.

Performing row reductions on A is now the same as doing a matrix multiplication PA, where $P \in \text{Mat}_{n \times n}(\mathbb{F})$ is a product of the elementary matrices. Note that such P are invertible and that P^{-1} is also a product of elementary matrices. The elementary 2×2 matrices look like.

$$I_{12} = \begin{bmatrix} 0 & 1 \\ 1 & 0 \end{bmatrix},$$

$$R_{12}(\alpha) = \begin{bmatrix} 1 & \alpha \\ 0 & 1 \end{bmatrix},$$

$$R_{21}(\alpha) = \begin{bmatrix} 1 & 0 \\ \alpha & 1 \end{bmatrix},$$

$$M_1(\alpha) = \begin{bmatrix} \alpha & 0 \\ 0 & 1 \end{bmatrix},$$

$$M_2(\alpha) = \begin{bmatrix} 1 & 0 \\ 0 & \alpha \end{bmatrix}.$$

If we multiply these matrices onto A from the left, we obtain the desired operations:

$$I_{12}A = \begin{bmatrix} 0 & 1 \\ 1 & 0 \end{bmatrix} \begin{bmatrix} \alpha_{11} & \alpha_{12} \\ \alpha_{21} & \alpha_{22} \end{bmatrix} = \begin{bmatrix} \alpha_{21} & \alpha_{22} \\ \alpha_{11} & \alpha_{12} \end{bmatrix}$$

$$R_{12}(\alpha)A = \begin{bmatrix} 1 & \alpha \\ 0 & 1 \end{bmatrix} \begin{bmatrix} \alpha_{11} & \alpha_{12} \\ \alpha_{21} & \alpha_{22} \end{bmatrix} = \begin{bmatrix} \alpha_{11} + \alpha\alpha_{21} & \alpha_{12} + \alpha\alpha_{22} \\ \alpha_{21} & \alpha_{22} \end{bmatrix}$$

$$R_{21}(\alpha)A = \begin{bmatrix} 1 & 0 \\ \alpha & 1 \end{bmatrix} \begin{bmatrix} \alpha_{11} & \alpha_{12} \\ \alpha_{21} & \alpha_{22} \end{bmatrix} = \begin{bmatrix} \alpha_{11} & \alpha_{12} \\ \alpha\alpha_{11} + \alpha_{21} & \alpha\alpha_{12} + \alpha_{22} \end{bmatrix}$$

$$M_1(\alpha)A = \begin{bmatrix} \alpha & 0 \\ 0 & 1 \end{bmatrix} \begin{bmatrix} \alpha_{11} & \alpha_{12} \\ \alpha_{21} & \alpha_{22} \end{bmatrix} = \begin{bmatrix} \alpha\alpha_{11} & \alpha\alpha_{12} \\ \alpha_{21} & \alpha_{22} \end{bmatrix}$$

$$M_2(\alpha)A = \begin{bmatrix} 1 & 0 \\ 0 & \alpha \end{bmatrix} \begin{bmatrix} \alpha_{11} & \alpha_{12} \\ \alpha_{21} & \alpha_{22} \end{bmatrix} = \begin{bmatrix} \alpha_{11} & \alpha_{12} \\ \alpha\alpha_{21} & \alpha\alpha_{22} \end{bmatrix}.$$

We can now move on to the important result mentioned above.

Theorem 1.13.4. (Uniqueness of Reduced Row Echelon Form) *The reduced row echelon form of an $n \times m$ matrix is unique.*

Proof. Let $A \in \mathrm{Mat}_{n \times m}(\mathbb{F})$ and assume that we have two reduced row echelon forms

$$PA = \begin{bmatrix} x_1 & \cdots & x_m \end{bmatrix},$$

$$QA = \begin{bmatrix} y_1 & \cdots & y_m \end{bmatrix},$$

where $P, Q \in \mathrm{Mat}_{n \times n}(\mathbb{F})$ are invertible. In particular, we have that

$$R \begin{bmatrix} x_1 & \cdots & x_m \end{bmatrix} = \begin{bmatrix} y_1 & \cdots & y_m \end{bmatrix},$$

where $R \in \mathrm{Mat}_{n \times n}(\mathbb{F})$ is invertible. We shall show that $x_i = y_i$, $i = 1, \ldots, m$ by induction on n.

First, observe that if $A = 0$, then there is nothing to prove. If $A \neq 0$, then both of the reduced row echelon forms have to be nontrivial. Then, we have that

$$x_{i_1} = e_1,$$
$$x_i = 0 \quad \text{for } i < i_1$$

and

$$y_{j_1} = e_1,$$
$$y_i = 0 \quad \text{for } i < j_1.$$

The relationship $Rx_i = y_i$ shows that $y_i = 0$ if $x_i = 0$. Thus, $j_1 \geq i_1$. Similarly, the relationship $y_i = R^{-1}x_i$ shows that $x_i = 0$ if $y_i = 0$. Hence, also $j_1 \leq i_1$. Thus, $i_1 = j_1$ and $x_{i_1} = e_1 = y_{j_1}$. This implies that $Re_1 = e_1$ and $R^{-1}e_1 = e_1$. In other words,

$$R = \begin{bmatrix} 1 & 0 \\ 0 & R' \end{bmatrix},$$

where $R' \in \mathrm{Mat}_{(n-1)\times(n-1)}(\mathbb{F})$ is invertible.

In the special case where $n = 1$, we are finished as we have shown that $R = [1]$. This anchors our induction. We can now assume the induction hypothesis: All $(n-1) \times m$ matrices have unique reduced row echelon forms.

If we define $x_i', y_i' \in \mathbb{F}^{n-1}$ as the last $n-1$ entries in x_i and y_i, i.e.,

$$x_i = \begin{bmatrix} \xi_{1i} \\ x_i' \end{bmatrix},$$

$$y_i = \begin{bmatrix} \upsilon_{1i} \\ y_i' \end{bmatrix},$$

then we see that $\begin{bmatrix} x_1' & \cdots & x_m' \end{bmatrix}$ and $\begin{bmatrix} y_1' & \cdots & y_m' \end{bmatrix}$ are still in reduced row echelon form. Moreover, the relationship

$$\begin{bmatrix} y_1 & \cdots & y_m \end{bmatrix} = R \begin{bmatrix} x_1 & \cdots & x_m \end{bmatrix}$$

now implies that

$$\begin{bmatrix} \upsilon_{11} & \cdots & \upsilon_{1m} \\ y_1' & \cdots & y_m' \end{bmatrix} = \begin{bmatrix} y_1 & \cdots & y_m \end{bmatrix}$$

$$= R \begin{bmatrix} x_1 & \cdots & x_m \end{bmatrix}$$

$$= \begin{bmatrix} 1 & 0 \\ 0 & R' \end{bmatrix} \begin{bmatrix} \xi_{11} & \cdots & \xi_{1m} \\ x_1' & \cdots & x_m' \end{bmatrix}$$

$$= \begin{bmatrix} \xi_{11} & \cdots & \xi_{1m} \\ R'x_1' & \cdots & R'x_m' \end{bmatrix}$$

Thus,

$$R' \begin{bmatrix} x_1' & \cdots & x_m' \end{bmatrix} = \begin{bmatrix} y_1' & \cdots & y_m' \end{bmatrix}.$$

The induction hypothesis now implies that $x_i' = y_i'$. This combined with

$$\begin{bmatrix} y_1 & \cdots & y_m \end{bmatrix} = \begin{bmatrix} \upsilon_{11} & \cdots & \upsilon_{1m} \\ y_1' & \cdots & y_m' \end{bmatrix}$$

$$= \begin{bmatrix} \xi_{11} & \cdots & \xi_{1m} \\ R'x_1' & \cdots & R'x_m' \end{bmatrix}$$

$$= \begin{bmatrix} x_1 & \cdots & x_m \end{bmatrix}$$

shows that $x_i = y_i$ for all $i = 1, \ldots, m$. $\square$

We are now ready to explain how the reduced row echelon form can be used to identify the kernel and image of a matrix. Along the way, we shall reprove some of our earlier results. Suppose that $A \in \mathrm{Mat}_{n \times m}(\mathbb{F})$ and

$$PA = A_{\mathrm{rref}}$$

$$= \begin{bmatrix} x_1 & \cdots & x_m \end{bmatrix},$$

where we can find $1 \leq j_1 < \cdots < j_k \leq m$, such that

$$x_{j_s} = e_s \text{ for } i = 1, \ldots, k$$

$$x_j = 0, \text{ if } j < j_1,$$

$$x_j \in \mathrm{span}\{e_1, \ldots, e_s\}, \text{ if } j_s < j < j_{s+1},$$

$$x_j \in \mathrm{span}\{e_1, \ldots, e_k\}, \text{ if } j_k < j.$$

Finally, let $i_1 < \cdots < i_{m-k}$ be the indices complementary to $j_1, .., j_k$, i.e.,

$$\{1, \ldots, m\} = \{j_1, .., j_k\} \cup \{i_1, \ldots, i_{m-k}\}.$$

We are first going to study the kernel of A. Since P is invertible, we see that $Ax = 0$ if and only if $A_{\mathrm{rref}}x = 0$. Thus we only need to study the equation $A_{\mathrm{rref}}x = 0$. If we let $x = (\xi_1, \ldots, \xi_m)$, then the nature of the equations $A_{\mathrm{rref}}x = 0$ will tell us that $(\xi_1, \ldots, \xi_m)$ are uniquely determined by $\xi_{i_1}, \ldots, \xi_{i_{m-k}}$. To see why this is, we note that if we have $A_{\mathrm{rref}} = \begin{bmatrix} \alpha_{ij} \end{bmatrix}$, then the reduced row echelon form tells us that

$$\xi_{j_1} + \alpha_{1i_1}\xi_{i_1} + \cdots + \alpha_{1i_{m-k}}\xi_{i_{m-k}} = 0,$$

$$\vdots \quad \vdots \quad \vdots$$

$$\xi_{j_k} + \alpha_{ki_1}\xi_{i_1} + \cdots + \alpha_{ki_{m-k}}\xi_{i_{m-k}} = 0.$$

Thus, $\xi_{j_1}, \ldots, \xi_{j_k}$ have explicit formulas in terms of $\xi_{i_1}, \ldots, \xi_{i_{m-k}}$. We actually get a bit more information: If we take $(\alpha_1, \ldots, \alpha_{m-k}) \in \mathbb{F}^{m-k}$ and construct the unique solution $x = (\xi_1, \ldots, \xi_m)$ such that $\xi_{i_1} = \alpha_1, \ldots, \xi_{i_{m-k}} = \alpha_{m-k}$, then we have actually constructed a map

$$\mathbb{F}^{m-k} \to \ker(A_{\mathrm{rref}})$$

$$(\alpha_1, \ldots, \alpha_{m-k}) \to (\xi_1, \ldots, \xi_m).$$

We have just seen that this map is onto. The construction also gives us explicit formulas for $\xi_{j_1}, \ldots, \xi_{j_k}$ that are linear in $\xi_{i_1} = \alpha_1, \ldots, \xi_{i_{m-k}} = \alpha_{m-k}$. Thus, the map is linear. Finally, if $(\xi_1, \ldots, \xi_m) = 0$, then we clearly also have $(\alpha_1, \ldots, \alpha_{m-k}) = 0$, so the map is one-to-one. All in all, it is a linear isomorphism.

This leads us to the following result.

Theorem 1.13.5. (Uniqueness of Dimension) *Let* $A \in \mathrm{Mat}_{n \times m}(\mathbb{F})$, *if* $n < m$, *then* $\ker(A) \neq \{0\}$. *Consequently,* $\mathbb{F}^n$ *and* $\mathbb{F}^m$ *are not isomorphic.*

Proof. Using the above notation, we have $k \leq n < m$. Thus, $m - k > 0$. From what we just saw, this implies $\ker(A) = \ker(A_{\mathrm{rref}}) \neq \{0\}$. In particular, it is not possible for A to be invertible. This shows that $\mathbb{F}^n$ and $\mathbb{F}^m$ cannot be isomorphic. $\square$

Having now shown that the dimension of a vector space is well defined, we can then establish the dimension formula. Part of the proof of this theorem is to identify a basis for the image of a matrix. Note that this proof does not depend on the result that subspaces of finite-dimensional vector spaces are finite-dimensional. In fact, for the subspaces under consideration, namely, the kernel and image, it is part of the proof to show that they are finite-dimensional.

Theorem 1.13.6. (The Dimension Formula) *Let* $A \in \mathrm{Mat}_{n \times m}(\mathbb{F})$, *then*

$$m = \dim(\ker(A)) + \dim(\mathrm{im}(A)).$$

Proof. We use the above notation. We just saw that $\dim(\ker(A)) = m - k$, so it remains to check why $\dim(\mathrm{im}(A)) = k$. If

$$A = \begin{bmatrix} y_1 & \cdots & y_m \end{bmatrix},$$

then we have $y_i = P^{-1} x_i$, where

$$A_{\mathrm{rref}} = \begin{bmatrix} x_1 & \cdots & x_m \end{bmatrix}.$$

We know that each

$$x_j \in \mathrm{span}\{e_1, \ldots, e_k\} = \mathrm{span}\{x_{j_1}, \ldots, x_{j_k}\};$$

thus, we have that

$$y_j \in \mathrm{span}\{y_{j_1}, \ldots, y_{j_k}\}.$$

Moreover, as P is invertible, we see that $y_{j_1}, \ldots, y_{j_k}$ must be linearly independent as $e_1, \ldots, e_k$ are linearly independent. This proves that $y_{j_1}, \ldots, y_{j_k}$ form a basis for im (A).   □

Corollary 1.13.7. (Subspace Theorem) *Let $M \subset \mathbb{F}^n$ be a subspace. Then, M is finite-dimensional and* dim $(M) \leq n$.

Proof. Recall from Sect. 1.10 that every subspace $M \subset \mathbb{F}^n$ has a complement. This means that we can construct a projection as in Sect. 1.11 that has M as kernel. This means that M is the kernel for some $A \in \mathrm{Mat}_{n \times n}(\mathbb{F})$. Thus, the previous theorem implies the claim.   □

It might help to see an example of how the above constructions work.

Example 1.13.8. Suppose that we have a 4×7 matrix

$$A = \begin{bmatrix} 0 & 1 & 4 & 1 & 0 & 3 & -1 \\ 0 & 0 & 0 & 1 & -2 & 5 & -4 \\ 0 & 0 & 0 & 0 & 0 & 0 & 1 \\ 0 & 0 & 0 & 0 & 0 & 0 & 1 \end{bmatrix}.$$

Then

$$A_{\mathrm{rref}} = \begin{bmatrix} 0 & 1 & 4 & 0 & 2 & -2 & 0 \\ 0 & 0 & 0 & 1 & -2 & 5 & 0 \\ 0 & 0 & 0 & 0 & 0 & 0 & 1 \\ 0 & 0 & 0 & 0 & 0 & 0 & 0 \end{bmatrix}.$$

Thus, $j_1 = 2$, $j_2 = 4$, and $j_3 = 7$. The complementary indices are $i_1 = 1$, $i_2 = 3$, $i_3 = 5$, and $i_4 = 6$. Hence,

$$\mathrm{im}\,(A) = \mathrm{span} \left\{ \begin{bmatrix} 1 \\ 0 \\ 0 \\ 0 \end{bmatrix}, \begin{bmatrix} 1 \\ 1 \\ 0 \\ 0 \end{bmatrix}, \begin{bmatrix} -1 \\ -4 \\ 1 \\ 1 \end{bmatrix} \right\}$$

and

$$\ker\,(A) = \left\{ \begin{bmatrix} \xi_1 \\ -4\xi_3 - 2\xi_5 + 2\xi_6 \\ \xi_3 \\ 2\xi_5 - 5\xi_6 \\ \xi_5 \\ \xi_6 \\ 0 \end{bmatrix} : \xi_1, \xi_3, \xi_5, \xi_6 \in \mathbb{F} \right\}.$$

Our method for finding a basis for the image of a matrix leads us to a different proof of the rank theorem. The *column rank* of a matrix is simply the dimension of the image, in other words, the maximal number of linearly independent column vectors. Similarly, the *row rank* is the maximal number of linearly independent rows. In other words, the row rank is the dimension of the image of the transposed matrix.

Theorem 1.13.9. (The Rank Theorem) *Any $n \times m$ matrix has the property that the row rank is equal to the column rank.*

Proof. We just saw that the column rank for A and A_{rref} is the same and equal to k with the above notation. Because of the row operations we use, it is clear that the rows of A_{rref} are linear combinations of the rows of A. As the process can be reversed, the rows of A are also linear combinations of the rows A_{rref}. Hence, A and A_{rref} also have the same row rank. Now, A_{rref} has k linearly independent rows and must therefore have row rank k.   □

Using the rank theorem together with the dimension formula leads to an interesting corollary.

Corollary 1.13.10. *Let $A \in \mathrm{Mat}_{n \times n}(\mathbb{F})$. Then,*

$$\dim\left(\ker\left(A\right)\right) = \dim\left(\ker\left(A^t\right)\right),$$

where $A^t \in \mathrm{Mat}_{n \times n}(\mathbb{F})$ is the transpose of A.

We are now going to clarify what type of matrices P occur when we do the row reduction to obtain $PA = A_{\mathrm{rref}}$. If we have an $n \times n$ matrix A with trivial kernel, then it must follow that $A_{\mathrm{rref}} = 1_{\mathbb{F}^n}$. Therefore, if we perform Gauss-Jordan elimination on the augmented matrix

$$A | 1_{\mathbb{F}^n},$$

then we end up with an answer that looks like

$$1_{\mathbb{F}^n} | B.$$

The matrix B evidently satisfies $BA = 1_{\mathbb{F}^n}$. To be sure that this is the inverse we must also check that $AB = 1_{\mathbb{F}^n}$. However, we know that A has an inverse A^{-1}. If we multiply the equation $BA = 1_{\mathbb{F}^n}$ by A^{-1} on the right we obtain $B = A^{-1}$. This settles the uncertainty.

Definition 1.13.11. The space of all invertible $n \times n$ matrices is called the *general linear group* and is denoted by

$$Gl_n(\mathbb{F}) = \left\{ A \in \mathrm{Mat}_{n \times n}(\mathbb{F}) \mid \exists\, A^{-1} \in \mathrm{Mat}_{n \times n}(\mathbb{F}) : AA^{-1} = A^{-1}A = 1_{\mathbb{F}^n} \right\}.$$

This space is a so-called *group*.

Definition 1.13.12. This means that we have a set G and a product operation $G \times G \to G$ denoted by $(g, h) \to gh$. This product operation must satisfy:

1. Associativity: $(g_1 g_2) g_3 = g_1 (g_2 g_3)$.
2. Existence of a unit $e \in G$ such that $eg = ge = g$.
3. Existence of inverses: For each $g \in G$, there is $g^{-1} \in G$ such that $gg^{-1} = g^{-1}g = e$.

If we use matrix multiplication in $Gl_n (\mathbb{F})$ and $1_{\mathbb{F}^n}$ as the unit, then it is clear that $Gl_n (\mathbb{F})$ is a group. Note that we do not assume that the product operation in a group is commutative, and indeed, it is not commutative in $Gl_n (\mathbb{F})$ unless $n = 1$.

Definition 1.13.13. If a possibly infinite subset $S \subset G$ of a group has the property that any element in G can be written as a product of elements in S, then we say that S *generates* G.

We can now prove,

Theorem 1.13.14. *The general linear group $Gl_n (\mathbb{F})$ is generated by the elementary matrices I_{kl}, $R_{kl} (\alpha)$, and $M_k (\alpha)$.*

Proof. We already observed that I_{kl}, $R_{kl} (\alpha)$, and $M_k (\alpha)$ are invertible and hence form a subset in $Gl_n (\mathbb{F})$. Let $A \in Gl_n (\mathbb{F})$, then we know that also $A^{-1} \in Gl_n (\mathbb{F})$. Now, observe that we can find $P \in Gl_n (\mathbb{F})$ as a product of elementary matrices such that $PA^{-1} = 1_{\mathbb{F}^n}$. This was the content of the Gauss-Jordan elimination process for finding the inverse of a matrix. This means that $P = A$ and hence A is a product of elementary matrices.   □

The row echelon representation of a matrix tells us:

Corollary 1.13.15. *Let $A \in \mathrm{Mat}_{n \times n} (\mathbb{F})$, then it is possible to find $P \in Gl_n (\mathbb{F})$ such that PA is upper triangular:*

$$PA = \begin{bmatrix} \beta_{11} & \beta_{12} & \cdots & \beta_{1n} \\ 0 & \beta_{22} & \cdots & \beta_{2n} \\ \vdots & \vdots & \ddots & \vdots \\ 0 & 0 & \cdots & \beta_{nn} \end{bmatrix}$$

Moreover,

$$\ker (A) = \ker (PA)$$

and $\ker (A) \neq \{0\}$ if and only if the product of the diagonal elements in PA is zero:

$$\beta_{11} \beta_{22} \cdots \beta_{nn} = 0.$$

We are now ready to see how the process of calculating A_{rref} using row operations can be interpreted as a change of basis in the image space.

Definition 1.13.16. Two matrices $A, B \in \mathrm{Mat}_{n \times m} (\mathbb{F})$ are said to be *row equivalent* if we can find $P \in Gl_n (\mathbb{F})$ such that $A = PB$.

Thus, row equivalent matrices are the matrices that can be obtained from each other via row operations. We can also think of row equivalent matrices as being different matrix representations of the same linear map with respect to different bases in $\mathbb{F}^n$. To see this, consider a linear map $L : \mathbb{F}^m \to \mathbb{F}^n$ that has matrix representation A with respect to the standard bases. If we perform a change of basis in $\mathbb{F}^n$ from the standard basis $f_1, \ldots, f_n$ to a basis $y_1, \ldots, y_n$ such that

$$\begin{bmatrix} y_1 \cdots y_n \end{bmatrix} = \begin{bmatrix} f_1 \cdots f_n \end{bmatrix} P,$$

i.e., the columns of P are regarded as a new basis for $\mathbb{F}^n$, then $B = P^{-1}A$ is simply the matrix representation for $L : \mathbb{F}^m \to \mathbb{F}^n$ when we have changed the basis in $\mathbb{F}^n$ according to P. This information can be encoded in the diagram

$$
\begin{array}{ccc}
\mathbb{F}^m & \xrightarrow{A} & \mathbb{F}^n \\
1_{\mathbb{F}^m} \downarrow & & \downarrow 1_{\mathbb{F}^n} \\
\mathbb{F}^m & \xrightarrow{L} & \mathbb{F}^n \\
1_{\mathbb{F}^m} \uparrow & & \uparrow P \\
\mathbb{F}^m & \xrightarrow{B} & \mathbb{F}^n.
\end{array}
$$

When we consider abstract matrices rather than systems of equations, we could equally well have performed column operations. This is accomplished by multiplying the elementary matrices on the right rather than the left. We can see explicitly what happens in the 2×2 case:

$$AI_{12} = \begin{bmatrix} \alpha_{11} & \alpha_{12} \\ \alpha_{21} & \alpha_{22} \end{bmatrix} \begin{bmatrix} 0 & 1 \\ 1 & 0 \end{bmatrix} = \begin{bmatrix} \alpha_{12} & \alpha_{11} \\ \alpha_{22} & \alpha_{21} \end{bmatrix}$$

$$AR_{12}(\alpha) = \begin{bmatrix} \alpha_{11} & \alpha_{12} \\ \alpha_{21} & \alpha_{22} \end{bmatrix} \begin{bmatrix} 1 & \alpha \\ 0 & 1 \end{bmatrix} = \begin{bmatrix} \alpha_{11} & \alpha\alpha_{11} + \alpha_{12} \\ \alpha_{21} & \alpha\alpha_{21} + \alpha_{22} \end{bmatrix}$$

$$AR_{21}(\alpha) = \begin{bmatrix} \alpha_{11} & \alpha_{12} \\ \alpha_{21} & \alpha_{22} \end{bmatrix} \begin{bmatrix} 1 & 0 \\ \alpha & 1 \end{bmatrix} = \begin{bmatrix} \alpha_{11} + \alpha\alpha_{12} & \alpha_{12} \\ \alpha_{21} + \alpha\alpha_{22} & \alpha_{22} \end{bmatrix}$$

$$AM_1(\alpha) = \begin{bmatrix} \alpha_{11} & \alpha_{12} \\ \alpha_{21} & \alpha_{22} \end{bmatrix} \begin{bmatrix} \alpha & 0 \\ 0 & 1 \end{bmatrix} = \begin{bmatrix} \alpha\alpha_{11} & \alpha_{12} \\ \alpha\alpha_{21} & \alpha_{22} \end{bmatrix}$$

$$AM_2(\alpha) = \begin{bmatrix} \alpha_{11} & \alpha_{12} \\ \alpha_{21} & \alpha_{22} \end{bmatrix} \begin{bmatrix} 1 & 0 \\ 0 & \alpha \end{bmatrix} = \begin{bmatrix} \alpha_{11} & \alpha\alpha_{12} \\ \alpha_{21} & \alpha\alpha_{22} \end{bmatrix}.$$

The only important and slightly confusing thing to be aware of is that, while $R_{kl}(\alpha)$ as a row operation multiplies row l by α and then adds it to row k, it now multiplies column k by α and adds it to column l as a column operation. This is because AE_{kl} is the matrix whose lth column is the kth column of A and whose other columns vanish.

Definition 1.13.17. Two matrices $A, B \in \mathrm{Mat}_{n \times m}(\mathbb{F})$ are said to be *column equivalent* if $A = BQ$ for some $Q \in Gl_m(\mathbb{F})$. According to the above interpretation, this corresponds to a change of basis in the domain space $\mathbb{F}^m$.

Definition 1.13.18. More generally, we say that $A, B \in \mathrm{Mat}_{n \times m}(\mathbb{F})$ are *equivalent* if $A = PBQ$, where $P \in Gl_n(\mathbb{F})$ and $Q \in Gl_m(\mathbb{F})$.

The diagram for the change of basis then looks like

$$
\begin{array}{ccc}
\mathbb{F}^m & \stackrel{A}{\longrightarrow} & \mathbb{F}^n \\
1_{\mathbb{F}^m} \downarrow & & \downarrow 1_{\mathbb{F}^n} \\
\mathbb{F}^m & \stackrel{L}{\longrightarrow} & \mathbb{F}^n \\
Q^{-1} \uparrow & & \uparrow P \\
\mathbb{F}^m & \stackrel{B}{\longrightarrow} & \mathbb{F}^n.
\end{array}
$$

In this way, we see that two matrices are equivalent if and only if they are matrix representations for the same linear map. Recall from Sect. 1.12 that any linear map between finite-dimensional spaces always has a matrix representation of the form

$$
\begin{bmatrix}
1 & \cdots & 0 & & & 0 \\
\vdots & \ddots & \vdots & & & \vdots \\
0 & \cdots & 1 & \vdots & & \vdots \\
& & & 0 & & 0 \\
\vdots & & \vdots & \vdots & \ddots & \vdots \\
0 & \cdots & 0 & 0 & \cdots & 0
\end{bmatrix},
$$

where there are k ones in the diagonal if the linear map has rank k. This implies

Corollary 1.13.19. (Characterization of Equivalent Matrices) $A, B \in \mathrm{Mat}_{n \times m}(\mathbb{F})$ *are equivalent if and only if they have the same rank. Moreover, any matrix of rank k is equivalent to a matrix that has k ones on the diagonal and zeros elsewhere.*

Exercises

1. Find bases for kernel and image for the following matrices:

   (a)
   $$
   \begin{bmatrix}
   1 & 3 & 5 & 1 \\
   2 & 0 & 6 & 0 \\
   0 & 1 & 7 & 2
   \end{bmatrix}
   $$

(b)

$$\begin{bmatrix} 1 & 2 \\ 0 & 3 \\ 1 & 4 \end{bmatrix}$$

(c)

$$\begin{bmatrix} 1 & 0 & 1 \\ 0 & 1 & 0 \\ 1 & 0 & 1 \end{bmatrix}$$

(d)

$$\begin{bmatrix} \alpha_{11} & 0 & \cdots & 0 \\ \alpha_{21} & \alpha_{22} & \cdots & 0 \\ \vdots & \vdots & \ddots & \vdots \\ \alpha_{n1} & \alpha_{n2} & \cdots & \alpha_{nn} \end{bmatrix}.$$

In this case, it will be necessary to discuss whether or not $\alpha_{ii} = 0$ for each $i = 1, \ldots, n$.

2. Find A^{-1} for each of the following matrices:

   (a)

   $$\begin{bmatrix} 0 & 0 & 0 & 1 \\ 0 & 0 & 1 & 0 \\ 0 & 1 & 0 & 0 \\ 1 & 0 & 0 & 0 \end{bmatrix}$$

   (b)

   $$\begin{bmatrix} 0 & 0 & 0 & 1 \\ 1 & 0 & 0 & 0 \\ 0 & 1 & 0 & 0 \\ 0 & 0 & 1 & 0 \end{bmatrix}$$

   (c)

   $$\begin{bmatrix} 0 & 1 & 0 & 1 \\ 1 & 0 & 0 & 0 \\ 0 & 0 & 1 & 0 \\ 0 & 0 & 0 & 1 \end{bmatrix}.$$

3. Let $A \in \mathrm{Mat}_{n \times m}(\mathbb{F})$. Show that we can find $P \in Gl_n(\mathbb{F})$ that is only a product of matrices of the types I_{ij} and $R_{ij}(\alpha)$ such that PA is upper triangular.

4. Let $A = \mathrm{Mat}_{n \times n}(\mathbb{F})$. We say that A has an LU decomposition if $A = LU$, where L is lower triangular with 1s on the diagonal and U is upper triangular. Show that A has an LU decomposition if all the *leading principal minors* are invertible. The leading principal $k \times k$ minor is the $k \times k$ submatrix gotten from A by eliminating the last $n - k$ rows and columns. Hint: Do Gauss elimination using only $R_{ij}(\alpha)$.

5. Assume that $A = PB$, where $P \in Gl_n (\mathbb{F})$.

   (a) Show that ker $(A) = $ ker (B).
   (b) Show that if the column vectors $y_{i_1}, \ldots, y_{i_k}$ of B form a basis for im (B), then the corresponding column vectors $x_{i_1}, \ldots, x_{i_k}$ for A form a basis for im (A).

6. Let $A \in \text{Mat}_{n \times m} (\mathbb{F})$.

   (a) Show that the $m \times m$ elementary matrices $I_{ij}, R_{ij} (\alpha), M_i (\alpha)$ when multiplied on the right correspond to column operations.
   (b) Show that we can find $Q \in Gl_m (\mathbb{F})$ such that AQ is lower triangular.
   (c) Use this to conclude that im $(A) = $ im (AQ) and describe a basis for im (A).
   (d) Use Q to find a basis for ker (A) given a basis for ker (AQ) and describe how you select a basis for ker (AQ).

7. Let $A \in \text{Mat}_{n \times n} (\mathbb{F})$ be upper triangular.

   (a) Show that dim $(\text{ker} (A)) \leq$ number of zero entries on the diagonal.
   (b) Give an example where dim $(\text{ker} (A)) <$ number of zero entries on the diagonal.

8. In this exercise, you are asked to show some relationships between the elementary matrices.

   (a) Show that $M_i (\alpha) = I_{ij} M_j (\alpha) I_{ji}$.
   (b) Show that $R_{ij} (\alpha) = M_j (\alpha^{-1}) R_{ij} (1) M_j (\alpha)$.
   (c) Show that $I_{ij} = R_{ij} (-1) R_{ji} (1) R_{ij} (-1) M_j (-1)$.
   (d) Show that $R_{kl} (\alpha) = I_{ki} I_{lj} R_{ij} (\alpha) I_{jl} I_{ik}$, where in case $i = k$ or $j = k$ we interpret $I_{kk} = I_{ll} = 1_{\mathbb{F}^n}$.

9. A matrix $A \in Gl_n (\mathbb{F})$ is a *permutation matrix* (see also Example 1.7.7) if $Ae_1 = e_{\sigma(i)}$ for some bijective map (permutation)

$$\sigma : \{1, \ldots, n\} \to \{1, \ldots, n\}.$$

   (a) Show that

$$A = \sum_{i=1}^{n} E_{\sigma(i)i}.$$

   (b) Show that A is a permutation matrix if and only if A has exactly one entry in each row and column which is 1 and all other entries are zero.
   (c) Show that A is a permutation matrix if and only if it is a product of the elementary matrices I_{ij}.

10. Assume that we have two fields $\mathbb{F} \subset \mathbb{L}$, such as $\mathbb{R} \subset \mathbb{C}$, and consider $A \in \text{Mat}_{n \times m} (\mathbb{F})$. Let $A_{\mathbb{L}} \in \text{Mat}_{n \times m} (\mathbb{L})$ be the matrix A thought of as

an element of $\mathrm{Mat}_{n \times m}(\mathbb{L})$. Show that $\dim_{\mathbb{F}} (\ker (A)) = \dim_{\mathbb{L}} (\ker (A_{\mathbb{L}}))$ and $\dim_{\mathbb{F}} (\mathrm{im}(A)) = \dim_{\mathbb{L}} (\mathrm{im}(A_{\mathbb{L}}))$. Hint: Show that A and $A_{\mathbb{L}}$ have the same reduced row echelon form.

11. Given $\alpha_{ij} \in \mathbb{F}$ for $i < j$ and $i, j = 1, \ldots, n$, we wish to solve

$$\frac{\xi_i}{\xi_j} = \alpha_{ij}.$$

(a) Show that this system either has no solutions or infinitely many solutions. Hint: Try $n = 2, 3$ first.
(b) Give conditions on α_{ij} that guarantee an infinite number of solutions.
(c) Rearrange this system into a linear system and explain the above results.

1.14 Dual Spaces*

Definition 1.14.1. For a vector space V over $\mathbb{F}$, we define the *dual vector space* $V' = \mathrm{Hom}(V, \mathbb{F})$ as the set of linear functions on V.

One often sees the notation V^* for V'. However, we have reserved V^* for the conjugate vector space to a complex vector space (see Exercise 6 in Sect. 1.4). When V is finite-dimensional we know that V and V' have the same dimension. In this section, we shall see how the dual vector space can be used as a substitute for an inner product on V in case V does not come with a natural inner product (see Chap. 3 for the theory on inner product spaces).

We have a natural dual pairing $V \times V' \to \mathbb{F}$ defined by $(x, f) = f(x)$ for $x \in V$ and $f \in V'$. We are going to think of (x, f) as a sort of inner product between x and f. Using this notation will enable us to make the theory virtually the same as for inner product spaces. Observe that this pairing is linear in both variables. Linearity in the first variable is a consequence of using linear functions in the second variable. Linearity in the second variable is completely trivial:

$$(\alpha x + \beta y, f) = f(\alpha x + \beta y)$$
$$= \alpha f(x) + \beta f(y)$$
$$= \alpha (x, f) + \beta (y, f),$$

$$(x, \alpha f + \beta g) = (\alpha f + \beta g)(x)$$
$$= \alpha f(x) + \beta g(x)$$
$$= \alpha (x, f) + \beta (x, g).$$

We start with our construction of a dual basis; these are similar to orthonormal bases. Let V have a basis $x_1, \ldots, x_n$, and define linear functions f_i by $f_i(x_j) = \delta_{ij}$. Thus, $(x_i, f_j) = f_j(x_i) = \delta_{ij}$.

Example 1.14.2. Recall that we defined $dx^i : \mathbb{R}^n \to \mathbb{R}$ as the linear function such that $dx^i (e_j) = \delta_{ij}$, where $e_1, \dots, e_n$ is the canonical basis for $\mathbb{R}^n$. Thus, dx^i is the dual basis to the canonical basis.

Proposition 1.14.3. *The vectors $f_1, \dots, f_n \in V'$ form a basis called the* dual basis *of $x_1, \dots, x_n$. Moreover, for $x \in V$ and $f \in V'$, we have the expansions*

$$x = (x, f_1) x_1 + \cdots + (x, f_n) x_n,$$
$$f = (x_1, f) f_1 + \cdots + (x_n, f) f_n.$$

Proof. Consider a linear combination $\alpha_1 f_1 + \cdots + \alpha_n f_n$. Then,

$$(x_i, \alpha_1 f_1 + \cdots + \alpha_n f_n) = \alpha_1 (x_i, f_1) + \cdots + \alpha_n (x_i, f_n)$$
$$= \alpha_i.$$

Thus, $\alpha_i = 0$ if $\alpha_1 f_1 + \cdots + \alpha_n f_n = 0$. Since V and V' have the same dimension, this shows that $f_1, \dots, f_n$ form a basis for V'. Moreover, if we have an expansion $f = \alpha_1 f_1 + \cdots + \alpha_n f_n$, then it follows that $\alpha_i = (x_i, f) = f(x_i)$.

Finally, assume that $x = \beta_1 x_1 + \cdots + \beta_n x_n$. Then,

$$(x, f_i) = (\beta_1 x_1 + \cdots + \beta_n x_n, f_i)$$
$$= \beta_1 (x_1, f_i) + \cdots + \beta_n (x_n, f_i)$$
$$= \beta_i,$$

which is what we wanted to prove. $\square$

Next, we define annihilators; these are counterparts to orthogonal complements.

Definition 1.14.4. Let $M \subset V$ be a subspace and define the *annihilator* to M in V as the subspace $M^o \subset V'$ given by

$$M^o = \{f \in V' : (x, f) = 0 \text{ for all } x \in M\}$$
$$= \{f \in V' : f(x) = 0 \text{ for all } x \in M\}$$
$$= \{f \in V' : f(M) = \{0\}\}$$
$$= \{f \in V' : f|_M = 0\}.$$

Using dual bases, we can get a slightly better grip on these annihilators.

Proposition 1.14.5. *If $M \subset V$ is a subspace of a finite-dimensional space and $x_1, \dots, x_n$ is a basis for V such that*

$$M = \text{span}\{x_1, \dots, x_m\},$$

then

$$M^o = \text{span}\{f_{m+1}, \dots, f_n\},$$

where $f_1, \dots, f_n$ is the dual basis. In particular, we have

$$\dim(M) + \dim(M^o) = \dim(V) = \dim(V').$$

Proof. If $M = \text{span}\{x_1, \dots, x_m\}$, then $f_{m+1}, \dots, f_n \in M^o$ by definition of the annihilator as each of $f_{m+1}, \dots, f_n$ vanish on the vectors $x_1, \dots, x_m$. Conversely, take $f \in M^o$ and expand it $f = \alpha_1 f_1 + \cdots + \alpha_n f_n$. If $1 \le i \le m$, then

$$0 = (x_i, f) = \alpha_i.$$

So $f = \alpha_{m+1} f_{m+1} + \cdots + \alpha_n f_n$ as desired.   □

We are now ready to discuss the *reflexive* property. This will allow us to go from V' back to V itself rather than to $(V')' = V''$. Thus, we have to find a natural identification $V \to V''$. There is, indeed, a natural linear map

$$\text{ev} : V \to V''$$

that takes each $x \in V$ to a linear function on V' defined by $\text{ev}_x(f) = (x, f) = f(x)$. To see that it is linear, observe that

$$(\alpha x + \beta y, f) = f(\alpha x + \beta y)$$
$$= \alpha f(x) + \beta f(y)$$
$$= \alpha(x, f) + \beta(y, f).$$

Evidently, we have defined ev_x in such a way that

$$(x, f) = (f, \text{ev}_x).$$

The map $x \to \text{ev}_x$ always has trivial kernel. To prove this in the finite-dimensional case, select a dual basis $f_1, \dots, f_n$ for V' and observe that since $\text{ev}_x(f_i) = (x, f_i)$ records the coordinates of x, it is not possible for x to be in the kernel unless it is zero. Finally, we use that $\dim(V) = \dim(V') = \dim(V'')$ to conclude that this map is an isomorphism. Thus, any element of V'' is of the form ev_x for a unique $x \in V$.

The first interesting observation we make is that if $f_1, \dots, f_n$ is dual to $x_1, \dots, x_n$, then $\text{ev}_{x_1}, \dots, \text{ev}_{x_n}$ is dual to $f_1, \dots, f_n$ as

$$\text{ev}_{x_i}(f_j) = (x_i, f_j) = \delta_{ij}.$$

If we agree to identify V'' with V, i.e., we think of x as identified with ev_x, then we can define the annihilator of a subspace $N \subset V'$ by

$$N^o = \{x \in V : (x, f) = 0 \text{ for all } f \in N\}$$
$$= \{x \in V : f(x) = 0 \text{ for all } f \in N\}.$$

The claim then is, that for $M \subset V$ and $N \subset V'$, we have $M^{oo} = M$ and $N^{oo} = N$. Both identities follow directly from the above proposition about the construction of a basis for the annihilator.

We now come to an interesting relationship between annihilators and the dual spaces of subspaces.

Proposition 1.14.6. *Assume that V is finite-dimensional. If $V = M \oplus N$, then $V' = M^o \oplus N^o$ and the restriction maps $V' \to M'$ and $V' \to N'$ give isomorphisms*

$$M^o \approx N',$$
$$N^o \approx M'.$$

Proof. Select a basis $x_1, \ldots, x_n$ for V such that

$$M = \text{span} \{x_1, \ldots, x_m\},$$
$$N = \text{span} \{x_{m+1}, \ldots, x_n\}.$$

Let $f_1, \ldots, f_n$ be the dual basis and observe that

$$M^o = \text{span} \{f_{m+1}, \ldots, f_n\},$$
$$N^o = \text{span} \{f_1, \ldots, f_m\}.$$

This proves that $V' = M^o \oplus N^o$. Next, we note that

$$\dim(M^o) = \dim(V) - \dim(M)$$
$$= \dim(N)$$
$$= \dim(N').$$

So at least M^o and N' have the same dimension. What is more, if we restrict $f_{m+1}, \ldots, f_n$ to N, then we still have that $(x_j, f_i) = \delta_{ij}$ for $j = m+1, \ldots, n$. As $N = \text{span} \{x_{m+1}, \ldots, x_n\}$, this means that $f_{m+1}|_N, \ldots, f_n|_N$ form a basis for N'. The proof that $N^o \approx M'$ is similar. $\square$

The main problem with using dual spaces rather than inner products is that while we usually have a good picture of what V is, we rarely get a good independent description of the dual space. Thus, the constructions mentioned here should be thought of as being theoretical and strictly auxiliary to the developments of the theory of linear operators on a fixed vector space V.

Below, we consider a few examples of constructions of dual spaces.

Example 1.14.7. Let $V = \mathrm{Mat}_{n \times m}(\mathbb{F})$, then we can identify $V' = \mathrm{Mat}_{m \times n}(\mathbb{F})$. For each $A \in \mathrm{Mat}_{m \times n}(\mathbb{F})$, the corresponding linear function is

$$f_A(X) = \mathrm{tr}(AX) = \mathrm{tr}(XA).$$

Example 1.14.8. If V is a finite-dimensional inner product space, then $f_y(x) = (x|y)$ defines a linear function, and we know that all linear functions are of that form. Thus, we can identify V' with V. Note, however, that in the complex case, $y \to f_y$ is not complex linear. It is in fact conjugate linear, i.e., $f_{\lambda y} = \bar{\lambda} f_y$. Thus, V' is identified with V^* (see Exercise 6 in Sect. 1.4). This conforms with the idea that the inner product defines a bilinear paring on $V \times V^*$ via $(x, y) \to (x|y)$ that is linear in both variables!

Example 1.14.9. If we think of V as $\mathbb{R}$ with $\mathbb{Q}$ as scalar multiplication, then it is not at all clear that we have any linear functions $f : \mathbb{R} \to \mathbb{Q}$. In fact, the axiom of choice has to be invoked in order to show that they exist.

Example 1.14.10. Finally, we have an exceedingly interesting infinite-dimensional example where the dual gets quite a bit bigger. Let $V = \mathbb{F}[t]$ be the vector space of polynomials. We have a natural basis $1, t, t^2, \ldots$. Thus, a linear map $f : \mathbb{F}[t] \to \mathbb{F}$ is determined by its values on this basis $\alpha_n = f(t^n)$. Conversely, given an infinite sequence $\alpha_0, \alpha_1, \alpha_2, \ldots \in \mathbb{F}$, we have a linear map such that $f(t^n) = \alpha_n$. So while V consists of finite sequences of elements from $\mathbb{F}$, the dual consists of infinite sequences of elements from $\mathbb{F}$. We can evidently identify $V' = \mathbb{F}[[t]]$ with power series by recording the values on the basis as coefficients:

$$\sum_{n=0}^{\infty} \alpha_n t^n = \sum_{n=0}^{\infty} f(t^n) t^n.$$

This means that V' inherits a product structure through taking products of power series. There is a large literature on this whole setup under the title *Umbral Calculus*. For more on this, see [Roman].

Definition 1.14.11. The dual space construction leads to a *dual map* $L' : W' \to V'$ for a linear map $L : V \to W$. This dual map is a generalization of the transpose of a matrix. The definition is quite simple:

$$L'(g) = g \circ L.$$

Thus, if $g \in W'$, then we get a linear function $g \circ L : V \to \mathbb{F}$ since $L : V \to W$. The dual to L is often denoted $L' = L^t$ as with matrices. This will be justified in the exercises to this section. Note that if we use the pairing (x, f) between V and V', then the dual map satisfies

$$(L(x), g) = (x, L'(g))$$

for all $x \in V$ and $g \in W'$. Thus, the dual map really is defined in a manner analogous to the adjoint.

The following properties follow almost immediately from the definition.

Proposition 1.14.12. *Let $L, \tilde{L} : V \to W$ and $K : W \to U$ be linear maps between finite-dimensional vector spaces; then:*

(1) $\left(\alpha L + \beta \tilde{L}\right)' = \alpha L' + \beta \tilde{L}'$.
(2) $(K \circ L)' = L' \circ K'$.
(3) $L'' = (L')' = L$ if we identify $V'' = V$ and $W'' = W$.
(4) If $M \subset V$ and $N \subset W$ are subspaces with $L(M) \subset N$, then $L'(N^o) \subset M^o$.

Proof. 1. Just note that

$$\left(\alpha L + \beta \tilde{L}\right)'(g) = g \circ \left(\alpha L + \beta \tilde{L}\right)$$
$$= \alpha g \circ L + \beta g \circ \tilde{L}$$
$$= \alpha L'(g) + \beta \tilde{L}'(g)$$

as g is linear.
2. This comes from

$$(K \circ L)'(h) = h \circ (K \circ L)$$
$$= (h \circ K) \circ L$$
$$= K'(h) \circ L$$
$$= L'\left(K'(h)\right).$$

3. Note that $L'' : V'' \to W''$. If we take $\mathrm{ev}_x \in V''$ and use $(x, f) = (f, \mathrm{ev}_x)$, then

$$\left(g, L''(\mathrm{ev}_x)\right) = \left(L'(g), \mathrm{ev}_x\right)$$
$$= \left(x, L'(g)\right)$$
$$= (L(x), g).$$

This shows that $L''(\mathrm{ev}_x)$ is identified with $L(x)$ as desired.
4. If $g \in V'$, then we have that $(x, L'(g)) = (L(x), g)$. So if $x \in M$, then we have $L(x) \in N$, and hence, $g(L(x)) = 0$ for $g \in N^o$. This means that $L'(g) \in M^o$. $\square$

Just like for adjoint maps, we have a type of Fredholm alternative for dual maps.

Theorem 1.14.13. (The Generalized Fredholm Alternative) *Let $L : V \to W$ be a linear map between finite-dimensional vector spaces. Then,*

$$\ker(L) = \mathrm{im}\left(L'\right),$$
$$\ker\left(L'\right) = \mathrm{im}\,(L)^o,$$

$$\ker{(L)}^o = \operatorname{im}(L'),$$
$$\ker{(L')}^o = \operatorname{im}(L).$$

Proof. We only need to prove the first statement as $L'' = L$ and $M^{oo} = M$,

$$\ker{(L)} = \{x \in V : Lx = 0\},$$
$$\operatorname{im}{(L')}^o = \{x \in V : (x, L'(g)) = 0 \text{ for all } g \in W\}.$$

Using that $(x, L'(g)) = (L(x), g)$, we note first that if $x \in \ker(L)$, then it must also belong to $\operatorname{im}(L')^o$. Conversely, if $0 = (x, L'(g)) = (L(x), g)$ for all $g \in W$, it must follow that $L(x) = 0$ and hence $x \in \ker(L)$. $\square$

As a corollary, we obtain a new version of the rank theorem (Theorem 1.12.11).

Corollary 1.14.14. (The Rank Theorem) *Let $L : V \to W$ be a linear map between finite-dimensional vector spaces. Then,*

$$\operatorname{rank}(L) = \operatorname{rank}(L').$$

Proof. The Fredholm alternative together with the dimension formula (Theorem 1.11.7) immediately shows:

$$\begin{aligned}
\operatorname{rank}(L) &= \dim V - \dim \ker(L) \\
&= \dim V - \dim \operatorname{im}{(L')}^o \\
&= \dim V - \dim V + \operatorname{im}(L') \\
&= \operatorname{rank}(L').
\end{aligned}$$

$\square$

Exercises

1. Let $x_1, \ldots, x_n$ be a basis for V and $f_1, \ldots, f_n$ a dual basis for V'. Show that the inverses to the isomorphisms

$$\begin{bmatrix} x_1 & \cdots & x_n \end{bmatrix} : \mathbb{F}^n \to V,$$
$$\begin{bmatrix} f_1 & \cdots & f_n \end{bmatrix} : \mathbb{F}^n \to V'$$

are given by

$$\begin{bmatrix} x_1 & \cdots & x_n \end{bmatrix}^{-1}(x) = \begin{bmatrix} f_1(x) \\ \vdots \\ f_n(x) \end{bmatrix},$$

$$[f_1 \cdots f_n]^{-1}(f) = \begin{bmatrix} f(x_1) \\ \vdots \\ f(x_n) \end{bmatrix}.$$

2. Let $L : V \to W$ with basis $x_1, \ldots, x_m$ for V, $y_1, \ldots, y_n$ for W and dual basis $g_1, \ldots, g_n$ for W'. Show that we have

$$L = [x_1 \cdots x_m][L][y_1 \cdots y_n]^{-1}$$

$$= [x_1 \cdots x_m][L]\begin{bmatrix} g_1 \\ \vdots \\ g_n \end{bmatrix},$$

where $[L]$ is the matrix representation for L with respect to the given bases.

3. Given the basis $1, t, t^2$ for P_2, identify P_2 with $\mathbb{C}^3$ (column vectors) and $(P_2)'$ with $\mathrm{Mat}_{1\times 3}(\mathbb{C})$ (row vectors).

   (a) Using these identifications, find a dual basis to $1, 1+t, 1+t+t^2$ in $(P_2)'$.
   (b) Using these identifications, find the matrix representation for $f \in (P_2)'$ defined by

   $$f(p) = p(t_0).$$

   (c) Using these identifications, find the matrix representation for $f \in (P_2)'$ defined by

   $$f(p) = \int_a^b p(t)\,dt.$$

   (d) Are all elements in $(P_2)'$ represented by the types of linear functions described in either (b) or (c)?

4. *(Lagrange Multiplier Construction)* Let $f, g \in V'$ and assume that $g \neq 0$. Show that $f = \lambda g$ for some $\lambda \in \mathbb{F}$ if and only if $\ker(f) \supset \ker(g)$.

5. Let $M \subset V$ be a subspace. Show that we have linear maps

$$M^o \xrightarrow[\to]{i} V' \xrightarrow[\to]{\pi} M',$$

where ι is one-to-one, π is onto, and $\mathrm{im}(i) = \ker(\pi)$. Conclude that V' is isomorphic to $M^o \times M'$.

6. Let V and W be finite-dimensional vector spaces. Exhibit an isomorphism between $V' \times W'$ and $(V \times W)'$ that does not depend on choosing bases for V and W.

7. Let $M, N \subset V$ be subspaces of a finite-dimensional vector space. Show that

$$M^o + N^o = (M \cap N)^o,$$
$$(M + N)^o = M^o \cap N^o.$$

8. Let $L : V \to W$ and assume that we have bases $x_1, \ldots, x_m$ for V, $y_1, \ldots, y_n$ for W and corresponding dual bases $f_1, \ldots, f_m$, for V' and $g_1, \ldots, g_n$ for W'. Show that if $[L]$ is the matrix representation for L with respect to the given bases, then $[L]^t = [L']$ with respect to the dual bases.

9. Assume that $L : V \to W$ is a linear map and that $L(M) \subset N$ for subspaces $M \subset V$ and $N \subset W$. Is there a relationship between $(L|_M)' : N' \to M'$ and $L'|_{N^o} : N^o \to M^o$?

10. (*The Rank Theorem*) This exercise is an abstract version of what happened in the proof of the Rank Theorem 1.12.11 in Sect. 1.12. Let $L : V \to W$ and $x_1, \ldots, x_k$ a basis for im (L).

   (a) Show that

   $$L(x) = (x, f_1) x_1 + \cdots + (x, f_k) x_k$$

   for suitable $f_1, \ldots, f_k \in V'$.

   (b) Show that

   $$L'(f) = (x_1, f) f_1 + \cdots + (x_k, f) f_k$$

   for $f \in W'$.

   (c) Conclude that rank $(L') \le$ rank (L).

   (d) Show that rank $(L') =$ rank (L).

11. Let $M \subset V$ be a finite-dimensional subspace of V and $x_1, \ldots, x_k$ a basis for M. Let

   $$L(x) = (x, f_1) x_1 + \cdots + (x, f_k) x_k$$

   for $f_1, \ldots, f_k \in V'$.

   (a) If $(x_j, f_i) = \delta_{ij}$, then L is a projection onto M, i.e., $L^2 = L$ and im $(L) = M$.

   (b) If E is a projection onto M, then

   $$E = (x, f_1) x_1 + \cdots + (x, f_k) x_k,$$

   with $(x_j, f_i) = \delta_{ij}$.

12. Let $M, N \subset V$ be subspaces of a finite-dimensional vector space and consider $L : M \times N \to V$ defined by $L(x, y) = x - y$.

   (a) Show that $L'(f)(x, y) = f(x) - f(y)$.

   (b) Show that ker (L') can be identified with both $M^o \cap N^o$ and $(M + N)^o$.

1.15 Quotient Spaces*

In Sect. 1.14, we saw that if $M \subset V$ is a subspace of a general vector space, then the annihilator subspace $M^o \subset V'$ can play the role of a canonical complement of M. One thing missing from this setup, however, is the projection whose kernel is M. In this section, we shall construct a different type of vector space that can substitute as a complement to M. It is called the *quotient space* of V over M and is denoted V/M. In this case, there is an onto linear map $P : V \to V/M$ whose kernel is M. The quotient space construction is somewhat abstract, but it is also quite general and can be developed with a minimum of information as we shall see. It is in fact quite fundamental and can be used to prove several of the important results mentioned in Sect. 1.11.

Similar to addition for subspaces in Sect. 1.10, we can in fact define addition for any subsets of a vector space.

Definition 1.15.1. If $S, T \subset V$ are subsets, then we define

$$S + T = \{x + y : x \in S \text{ and } y \in T\}.$$

It is immediately clear that this addition on subsets is associative and commutative. In case one of the sets contains only one element, we simplify the notation by writing

$$S + \{x_0\} = S + x_0 = \{x + x_0 : x \in S\},$$

and we call $S + x_0$ a *translate* of S. Geometrically, all of the sets $S + x_0$ appear to be parallel pictures of S (see Fig. 1.4) that are translated in V as we change x_0. We also say that S and T are *parallel* and denote it $S \parallel T$ if $T = S + x_0$ for some $x_0 \in V$.

It is also possible to scale subsets

$$\alpha S = \{\alpha x : x \in S\}.$$

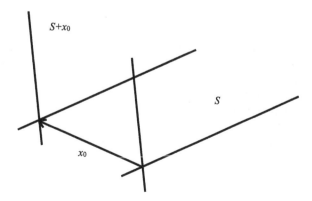

Fig. 1.4 Parallel subspaces

This scalar multiplication satisfies some of the usual properties of scalar multiplication

$$(\alpha\beta)\, S = \alpha\,(\beta S),$$

$$1S = S,$$

$$\alpha\,(S + T) = \alpha S + \alpha T.$$

However, the other distributive law can fail

$$(\alpha + \beta)\, S \stackrel{?}{=} \alpha S + \beta S$$

since it may not be true that

$$2S \stackrel{?}{=} S + S.$$

Certainly, $2S \subset S + S$, but elements $x + y$ do not have to belong to $2S$ if $x, y \in S$ are distinct. Take, e.g., $S = \{x, -x\}$, where $x \neq 0$. Then, $2S = \{2x, -2x\}$, while $S + S = \{2x, 0, -2x\}$.

Definition 1.15.2. Our picture of the quotient space V/M, when $M \subset V$ is a subspace, is the set of all translates $M + x_0$ for $x_0 \in V$

$$V/M = \{M + x_0 : x_0 \in V\}$$

Several of these translates are in fact equal as

$$x_1 + M = x_2 + M$$

precisely when $x_1 - x_2 \in M$. To see why this is, note that if $z \in M$, then $z + M = M$ since M is a subspace. Thus, $x_1 - x_2 \in M$ implies that

$$x_1 + M = x_2 + (x_1 - x_2) + M$$

$$= x_2 + M.$$

Conversely, if $x_1 + M = x_2 + M$, then $x_1 = x_2 + x$ for some $x \in M$ implying that $x_1 - x_2 \in M$.

We see that in the trivial case where $M = \{0\}$, the translates of $\{0\}$ can be identified with V itself. Thus, $V/\{0\} \approx V$. In the other trivial case where $M = V$, all the translates are simply V itself. So V/V is the one element set $\{V\}$ whose element is the vector space V.

We now need to see how addition and scalar multiplication works on V/M. The important property that simplifies calculations and will turn V/M into a vector space is the fact that M is a subspace, i.e., for all scalars $\alpha, \beta \in \mathbb{F}$,

$$\alpha M + \beta M = M.$$

This implies that addition and scalar multiplication is considerably simplified.

$$\alpha\,(M + x) + \beta\,(M + y) = \alpha M + \beta M + \alpha x + \beta y$$
$$= M + \alpha x + \beta y.$$

With this in mind, we can show that V/M is a vector space. The zero element is M since $M + (M + x_0) = M + x_0$. The negative of $M + x_0$ is the translate $M - x_0$. Finally, the important distributive law that was not true in general also holds because

$$(\alpha + \beta)\,(M + x_0) = M + (\alpha + \beta)\,x_0$$
$$= M + \alpha x_0 + \beta x_0$$
$$= (M + \alpha x_0) + (M + \beta x_0)$$
$$= \alpha\,(M + x_0) + \beta\,(M + x_0)\,.$$

The "projection" $P : V \to V/M$ is now defined by

$$P\,(x) = M + x.$$

Clearly, P is onto and $P\,(x) = 0$ if and only if $x \in M$. The fact that P is linear follows from the way we add elements in V/M,

$$P\,(\alpha x + \beta y) = M + \alpha x + \beta y$$
$$= \alpha\,(M + x) + \beta\,(M + y)$$
$$= \alpha P\,(x) + \beta P\,(y)\,.$$

This projection can be generalized to the setting where $M \subset N \subset V$. Here we get $V/M \to V/N$ by mapping $x + M$ to $x + N$.

If $L : V \to W$ and $M \subset V$, $L\,(M) \subset N \subset W$, then we get an induced map $L : V/M \to W/N$ by sending $x + M$ to $L\,(x) + N$. We need to check that this indeed gives a well-defined map. Assuming that $x_1 + M = x_2 + M$, we have to show that $L\,(x_1) + N = L\,(x_2) + N$. The first condition is equivalent to $x_1 - x_2 \in M$; thus,

$$L\,(x_1) - L\,(x_2) = L\,(x_1 - x_2)$$
$$\in L\,(M) \subset N,$$

implying that $L\,(x_1) + N = L\,(x_2) + N$.

We are now going to investigate how the quotient space can be used to understand some of the developments from Sect. 1.11. For any linear map, we have that $L\,(\ker\,(L)) = \{0\}$. Thus, L induces a linear map

$$V/\,(\ker\,(L)) \to W/\,\{0\} \approx W.$$

Since the image of $\ker(L) + x$ is $\{0\} + L(x) \approx L(x)$, we see that the induced map has trivial kernel. This implies that we have an isomorphism

$$V/(\ker(L)) \to \mathrm{im}(L).$$

We can put all of this into a commutative diagram:

$$
\begin{array}{ccc}
V & \xrightarrow{\;L\;} & W \\
P \downarrow & & \uparrow \\
V/(\ker(L)) & \xrightarrow{\approx} & \mathrm{im}(L)
\end{array}
$$

Note that, as yet, we have not used any of the facts we know about finite-dimensional spaces. The two facts we shall use are that the dimension of a vector space is well defined (Theorem 1.8.4) and that any subspace in a finite-dimensional vector space has a finite-dimensional complement (Theorem 1.10.20 and Corollary 1.12.6). We start by considering subspaces.

Theorem 1.15.3. (The Subspace Theorem) *Let V be a finite-dimensional vector space. If $M \subset V$ is a subspace, then both M and V/M are finite-dimensional and*

$$\dim V = \dim M + \dim(V/M).$$

Proof. We start by selecting a finite-dimensional subspace $N \subset V$ that is complementary to M (see Corollary 1.12.6). If we restrict the projection $P : V \to V/M$ to $P|_N : N \to V/M$, then it has trivial kernel as $M \cap N = \{0\}$, so $P|_N$ is one-to-one. $P|_N$ is also onto since any $z \in V$ can be written as $z = x + y$ where $x \in M$ and $y \in N$, so it follows that

$$M + z = M + x + y$$
$$= M + y$$
$$= P(y).$$

Thus, $P|_N : N \to V/M$ is an isomorphism. This shows that V/M is finite-dimensional. In the same way, we see that the projection $Q : V \to V/N$ restricts to an isomorphism $Q|_M : M \to V/N$. By selecting a finite-dimensional complement for $N \subset V$, we also get that V/N is finite-dimensional. This in turn shows that M is finite-dimensional.

We can now use that $V = M \oplus N$ to show that

$$\dim V = \dim M + \dim N$$
$$= \dim M + \dim(V/M).$$

$\square$

The dimension formula now follows from our observations above.

Corollary 1.15.4. (The Dimension Formula) *Let V be a finite-dimensional vector space. If $L : V \to W$ is a linear map, then*

$$\dim V = \dim(\ker(L)) + \dim(\operatorname{im}(L)).$$

Proof. We just saw that

$$\dim V = \dim(\ker(L)) + \dim(V/(\ker(L))).$$

In addition, we have an isomorphism

$$V/(\ker(L)) \longrightarrow \operatorname{im}(L).$$

This proves the claim.   □

Exercises

1. An affine subspace $A \subset V$ is a subset such that if $x_1, \ldots, x_k \in A$, $\alpha_1, \ldots, \alpha_k \in \mathbb{F}$, and $\alpha_1 + \cdots + \alpha_k = 1$, then $\alpha_1 x_1 + \cdots + \alpha_k x_k \in A$. Show that V/M consists of all of the affine subspaces parallel to M.

2. Find an example of a nontrivial linear operator $L : V \to V$ and a subspace $M \subset V$ such that $L|_M = 0$ and the induced map $L : V/M \to V/M$ is also zero.

3. This exercise requires knowledge of the characteristic polynomial (see Sects. 2.3, 2.7, or 5.7). Let $L : V \to V$ be a linear operator with an invariant subspace $M \subset V$. Show that $\chi_L(t)$ is the product of the characteristic polynomials of $L|_M$ and the induced map $L : V/M \to V/M$.

4. Let $M \subset V$ be a subspace and assume that we have $x_1, \ldots, x_n \in V$ such that $x_1, \ldots, x_k$ form a basis for M and $x_{k+1} + M, \ldots, x_n + M$ form a basis for V/M. Show that $x_1, \ldots, x_n$ is a basis for V.

5. Let $L : V \to W$ be a linear map and assume that $L(M) \subset N$. How does the induced map $L : V/M \to W/N$ compare to the dual maps constructed in Exercise 2 in Sect. 1.14?

6. Let $M \subset V$ be a subspace. Show that there is a natural or canonical isomorphism $M^o \to (V/M)'$, i.e., an isomorphism that does not depend on a choice of basis for the spaces.

Chapter 2
Linear Operators

In this chapter, we are going to present all of the results that relate to linear operators on abstract finite-dimensional vector spaces. Aside from a section on polynomials, we start with a section on linear differential equations in order to motivate both some material from Chap. 1 and also give a reason for why it is desirable to study matrix representations. Eigenvectors and eigenvalues are first introduced in the context of differential equations where they are used to solve such equations. It is, however, possible to start with the Sect. 2.3 and ignore the discussion on differential equations. The material developed in Chap. 1 on Gauss elimination is used to calculate eigenvalues and eigenvectors and to give a "weak" definition of the characteristic polynomial. We also introduce the minimal polynomial and use it to characterize diagonalizable maps. We then move on to cyclic subspaces leading us to fairly simple proofs of the Cayley–Hamilton theorem and the cyclic subspace decomposition. This in turn gives us a nice proof of the Frobenius canonical form as well as the Jordan canonical form. We finish the chapter with the Smith normal form. This result gives a direct method for calculating the Frobenius canonical form as well as a complete set of similarity invariants for a matrix. It also shows how a system of higher order differential equations (or recurrence equations) can be decoupled and solved as independent higher order equations.

Various properties of polynomials are used quite a bit in this chapter. Most of these properties are probably already known to the student and in any case are certainly well known from arithmetic of integers nevertheless, we have chosen to collect some of these facts in an optional section at the beginning of this chapter.

It is possible to simply cover Sects. 2.3 and 2.5 and then move on to the chapters on inner product spaces. In fact, it is possible to skip this chapter entirely as it is not absolutely necessary in the theory of inner product spaces.

P. Petersen, *Linear Algebra*, Undergraduate Texts in Mathematics,
DOI 10.1007/978-1-4614-3612-6_2, © Springer Science+Business Media New York 2012

2.1 Polynomials*

The space of polynomials with coefficients in the field $\mathbb{F}$ is denoted $\mathbb{F}[t]$. This space consists of expressions of the form

$$\alpha_0 + \alpha_1 t + \cdots + \alpha_k t^k,$$

where $\alpha_0, \ldots, \alpha_k \in \mathbb{F}$ and k is a nonnegative integer. One can think of these expressions as functions on $\mathbb{F}$, but in this section, we shall only use the formal algebraic structure that comes from writing polynomials in the above fashion. Recall that integers are written in a similar way if we use the standard positional base 10 system (or any other base for that matter):

$$a_k \cdots a_0 = a_k 10^k + a_{k-1} 10^{k-1} + \cdots + a_1 10 + a_0.$$

Indeed, there are many basic number theoretic similarities between integers and polynomials as we shall see below.

Addition is defined by adding term by term:

$$\left(\alpha_0 + \alpha_1 t + \alpha_2 t^2 + \cdots\right) + \left(\beta_0 + \beta_1 t + \beta_2 t^2 + \cdots\right)$$
$$= (\alpha_0 + \beta_0) + (\alpha_1 + \beta_1) t + (\alpha_2 + \beta_2) t^2 + \cdots$$

Multiplication is a bit more complicated but still completely naturally defined by multiplying all the different terms and then collecting according to the powers of t:

$$\left(\alpha_0 + \alpha_1 t + \alpha_2 t^2 + \cdots\right) \cdot \left(\beta_0 + \beta_1 t + \beta_2 t^2 + \cdots\right)$$
$$= \alpha_0 \cdot \beta_0 + (\alpha_0 \beta_1 + \alpha_1 \beta_0) t + (\alpha_0 \beta_2 + \alpha_1 \beta_1 + \alpha_2 \beta_0) t^2 + \cdots$$

Note that in "addition," the indices match the power of t, while in "multiplication," each term has the property that the sum of the indices matches the power of t.

The *degree* of a polynomial $\alpha_0 + \alpha_1 t + \cdots + \alpha_n t^n$ is the largest k such that $\alpha_k \neq 0$. In particular,

$$\alpha_0 + \alpha_1 t + \cdots + \alpha_k t^k + \cdots + \alpha_n t^n = \alpha_0 + \alpha_1 t + \cdots + \alpha_k t^k,$$

where k is the degree of the polynomial. We also write $\deg(p) = k$. The degree satisfies the following elementary properties:

$$\deg(p + q) \leq \max\{\deg(p), \deg(q)\},$$
$$\deg(pq) = \deg(p) + \deg(q).$$

Note that if $\deg(p) = 0$, then $p(t) = \alpha_0$ is simply a scalar.

It is often convenient to work with *monic* polynomials. These are the polynomials of the form

$$\alpha_0 + \alpha_1 t + \cdots + 1 \cdot t^k.$$

Note that any polynomial can be made into a monic polynomial by diving by the scalar that appears in front of the term of highest degree. Working with monic polynomials is similar to working with positive integers rather than all integers.

If $p, q \in \mathbb{F}[t]$, then we say that p *divides* q if $q = pd$ for some $d \in \mathbb{F}[t]$. Note that if p divides q, then it must follow that $\deg(p) \leq \deg(q)$. The converse is of course not true, but polynomial long division gives us a very useful partial answer to what might happen.

Theorem 2.1.1. (The Euclidean Algorithm) *If $p, q \in \mathbb{F}[t]$ and $\deg(p) \leq \deg(q)$, then $q = pd + r$, where $\deg(r) < \deg(p)$.*

Proof. The proof is along the same lines as how we do long division with remainder. The idea of the Euclidean algorithm is that whenever $\deg(p) \leq \deg(q)$, it is possible to find d_1 and r_1 such that

$$q = pd_1 + r_1,$$

$$\deg(r_1) < \deg(q).$$

To establish, this assume

$$q = \alpha_n t^n + \alpha_{n-1} t^{n-1} + \cdots + \alpha_0,$$

$$p = \beta_m t^m + \beta_{m-1} t^{m-1} + \cdots + \beta_0,$$

where $\alpha_n, \beta_m \neq 0$. Then, define $d_1 = \frac{\alpha_n}{\beta_m} t^{n-m}$ and

$$
\begin{aligned}
r_1 &= q - pd_1 \\
&= \left(\alpha_n t^n + \alpha_{n-1} t^{n-1} + \cdots + \alpha_0\right) \\
&\quad - \left(\beta_m t^m + \beta_{m-1} t^{m-1} + \cdots + \beta_0\right) \frac{\alpha_n}{\beta_m} t^{n-m} \\
&= \left(\alpha_n t^n + \alpha_{n-1} t^{n-1} + \cdots + \alpha_0\right) \\
&\quad - \left(\alpha_n t^n + \beta_{m-1}\frac{\alpha_n}{\beta_m} t^{n-1} + \cdots + \beta_0 \frac{\alpha_n}{\beta_m} t^{n-m}\right) \\
&= 0 \cdot t^n + \left(\alpha_{n-1} - \beta_{m-1}\frac{\alpha_n}{\beta_m}\right) t^{n-1} + \cdots.
\end{aligned}
$$

Thus, $\deg(r_1) < n = \deg(q)$.

If $\deg(r_1) < \deg(p)$, we are finished; otherwise, we use the same construction to get

$$r_1 = pd_2 + r_2,$$

$$\deg(r_2) < \deg(r_1).$$

We then continue this process and construct

$$r_k = pd_{k+1} + r_{k+1},$$

$$\deg(r_{k+1}) < \deg(r_k).$$

Eventually, we must arrive at a situation where $\deg(r_k) \geq \deg(p)$ while $\deg(r_{k+1}) < \deg(p)$.

Collecting each step in this process, we see that

$$\begin{aligned}
q &= pd_1 + r_1 \\
&= pd_1 + pd_2 + r_2 \\
&= p(d_1 + d_2) + r_2 \\
&\ \ \vdots \\
&= p(d_1 + d_2 + \cdots + d_{k+1}) + r_{k+1}.
\end{aligned}$$

This proves the theorem.   □

The Euclidean algorithm is the central construction that makes all of the following results work.

Proposition 2.1.2. *Let $p \in \mathbb{F}[t]$ and $\lambda \in \mathbb{F}$. $(t - \lambda)$ divides p if and only if λ is a root of p, i.e., $p(\lambda) = 0$.*

Proof. If $(t - \lambda)$ divides p, then $p = (t - \lambda)q$. Hence, $p(\lambda) = 0 \cdot q(\lambda) = 0$.

Conversely, use the Euclidean algorithm to write

$$p = (t - \lambda)q + r,$$

$$\deg(r) < \deg(t - \lambda) = 1.$$

This means that $r = \beta \in \mathbb{F}$. Now, evaluate this at λ

$$\begin{aligned}
0 &= p(\lambda) \\
&= (\lambda - \lambda)q(\lambda) + r \\
&= r \\
&= \beta.
\end{aligned}$$

Thus, $r = 0$ and $p = (t - \lambda)q$.   □

This gives us an important corollary.

Corollary 2.1.3. *Let $p \in \mathbb{F}[t]$. If $\deg(p) = k$, then p has no more than k roots.*

Proof. We prove this by induction. When $k = 0$ or 1, there is nothing to prove. If p has a root $\lambda \in \mathbb{F}$, then $p = (t - \lambda)q$, where $\deg(q) < \deg(p)$. Thus, q has no more than $\deg(q)$ roots. In addition, we have that $\mu \neq \lambda$ is a root of p if and only if it is a root of q. Thus p, cannot have more than $1 + \deg(q) \leq \deg(p)$ roots. $\quad\square$

In the next proposition, we show that two polynomials always have a *greatest common divisor*.

Proposition 2.1.4. *Let* $p, q \in \mathbb{F}[t]$, *then there is a unique monic polynomial* $d = \gcd\{p, q\}$ *with the property that if* d_1 *divides both* p *and* q, *then* d_1 *divides* d. *Moreover, there exist* $r, s \in \mathbb{F}[t]$ *such that* $d = pr + qs$.

Proof. Let d be a monic polynomial of smallest degree such that $d = ps_1 + qs_2$. It is clear that any polynomial d_1 that divides p and q must also divide d. So we must show that d divides p and q. We show more generally that d divides all polynomials of the form $d' = ps'_1 + qs'_2$. For such a polynomial, we have $d' = du + r$ where $\deg(r) < \deg(d)$. This implies

$$
\begin{aligned}
r &= d' - du \\
&= p\left(s'_1 - us_1\right) + q\left(s'_2 - us_2\right).
\end{aligned}
$$

It must follow that $r = 0$ as we could otherwise find a monic polynomial of the form $ps''_1 + qs''_2$ of degree $< \deg(d)$. Thus, d divides d'. In particular, d must divide $p = p \cdot 1 + q \cdot 0$ and $q = p \cdot 0 + q \cdot 1$.

To check uniqueness, assume d_1 is a monic polynomial with the property that any polynomial that divides p and q also divides d_1. This means that d divides d_1 and also that d_1 divides d. Since both polynomials are monic, this shows that $d = d_1$.

$\quad\square$

We can more generally show that for any finite collection $p_1, \ldots, p_n$ of polynomials, there is a *greatest common divisor*

$$
d = \gcd\{p_1, \ldots, p_n\}.
$$

As in the above proposition, the polynomial d is a monic polynomial of smallest degree such that

$$
d = p_1 s_1 + \cdots + p_n s_n.
$$

Moreover, it has the property that any polynomial that divides $p_1, \ldots, p_n$ also divides d. The polynomials $p_1, \ldots, p_n \in \mathbb{F}[t]$ are said to be *relatively prime* or have *no common factors* if the only monic polynomial that divides $p_1, \ldots, p_n$ is 1. In other words, $\gcd\{p_1, \ldots, p_n\} = 1$.

We can also show that two polynomials have a *least common multiple*.

Proposition 2.1.5. *Let* $p, q \in \mathbb{F}[t]$, *then there is a unique monic polynomial* $m = \operatorname{lcm}\{p, q\}$ *with the property that if* p *and* q *divide* m_1, *then* m *divides* m_1.

Proof. Let m be the monic polynomial of smallest degree that is divisible by both p and q. Note that such polynomials exist as pq is divisible by both p and q. Next, suppose that p and q divide m_1. Since $\deg(m_1) \geq \deg(m)$, we have that $m_1 = sm + r$ with $\deg(r) < \deg(m)$. Since p and q divide m_1 and m, they must also divide $m_1 - sm = r$. As m has the smallest degree with this property, it must follow that $r = 0$. Hence, m divides m_1.   □

A monic polynomial $p \in \mathbb{F}[t]$ of degree ≥ 1 is said to be *prime* or *irreducible* if the only monic polynomials from $\mathbb{F}[t]$ that divide p are 1 and p. The simplest irreducible polynomials are the linear ones $t - \alpha$. If the field $\mathbb{F} = \mathbb{C}$, then all irreducible polynomials are linear. While if the field $\mathbb{F} = \mathbb{R}$, then the only other irreducible polynomials are the quadratic ones $t^2 + \alpha t + \beta$ with negative discriminant $D = \alpha^2 - 4\beta < 0$. These two facts are not easy to prove and depend on the Fundamental Theorem of Algebra, which we discuss below.

In analogy with the prime factorization of integers, we also have a prime factorization of polynomials. Before establishing this decomposition, we need to prove a very useful property for irreducible polynomials.

Lemma 2.1.6. *Let $p \in \mathbb{F}[t]$ be irreducible. If p divides $q_1 \cdot q_2$, then p divides either q_1 or q_2.*

Proof. Let $d_1 = \gcd(p, q_1)$. Since d_1 divides p, it follows that $d_1 = 1$ or $d_1 = p$. In the latter case, $d_1 = p$ divides q_1 so we are finished. If $d_1 = 1$, then we can write $1 = pr + q_1 s$. In particular,

$$q_2 = q_2 pr + q_2 q_1 s.$$

Here we have that p divides $q_2 q_1$ and p. Thus, it also divides

$$q_2 = q_2 pr + q_2 q_1 s.$$ □

Theorem 2.1.7. (Unique Factorization of Polynomials) *Let $p \in \mathbb{F}[t]$ be a monic polynomial, then $p = p_1 \cdots p_k$ is a product of irreducible polynomials. Moreover, except for rearranging these polynomials this factorization is unique.*

Proof. We can prove this result by induction on $\deg(p)$. If p is only divisible by 1 and p, then p is irreducible and we are finished. Otherwise, $p = q_1 \cdot q_2$, where q_1 and q_2 are monic polynomials with $\deg(q_1), \deg(q_2) < \deg(p)$. By assumption, each of these two factors can be decomposed into irreducible polynomials; hence, we also get such a decomposition for p.

For uniqueness, assume that $p = p_1 \cdots p_k = q_1 \cdots q_l$ are two decompositions of p into irreducible factors. Using induction again, we see that it suffices to show that $p_1 = q_i$ for some i. The previous lemma now shows that p_1 must divide q_1 or $q_2 \cdots q_l$. In the former case, it follows that $p_1 = q_1$ as q_1 is irreducible. In the latter case, we get again that p_1 must divide q_2 or $q_3 \cdots q_l$. Continuing in this fashion, it must follow that $p_1 = q_i$ for some i.   □

If all the irreducible factors of a monic polynomial $p \in \mathbb{F}[t]$ are linear, then we say that that p *splits*. Thus, p splits if and only if

$$p(t) = (t - \alpha_1) \cdots (t - \alpha_k)$$

for $\alpha_1, \ldots, \alpha_k \in \mathbb{F}$.

Finally, we show that all complex polynomials have a root. It is curious that while this theorem is algebraic in nature, the proof is analytic. There are many completely different proofs of this theorem including ones that are far more algebraic. The one presented here, however, seems to be the most elementary.

Theorem 2.1.8. (The Fundamental Theorem of Algebra) *Any complex polynomial of degree ≥ 1 has a root.*

Proof. Let $p(z) \in \mathbb{C}[z]$ have degree $n \geq 1$. Our first claim is that we can find $z_0 \in \mathbb{C}$ such that $|p(z)| \geq |p(z_0)|$ for all $z \in \mathbb{C}$. To see why $|p(z)|$ has to have a minimum, we first observe that

$$
\begin{aligned}
\frac{p(z)}{z^n} &= \frac{a_n z^n + a_{n-1} z^{n-1} + \cdots + a_1 z + a_0}{z^n} \\
&= a_n + a_{n-1} \frac{1}{z} + \cdots + a_1 \frac{1}{z^{n-1}} + a_0 \frac{1}{z^n} \\
&\rightarrow a_n \text{ as } z \rightarrow \infty.
\end{aligned}
$$

Since $a_n \neq 0$, we can therefore choose $R > 0$ so that

$$
|p(z)| \geq \frac{|a_n|}{2} |z|^n \text{ for } |z| \geq R.
$$

By possibly increasing R further, we can also assume that

$$
\frac{|a_n|}{2} |R|^n \geq |p(0)|.
$$

On the compact set $\bar{B}(0, R) = \{z \in \mathbb{C} : |z| \leq R\}$, we can now find z_0 such that $|p(z)| \geq |p(z_0)|$ for all $z \in \bar{B}(0, R)$. By our assumptions, this also holds when $|z| \geq R$ since in that case

$$
\begin{aligned}
|p(z)| &\geq \frac{|a_n|}{2} |z|^n \\
&\geq \frac{|a_n|}{2} |R|^n \\
&\geq |p(0)| \\
&\geq |p(z_0)|.
\end{aligned}
$$

Thus, we have found our global minimum for $|p(z)|$.

If $|p(z_0)| = 0$, then z_0 is a root and we are finished. Otherwise we can define a new polynomial of degree $n \geq 1$:

$$
q(z) = \frac{p(z + z_0)}{p(z_0)}.
$$

This polynomial satisfies

$$q(0) = \frac{p(z_0)}{p(z_0)} = 1,$$

$$|q(z)| = \left| \frac{p(z+z_0)}{p(z_0)} \right|$$

$$\geq \left| \frac{p(z_0)}{p(z_0)} \right|$$

$$= 1$$

Thus,

$$q(z) = 1 + b_k z^k + \cdots + b_n z^n,$$

where $b_k \neq 0$. We can now investigate what happens to $q(z)$ for small z. We first note that

$$q(z) = 1 + b_k z^k + b_{k+1} z^{k+1} + \cdots + b_n z^n$$
$$= 1 + b_k z^k + \left(b_{k+1} z + \cdots + b_n z^{n-k} \right) z^k,$$

where

$$\left(b_{k+1} z + \cdots + b_n z^{n-k} \right) \to 0 \text{ as } z \to 0.$$

If we write $z = re^{i\theta}$ and fix θ so that

$$b_k e^{ik\theta} = -|b_k|,$$

then

$$|q(z)| = \left| 1 + b_k z^k + \left(b_{k+1} z + \cdots + b_n z^{n-k} \right) z^k \right|$$
$$= \left| 1 - |b_k| \, r^k + \left(b_{k+1} z + \cdots + b_n z^{n-k} \right) r^k e^{ik\theta} \right|$$
$$\leq 1 - |b_k| \, r^k + \left| \left(b_{k+1} z + \cdots + b_n z^{n-k} \right) r^k e^{ik\theta} \right|$$
$$= 1 - |b_k| \, r^k + \left| b_{k+1} z + \cdots + b_n z^{n-k} \right| r^k$$
$$\leq 1 - \frac{|b_k|}{2} r^k$$

as long as r is chosen so small that $1 - |b_k| \, r^k > 0$ and $\left| b_{k+1} z \cdots + b_n z^{n-k} \right| \leq \frac{|b_k|}{2}$. This, however, implies that $\left| q\left(re^{i\theta} \right) \right| < 1$ for small r. We have therefore arrived at a contradiction. $\qquad\qquad\square$

2.2   Linear Differential Equations*

In this section, we shall study linear differential equations. Everything we have learned about linear independence, bases, special matrix representations, etc. will be extremely useful when trying to solve such equations. In fact, we shall in several sections of this text see that virtually every development in linear algebra can be used to understand the structure of solutions to linear differential equations. It is, however, possible to skip this section if one does not want to be bothered by differential equations while learning linear algebra.

We start with systems of differential equations:

$$\dot{x}_1 = a_{11}x_1 + \cdots + a_{1m}x_m + b_1$$
$$\vdots \qquad\qquad \vdots$$
$$\dot{x}_m = a_{n1}x_1 + \cdots + a_{nm}x_m + b_n,$$

where $a_{ij}, b_i \in C^\infty([a,b], \mathbb{C})$ (or just $C^\infty([a,b], \mathbb{R})$) and the functions $x_j :$ $[a,b] \to \mathbb{C}$ are to be determined. We can write the system in matrix form and also rearrange it a bit to make it look like we are solving $L(x) = b$. To do this, we use

$$x = \begin{bmatrix} x_1 \\ \vdots \\ x_m \end{bmatrix}, \ b = \begin{bmatrix} b_1 \\ \vdots \\ b_n \end{bmatrix}, \ A = \begin{bmatrix} a_{11} & \cdots & a_{1m} \\ \vdots & \ddots & \vdots \\ a_{n1} & \cdots & a_{nm} \end{bmatrix}$$

and define

$$L : C^\infty([a,b], \mathbb{C}^m) \to C^\infty([a,b], \mathbb{C}^n)$$
$$L(x) = \dot{x} - Ax.$$

The equation $L(x) = 0$ is called the *homogeneous system*. We note that the following three properties can be used as a general outline for what to do:

1. $L(x) = b$ can be solved if and only if $b \in \text{im}(L)$.
2. If $L(x_0) = b$ and $x \in \text{ker}(L)$, then $L(x + x_0) = b$.
3. If $L(x_0) = b$ and $L(x_1) = b$, then $x_0 - x_1 \in \text{ker}(L)$.

The specific implementation of actually solving the equations, however, is quite different from what we did with systems of (algebraic) equations.

First of all, we only consider the case where $n = m$. This implies that for given $t_0 \in [a,b]$ and $x_0 \in \mathbb{C}^n$, the *initial value problem*

$$L(x) = b,$$
$$x(t_0) = a_0$$

has a unique solution $x \in C^{\infty}\left([a,b],\mathbb{C}^n\right)$. We shall not prove this result in this generality, but we shall eventually see why this is true when the matrix A has entries that are constants rather than functions (see Sect. 3.7). As we learn more about linear algebra, we shall revisit this problem and slowly try to gain a better understanding of it. For now, let us just note an important consequence.

Theorem 2.2.1. *The complete collection of solutions to*

$$\dot{x}_1 = a_{11}x_1 + \cdots + a_{1n}x_n + b_1$$
$$\vdots \qquad\qquad \vdots$$
$$\dot{x}_n = a_{n1}x_1 + \cdots + a_{nn}x_n + b_n$$

can be found by finding one solution x_0 and then adding it to the solutions of the homogeneous equation $L(z) = 0$, i.e.,

$$x = z + x_0,$$
$$L(z) = 0;$$

moreover, $\dim\left(\ker\left(L\right)\right) = n.$

Some particularly interesting and important linear equations are the nth order equations

$$D^n x + a_{n-1}D^{n-1}x + \cdots + a_1 Dx + a_0 x = b,$$

where $D^k x$ is the kth order derivative of x. If we assume that $a_{n-1}, \ldots, a_0, b \in C^{\infty}\left([a,b],\mathbb{C}\right)$ and define

$$L : C^{\infty}\left([a,b],\mathbb{C}\right) \to C^{\infty}\left([a,b],\mathbb{C}\right)$$
$$L(x) = \left(D^n + a_{n-1}D^{n-1} + \cdots + a_1 D + a_0\right)(x),$$

then we have a nice linear problem just as in the previous cases of linear systems of differential or algebraic equations. The problem of solving $L(x) = b$ can also be reinterpreted as a linear system of differential equations by defining

$$x_1 = x, x_2 = Dx, \ldots, x_n = D^{n-1}x$$

and then considering the system

$$\dot{x}_1 = \qquad\qquad x_2$$
$$\dot{x}_2 = \qquad\qquad x_3$$
$$\vdots \qquad\qquad\qquad \vdots$$
$$\dot{x}_n = -a_{n-1}x_n - \cdots - a_1 x_2 - a_0 x_1 + b_n.$$

This will not help us in solving the desired equation, but it does tells us that the initial value problem

$$L(x) = b,$$
$$x(t_0) = c_0, Dx(t_0) = c_1, \ldots, D^{n-1}x(t_0) = c_{n-1},$$

has a unique solution, and hence, the above theorem can be paraphrased.

Theorem 2.2.2. *The complete collection of solutions to*

$$D^n x + a_{n-1} D^{n-1} x + \cdots + a_1 Dx + a_0 x = b$$

can be found by finding one solution x_0 and then adding it to the solutions of the homogeneous equation $L(z) = 0$, i.e.,

$$x = z + x_0,$$
$$L(z) = 0.$$

Moreover, $\dim(\ker(L)) = n$.

It is not hard to give a complete account of how to solve the homogeneous problem $L(x) = 0$ when $a_0, \ldots, a_{n-1} \in \mathbb{C}$ are constants. Let us start with $n = 1$. Then we are trying to solve

$$Dx + a_0 x = \dot{x} + a_0 x = 0.$$

Clearly, $x = \exp(-a_0 t)$, is a solution and the complete set of solutions is

$$x = c \exp(-a_0 t), c \in \mathbb{C}.$$

The initial value problem

$$\dot{x} + a_0 x = 0,$$
$$x(t_0) = c_0$$

has the solution

$$x = c_0 \exp(-a_0 (t - t_0)).$$

The trick to solving the higher order case is to note that we can rewrite L as

$$L = D^n + a_{n-1} D^{n-1} + \cdots + a_1 D + a_0$$
$$= p(D).$$

This makes L look like a polynomial where D is the variable. The corresponding polynomial

$$p(t) = t^n + a_{n-1} t^{n-1} + \cdots + a_1 t + a_0$$

is called the *characteristic polynomial*. The idea behind solving these equations comes from

Proposition 2.2.3. (The Reduction Principle) *If $q(t) = t^m + b_{m-1}t^{m-1} + \cdots + b_0$ is a polynomial that divides $p(t) = t^n + a_{n-1}t^{n-1} + \cdots + a_1 t + a_0$, then any solution to $q(D)(x) = 0$ is also a solution to $p(D)(x) = 0$.*

Proof. This simply hinges of observing that $p(t) = r(t)q(t)$, then $p(D) = r(D)q(D)$. So by evaluating the latter on x, we get $p(D)(x) = r(D)(q(D)(x)) = 0$. $\square$

The simplest factors are, of course, the linear factors $t - \lambda$, and we know that the solutions to

$$(D - \lambda)(x) = Dx - \lambda x = 0$$

are given by $x(t) = C \exp(\lambda t)$. This means that we should be looking for roots to $p(t)$. These roots are called *eigenvalues* or *characteristic values*. The Fundamental Theorem of Algebra asserts that any polynomial $p \in \mathbb{C}[t]$ can be factored over the complex numbers

$$p(t) = t^n + a_{n-1}t^{n-1} + \cdots + a_1 t + a_0$$
$$= (t - \lambda_1)^{k_1} \cdots (t - \lambda_m)^{k_m}.$$

Here the roots $\lambda_1, \ldots, \lambda_m$ are assumed to be distinct, each occurs with multiplicity $k_1, \ldots, k_m$, and $k_1 + \cdots + k_m = n$.

The original equation

$$L = D^n + a_{n-1}D^{n-1} + \cdots + a_1 D + a_0,$$

then factors

$$L = D^n + a_{n-1}D^{n-1} + \cdots + a_1 D + a_0$$
$$= (D - \lambda_1)^{k_1} \cdots (D - \lambda_m)^{k_m}.$$

Thus, the original problem has been reduced to solving the equations

$$(D - \lambda_1)^{k_1}(x) = 0,$$

$$\vdots \quad \vdots$$

$$(D - \lambda_m)^{k_m}(x) = 0.$$

Note that if we had not insisted on using the more abstract and less natural complex numbers, we would not have been able to make the reduction so easily. If we are in a case where the differential equation is real and there is a good physical reason for keeping solutions real as well, then we can still solve it as if it were complex and then take real and imaginary parts of the complex solutions to get real ones. It would seem that the n complex solutions would then lead to $2n$ real ones. This is not really the case. First, observe that each real eigenvalue λ only gives rise to a one parameter family of real solutions $c \exp(\lambda(t - t_0))$. As for complex eigenvalues, we know that real polynomials have the property that complex roots come in conjugate pairs. Then, we note that $\exp(\lambda(t-t_0))$ and $\exp(\bar{\lambda}(t-t_0))$ up to sign have the same real and imaginary parts, and so these pairs of eigenvalues only lead to a two-parameter family of real solutions which if $\lambda = \lambda_1 + i\lambda_2$ looks like

$$c \exp(\lambda_1(t-t_0))\cos(\lambda_2(t-t_0)) + d \exp(\lambda_1(t-t_0))\sin(\lambda_2(t-t_0)).$$

Let us return to the complex case again. If $m = n$ and $k_1 = \cdots = k_m = 1$, we simply get n first-order equations, and we see that the complete set of solutions to $L(x) = 0$ is given by

$$x = c_1 \exp(\lambda_1 t) + \cdots + c_n \exp(\lambda_n t).$$

It should be noted that we need to show that $\exp(\lambda_1 t),\ldots,\exp(\lambda_n t)$ are linearly independent in order to show that we have found all solutions. This was discussed in Example 1.12.15 and will also be established in Sect. 2.5.

With a view towards solving the initial value problem, we rewrite the solution as

$$x = d_1 \exp(\lambda_1(t-t_0)) + \cdots + d_n \exp(\lambda_n(t-t_0)).$$

To solve the initial value problem requires differentiating this expression several times and then solving

$$x(t_0) = d_1 + \cdots + d_n,$$
$$Dx(t_0) = \lambda_1 d_1 + \cdots + \lambda_n d_n,$$

$$\vdots \quad \vdots$$

$$D^{n-1}x(t_0) = \lambda_1^{n-1} d_1 + \cdots + \lambda_n^{n-1} d_n$$

for $d_1,\ldots,d_n$. In matrix form, this becomes

$$\begin{bmatrix} 1 & \cdots & 1 \\ \lambda_1 & \cdots & \lambda_n \\ \vdots & \ddots & \vdots \\ \lambda_1^{n-1} & \cdots & \lambda_n^{n-1} \end{bmatrix} \begin{bmatrix} d_1 \\ \vdots \\ d_n \end{bmatrix} = \begin{bmatrix} x(t_0) \\ \dot{x}(t_0) \\ \vdots \\ x^{(n-1)}(t_0) \end{bmatrix}.$$

In Example 1.12.12, we saw that this matrix has rank n if $\lambda_1,\ldots,\lambda_n$ are distinct. Thus, we can solve for the ds in this case.

When roots have multiplicity, things get a little more complicated. We first need to solve the equation

$$(D-\lambda)^k(x) = 0.$$

One can check that the k functions $\exp(\lambda t), t\exp(\lambda t),\ldots,t^{k-1}\exp(\lambda t)$ are solutions to this equation. One can also prove that they are linearly independent using that $1, t,\ldots,t^{k-1}$ are linearly independent. This will lead us to a complete set of solutions to $L(x) = 0$ even when we have multiple roots. The issue of solving the initial value is somewhat more involved due to the problem of taking derivatives of $t^l \exp(\lambda t)$. This can be simplified a little by considering the solutions $\exp(\lambda(t - t_0)), (t - t_0)\exp(\lambda(t - t_0)), \ldots, (t - t_0)^{k-1}\exp(\lambda(t - t_0))$.

For the sake of illustration, let us consider the simplest case of trying to solve $(D - \lambda)^2(x) = 0$. The complete set of solutions can be parametrized as

$$x = d_1 \exp(\lambda(t - t_0)) + d_2(t - t_0)\exp(\lambda(t - t_0)).$$

Then,

$$Dx = \lambda d_1 \exp(\lambda(t - t_0)) + (1 + \lambda(t - t_0))d_2 \exp(\lambda(t - t_0)).$$

Thus, we have to solve

$$x(t_0) = d_1,$$
$$Dx(t_0) = \lambda d_1 + d_2.$$

This leads us to the system

$$\begin{bmatrix} 1 & 0 \\ \lambda & 1 \end{bmatrix} \begin{bmatrix} d_1 \\ d_2 \end{bmatrix} = \begin{bmatrix} x(t_0) \\ Dx(t_0) \end{bmatrix}.$$

If $\lambda = 0$, we are finished. Otherwise, we can multiply the first equation by λ and subtract it from the second to obtain

$$\begin{bmatrix} 1 & 0 \\ 0 & 1 \end{bmatrix} \begin{bmatrix} d_1 \\ d_2 \end{bmatrix} = \begin{bmatrix} x(t_0) \\ Dx(t_0) - \lambda x(t_0) \end{bmatrix}.$$

Thus, the solution to the initial value problem is

$$x = x(t_0)\exp(\lambda(t - t_0)) + (Dx(t_0) - \lambda x(t_0))(t - t_0)\exp(\lambda(t - t_0)).$$

A similar method of finding a characteristic polynomial and its roots can also be employed in solving linear systems of equations as well as homogeneous systems of linear differential with constant coefficients. The problem lies in deciding what the characteristic polynomial should be and what its roots mean for the system. This will be studied in subsequent sections and chapters. In Sects. 2.6 and 2.7, we shall also see that systems of first-order differential equations can be solved using our knowledge of higher order equations.

For now, let us see how one can approach systems of linear differential equations from the point of view of first trying to define the eigenvalues. We are considering the homogeneous problem

$$L(x) = \dot{x} - Ax = 0,$$

where A is an $n \times n$ matrix with real or complex numbers as entries. If the system is *decoupled*, i.e., $\dot{x}_i$, depends only on x_i then we have n first-order equations that can be solved as above. In this case, the entries that are not on the diagonal of A are zero. A particularly simple case occurs when $A = \lambda 1_{\mathbb{C}^n}$ for some λ. In this case, the general solution is given by

$$x = a_0 \exp\left(\lambda\left(t - t_0\right)\right).$$

We now observe that for fixed a_0, this is still a solution to the general equation $\dot{x} = Ax$ provided only that $Aa_0 = \lambda a_0$. Thus, we are lead to seek pairs of scalars λ and vectors a_0 such that $Aa_0 = \lambda a_0$. If we can find such pairs where $a_0 \neq 0$, then we call λ an *eigenvalue* for A and a_0 and *eigenvector* for λ. Therefore, if we can find a basis $v_1, \ldots, v_n$ for $\mathbb{R}^n$ or $\mathbb{C}^n$ of eigenvectors with $Av_1 = \lambda_1 v_1, \ldots, Av_n = \lambda_n v_x$, then we have that the complete solution must be

$$x = v_1 \exp\left(\lambda_1\left(t - t_0\right)\right)c_1 + \cdots + v_n \exp\left(\lambda_n\left(t - t_0\right)\right)c_n.$$

The initial value problem $L(x) = 0$, $x(t_0) = x_0$ is then handled by solving

$$v_1 c_1 + \cdots + v_n c_n = \begin{bmatrix} v_1 \cdots v_n \end{bmatrix} \begin{bmatrix} c_1 \\ \vdots \\ c_n \end{bmatrix} = x_0.$$

Since $v_1, \ldots, v_n$ was assumed to be a basis, we know that this system can be solved. Gauss elimination can then be used to find $c_1, \ldots, c_n$.

What we accomplished by this change of basis was to decouple the system in a different coordinate system. One of the goals in the study of linear operators is to find a basis that makes the matrix representation of the operator as simple as possible. As we have just seen, this can then be used to great effect in solving what might appear to be a rather complicated problem. Even so, it might not be possible to find the desired basis of eigenvectors. This happens if we consider the second-order equation $(D - \lambda)^2 = 0$ and convert it to a system

$$\begin{bmatrix} \dot{x}_1 \\ \dot{x}_2 \end{bmatrix} = \begin{bmatrix} 0 & 1 \\ -\lambda^2 & 2\lambda \end{bmatrix} \begin{bmatrix} x_1 \\ x_2 \end{bmatrix}.$$

Here the general solution to $(D - \lambda)^2 = 0$ is of the form

$$x = x_1 = c_1 \exp\left(\lambda t\right) + c_2 t \exp\left(\lambda t\right)$$

so,

$$x_2 = \dot{x}_1 = c_1 \lambda \exp\left(\lambda t\right) + c_2 \left(\lambda t + 1\right) \exp\left(\lambda t\right).$$

This means that

$$\begin{bmatrix} x_1 \\ x_2 \end{bmatrix} = c_1 \begin{bmatrix} 1 \\ \lambda \end{bmatrix} \exp(\lambda t) + c_2 \begin{bmatrix} t \\ \lambda t + 1 \end{bmatrix} \exp(\lambda t).$$

Since we cannot write this in the form

$$\begin{bmatrix} x_1 \\ x_2 \end{bmatrix} = c_1 v_1 \exp(\lambda_1 t) + c_2 v_2 \exp(\lambda_2 t),$$

there cannot be any reason to expect that a basis of eigenvectors can be found even for the simple matrix

$$A = \begin{bmatrix} 0 & 1 \\ 0 & 0 \end{bmatrix}.$$

In Sect. 2.3 we shall see that any square matrix and indeed any linear operator on a finite-dimensional vector space has a characteristic polynomial whose roots are the eigenvalues of the map. Having done that, we shall in Sect. 2.4 and especially Sect. 2.5 try to determine exactly what properties of the linear map further guarantee that it admits a basis of eigenvectors. In Sects. 2.6–2.8, we shall show that any system of equations can be transformed into a new system that looks like several uncoupled higher order equations.

There is another rather intriguing way of solving linear differential equations by reducing them to *recurrences*. We will emphasize higher order equations, but it works equally well with systems. The goal is to transform the differential equation:

$$D^n x + a_{n-1} D^{n-1} x + \cdots + a_1 D x + a_0 x = p(D)(x) = 0$$

into something that can be solved using combinatorial methods.

Assume that x is given by its MacLaurin expansion

$$x(t) = \sum_{k=0}^{\infty} (D^k x)(0) \frac{t^k}{k!}$$

$$= \sum_{k=0}^{\infty} c_k \frac{t^k}{k!}.$$

The derivative is then given by

$$Dx = \sum_{k=1}^{\infty} c_k \frac{t^{k-1}}{(k-1)!}$$

$$= \sum_{k=0}^{\infty} c_{k+1} \frac{t^k}{k!},$$

and more generally

$$D^l x = \sum_{k=0}^{\infty} c_{k+l} \frac{t^k}{k!}.$$

Thus, the derivative of x is simply a shift in the index for the sequence (c_k). The differential equation now gets to look like

$$D^n x + a_{n-1} D^{n-1} x + \cdots + a_1 Dx + a_0 x$$

$$= \sum_{k=0}^{\infty} (c_{k+n} + a_{n-1} c_{k+n-1} + \cdots + a_1 c_{k+1} + a_0 c_k) \frac{t^k}{k!}.$$

From this we can conclude that x is a solution if and only if the sequence c_k solves the *linear nth-order recurrence*

$$c_{k+n} + a_{n-1} c_{k+n-1} + \cdots + a_1 c_{k+1} + a_0 c_k = 0$$

or

$$c_{k+n} = -(a_{n-1} c_{k+n-1} + \cdots + a_1 c_{k+1} + a_0 c_k).$$

For such a sequence, it is clear that we need to know the initial values $c_0, \ldots, c_{n-1}$ in order to find the whole sequence. This corresponds to the initial value problem for the corresponding differential equation as $c_k = (D^k x)(0)$.

The correspondence between systems $\dot{x} = Ax$ and recurrences of vectors $c_{n+1} = Ac_n$ comes about by assuming that the solution to the differential equation looks like

$$x(t) = \sum_{n=0}^{\infty} c_n \frac{t^n}{n!},$$

$$c_n \in \mathbb{C}^n.$$

Finally, we point out that in Sect. 2.9, we offer an explicit algorithm for reducing systems of possibly higher order equations to independent higher order equations.

Exercises

1. Find the solution to the differential equations with the general initial values: $x(t_0) = x_0$, $\dot{x}(t_0) = \dot{x}_0$, and $\ddot{x}(t_0) = \ddot{x}_0$.

   (a) $\dddot{x} - 3\ddot{x} + 3\dot{x} - x = 0$.
   (b) $\dddot{x} - 5\ddot{x} + 8\dot{x} - 4x = 0$.
   (c) $\dddot{x} + 6\ddot{x} + 11\dot{x} + 6x = 0$.

2. Find the complete solution to the initial value problems.

(a) $\begin{bmatrix} \dot{x} \\ \dot{y} \end{bmatrix} = \begin{bmatrix} 0 & 2 \\ 1 & 3 \end{bmatrix} \begin{bmatrix} x \\ y \end{bmatrix}$, where $\begin{bmatrix} x(t_0) \\ y(t_0) \end{bmatrix} = \begin{bmatrix} x_0 \\ y_0 \end{bmatrix}$.

(b) $\begin{bmatrix} \dot{x} \\ \dot{y} \end{bmatrix} = \begin{bmatrix} 0 & 1 \\ 1 & 2 \end{bmatrix} \begin{bmatrix} x \\ y \end{bmatrix}$, where $\begin{bmatrix} x(t_0) \\ y(t_0) \end{bmatrix} = \begin{bmatrix} x_0 \\ y_0 \end{bmatrix}$.

3. Find the real solution to the differential equations with the general initial values: $x(t_0) = x_0$, $\dot{x}(t_0) = \dot{x}_0$, and $\ddot{x}(t_0) = \ddot{x}_0$ in the third-order cases.

(a) $\ddot{x} + x = 0$.

(b) $\ddot{x} + \dot{x} = 0$.

(c) $\ddot{x} - 6\dot{x} + 25x = 0$.

(d) $\dddot{x} - 5\ddot{x} + 19\dot{x} + 25 = 0$.

4. Consider the vector space $C^\infty([a,b], \mathbb{C}^n)$ of infinitely differentiable curves in $\mathbb{C}^n$ and let $z_1, \ldots, z_n \in C^\infty([a,b], \mathbb{C}^n)$.

(a) Show that if we can find $t_0 \in [a,b]$ so that the vectors $z_1(t_0), \ldots, z_n(t_0) \in \mathbb{C}^n$ are linearly independent, then the functions $z_1, \ldots, z_n \in C^\infty([a,b], \mathbb{C}^n)$ are also linearly independent.

(b) Find a linearly independent pair $z_1, z_2 \in C^\infty([a,b], \mathbb{C}^2)$ so that $z_1(t), z_2(t) \in \mathbb{C}^2$ are linearly dependent for all $t \in [a,b]$.

(c) Assume now that each $z_1, \ldots, z_n$ solves the linear differential equation $\dot{x} = Ax$. Show that if $z_1(t_0), \ldots, z_n(t_0) \in \mathbb{C}^n$ are linearly dependent for some t_0, then $z_1, \ldots, z_n \in C^\infty([a,b], \mathbb{C}^n)$ are linearly dependent as well.

5. Let $p(t) = (t - \lambda_1) \cdots (t - \lambda_n)$, where we allow multiplicities among the roots.

(a) Show that $(D - \lambda)(x) = f(t)$ has

$$x = \exp(\lambda t) \int_0^t \exp(-\lambda s) f(s) \, ds$$

as a solution.

(b) Show that a solution x to $p(D)(x) = f$ can be found by successively solving

$$(D - \lambda_1)(z_1) = f,$$
$$(D - \lambda_2)(z_2) = z_1,$$

$$\vdots \quad \vdots$$

$$(D - \lambda_n)(z_n) = z_{n-1}.$$

6. Show that the initial value problem

$$\dot{x} = Ax,$$

$$x\,(t_0) = x_0$$

can be solved "explicitly" if A is upper (or lower) triangular. This holds even in the case where the entries of A and b are functions of t.

7. Assume that $x\,(t)$ is a solution to $\dot{x} = Ax$, where $A \in \mathrm{Mat}_{n \times n}\,(\mathbb{C})$.

   (a) Show that the phase shifts $x_\omega\,(t) = x\,(t + \omega)$ are also solutions.
   (b) Show that if the vectors $x\,(\omega_1), \ldots, x\,(\omega_n)$ form a basis for $\mathbb{C}^n$, then all solutions to $\dot{x} = Ax$ are linear combinations of the phase-shifted solutions $x_{\omega_1}, \ldots, x_{\omega_n}$.

8. Assume that x is a solution to $p\,(D)\,(x) = 0$, where $p\,(D) = D^n + \cdots + a_1 D + a_0$.

   (a) Show that the phase shifts $x_\omega\,(t) = x\,(t + \omega)$ are also solutions.
   (b) Show that, if the vectors

$$
\begin{bmatrix} x\,(\omega_1) \\ Dx\,(\omega_1) \\ \vdots \\ D^{n-1}x\,(\omega_1) \end{bmatrix}, \ldots,
\begin{bmatrix} x\,(\omega_n) \\ Dx\,(\omega_n) \\ \vdots \\ D^{n-1}x\,(\omega_n) \end{bmatrix}
$$

   form a basis for $\mathbb{C}^n$, then all solutions to $p\,(D)\,(x) = 0$ are linear combinations of the phase shifted solutions $x_{\omega_1}, \ldots, x_{\omega_n}$.

9. Let $p\,(t) = (t - \lambda_1) \cdots (t - \lambda_n)$. Show that the higher order equation $L\,(y) = p\,(D)\,(y) = 0$ can be made into a system of equations $\dot{x} - Ax = 0$, where

$$
A = \begin{bmatrix}
\lambda_1 & 1 & & 0 \\
0 & \lambda_2 & \ddots & \\
& & \ddots & 1 \\
0 & & & \lambda_n
\end{bmatrix}
$$

   by choosing

$$
x = \begin{bmatrix}
y \\
(D - \lambda_1)\,y \\
\vdots \\
(D - \lambda_1) \cdots (D - \lambda_{n-1})\,y
\end{bmatrix}.
$$

10. Show that $p(t) \exp(\lambda t)$ solves $(D - \lambda)^k x = 0$ if $p(t) \in \mathbb{C}[t]$ and $\deg(p) \leq k - 1$. Conclude that $\ker((D - \lambda)^k)$ contains a k-dimensional subspace.

11. Let $V = \mathrm{span}\{\exp(\lambda_1 t), \ldots, \exp(\lambda_n t)\}$, where $\lambda_1, \ldots, \lambda_n \in \mathbb{C}$ are distinct.

   (a) Show that $\exp(\lambda_1 t), \ldots, \exp(\lambda_n t)$ form a basis for V. Hint: One way of doing this is to construct a linear isomorphism

$$L : V \to \mathbb{C}^n$$
$$L(f) = (f(t_1), \ldots, f(t_n))$$

   by selecting suitable points $t_1, \ldots, t_n \in \mathbb{R}$ depending on $\lambda_1, \ldots, \lambda_n \in \mathbb{C}$ such that $L(\exp(\lambda_i t))$, $i = 1, \ldots, n$ form a basis.

   (b) Show that if $x \in V$, then $Dx \in V$.

   (c) Compute the matrix representation for the linear operator $D : V \to V$ with respect to $\exp(\lambda_1 t), \ldots, \exp(\lambda_n t)$.

   (d) More generally, show that $p(D) : V \to V$, where $p(D) = a_k D^k + \cdots + a_1 D + a_0 1_V$.

   (e) Show that $p(D) = 0$ if and only if $\lambda_1, \ldots, \lambda_n$ are all roots of $p(t)$.

12. Let $p \in \mathbb{C}[t]$ and consider $\ker(p(D)) = \{x : p(D)(x) = 0\}$, i.e., it is the space of solutions to $p(D) = 0$.

   (a) Assuming unique solutions to initial value problems, show that

$$\dim_{\mathbb{C}} \ker(p(D)) = \deg p = n.$$

   (b) Show that $D : \ker(p(D)) \to \ker(p(D))$ (see also Exercise 3 in Sect. 1.11).

   (c) Show that $q(D) : \ker(p(D)) \to \ker(p(D))$ for any polynomial $q(t) \in \mathbb{C}[t]$.

   (d) Show that $\ker(p(D))$ has a basis for the form $x, Dx, \ldots, D^{n-1}x$. Hint: Let x be the solution to $p(D)(x) = 0$ with the initial values $x(0) = Dx(0) = \cdots = D^{n-2}x(0) = 0$ and $D^{n-1}x(0) = 1$.

13. Let $p \in \mathbb{R}[t]$ and consider

$$\ker_{\mathbb{R}}(p(D)) = \{x : \mathbb{R} \to \mathbb{R} : p(D)(x) = 0\},$$
$$\ker_{\mathbb{C}}(p(D)) = \{z : \mathbb{R} \to \mathbb{C} : p(D)(z) = 0\}$$

   i.e., the real-valued, respectively, complex-valued solutions.

   (a) Show that $x \in \ker_{\mathbb{R}}(p(D))$ if and only if $x = \mathrm{Re}(z)$ where $z \in \ker_{\mathbb{C}}(p(D))$.

   (b) Show that $\dim_{\mathbb{C}} \ker(p(D)) = \deg p = \dim_{\mathbb{R}} \ker(p(D))$.

2.3 Eigenvalues

We are now ready to give the abstract definitions for eigenvalues and eigenvectors.

Definition 2.3.1. Consider a linear operator $L : V \to V$ on a vector space over $\mathbb{F}$. If we have a scalar $\lambda \in \mathbb{F}$ and a vector $x \in V - \{0\}$ such that $L(x) = \lambda x$, then we say that λ is an *eigenvalue* of L and x is an *eigenvector* for λ. If we add the zero vector to the space of eigenvectors for λ, then it can be identified with the subspace

$$\ker(L - \lambda 1_V) = \{x \in V : L(x) - \lambda x = 0\} \subset V.$$

This is also called the *eigenspace* for λ. In many texts, this space is often denoted

$$E_\lambda = \ker(L - \lambda 1_V).$$

Eigenvalues are also called proper values or characteristic values in some texts. "Eigen" is a German adjective that often is translated as "own" or "proper" (think "property").

For linear operators defined by $n \times n$ matrices, we can give a procedure for computing the eigenvalues/vectors using Gauss elimination. The more standard method that employs determinants can be found in virtually every other book on linear algebra and will be explained in Sect. 5.7. We start by considering a matrix $A \in \mathrm{Mat}_{n \times n}(\mathbb{F})$. If we wish to find an eigenvalue λ for A, then we need to determine when there is a nontrivial solution to $(A - \lambda 1_{\mathbb{F}^n})(x) = 0$. In other words, the augmented system

$$\begin{bmatrix} \alpha_{11} - \lambda & \cdots & \alpha_{1n} & 0 \\ \vdots & \ddots & \vdots & \vdots \\ \alpha_{n1} & \cdots & \alpha_{nn} - \lambda & 0 \end{bmatrix}$$

should have a nontrivial solution. This is something we know how to deal with using Gauss elimination. The only complication is that if λ is simply an abstract number, then it can be a bit tricky to decide when we are allowed to divide by expressions that involve λ.

Note that we do not necessarily need to carry the last column of zeros through the calculations as row reduction will never change those entries. Thus, we only need to do row reduction on $A - \lambda 1_{\mathbb{F}^n}$ or if convenient $\lambda 1_{\mathbb{F}^n} - A$.

Example 2.3.2. Assume that $\mathbb{F} = \mathbb{C}$ and let

$$A = \begin{bmatrix} 0 & 1 & 0 & 0 \\ -1 & 0 & 0 & 0 \\ 0 & 0 & 0 & 1 \\ 0 & 0 & 1 & 0 \end{bmatrix}.$$

Row reduction tells us the augmented system $[A - \lambda 1_{\mathbb{C}^4}|0]$ becomes

$$\begin{bmatrix} -\lambda & 1 & 0 & 0 & 0 \\ -1 & -\lambda & 0 & 0 & 0 \\ 0 & 0 & -\lambda & 1 & 0 \\ 0 & 0 & 1 & -\lambda & 0 \end{bmatrix}$$ Interchange rows 1 and 2.

    Interchange rows 3 and 4.

$$\begin{bmatrix} -1 & -\lambda & 0 & 0 & 0 \\ -\lambda & 1 & 0 & 0 & 0 \\ 0 & 0 & 1 & -\lambda & 0 \\ 0 & 0 & -\lambda & 1 & 0 \end{bmatrix}$$ Use row 1 to eliminate $-\lambda$ in row 2.

    Use row 3 to eliminate $-\lambda$ in row 4.

$$\begin{bmatrix} -1 & -\lambda & 0 & 0 & 0 \\ 0 & 1+\lambda^2 & 0 & 0 & 0 \\ 0 & 0 & 1 & -\lambda & 0 \\ 0 & 0 & 0 & 1-\lambda^2 & 0 \end{bmatrix}$$

$$\begin{bmatrix} 1 & \lambda & 0 & 0 & 0 \\ 0 & 1+\lambda^2 & 0 & 0 & 0 \\ 0 & 0 & 1 & -\lambda & 0 \\ 0 & 0 & 0 & 1-\lambda^2 & 0 \end{bmatrix}$$ Multiply the first row by -1.

Thus, $(A - \lambda 1_{\mathbb{C}^4})(x) = 0$ has nontrivial solutions precisely when $1 + \lambda^2 = 0$ or $1 - \lambda^2 = 0$. Therefore, the eigenvalues, are $\lambda = \pm i$ and $\lambda = \pm 1$. Note that the two conditions can be multiplied into one *characteristic equation* of degree 4: $(1 + \lambda^2)(1 - \lambda^2) = 0$. Having found the eigenvalues we then need to insert them into the augmented system and find the eigenvectors. Since the system has already been reduced, this is quite simple. First, let $\lambda = \pm i$ so that the augmented system is

$$\begin{bmatrix} 1 & \pm i & 0 & 0 & 0 \\ 0 & 0 & 0 & 0 & 0 \\ 0 & 0 & 1 & \mp i & 0 \\ 0 & 0 & 0 & 2 & 0 \end{bmatrix}$$

Thus, we get

$$\begin{bmatrix} 1 \\ i \\ 0 \\ 0 \end{bmatrix} \leftrightarrow \lambda = i \text{ and } \begin{bmatrix} i \\ 1 \\ 0 \\ 0 \end{bmatrix} \leftrightarrow \lambda = -i$$

Next, we let $\lambda = \pm 1$ and consider

$$\begin{bmatrix} 1 & \pm 1 & 0 & 0 & 0 \\ 0 & 2 & 0 & 0 & 0 \\ 0 & 0 & 1 & \mp 1 & 0 \\ 0 & 0 & 0 & 0 & 0 \end{bmatrix}$$

to get

$$\begin{bmatrix} 0 \\ 0 \\ 1 \\ 1 \end{bmatrix} \leftrightarrow 1 \text{ and } \begin{bmatrix} 0 \\ 0 \\ -1 \\ 1 \end{bmatrix} \leftrightarrow -1$$

Example 2.3.3. Let

$$A = \begin{bmatrix} \alpha_{11} & \cdots & \alpha_{1n} \\ \vdots & \ddots & \vdots \\ 0 & \cdots & \alpha_{nn} \end{bmatrix}$$

be upper triangular, i.e., all entries below the diagonal are zero: $\alpha_{ij} = 0$ if $i > j$. Then, we are looking at

$$\begin{bmatrix} \alpha_{11} - \lambda & \cdots & \alpha_{1n} & 0 \\ \vdots & \ddots & \vdots & \vdots \\ 0 & \cdots & \alpha_{nn} - \lambda & 0 \end{bmatrix}.$$

Note again that we do not perform any divisions so as to make the diagonal entries 1. This is because if they are zero, we evidently have a nontrivial solution and that is what we are looking for. Therefore, the eigenvalues are $\lambda = \alpha_{11}, \ldots, \alpha_{nn}$. Note that the eigenvalues are precisely the roots of the polynomial that we get by multiplying the diagonal entries. This polynomial is going to be proportional to the *characteristic polynomial* of A.

The next examples show what we have to watch out for when performing the elementary row operations.

Example 2.3.4. Let

$$A = \begin{bmatrix} 1 & 2 & 4 \\ -1 & 0 & 2 \\ 3 & -1 & 5 \end{bmatrix},$$

and perform row operations on

$$\begin{bmatrix} 1-\lambda & 2 & 4 & 0 \\ -1 & -\lambda & 2 & 0 \\ 3 & -1 & 5-\lambda & 0 \end{bmatrix} \quad \begin{array}{l} \text{Change sign in row 2.} \\ \text{Interchange rows 1 and 2.} \end{array}$$

$$\begin{bmatrix} 1 & \lambda & -2 & 0 \\ 1-\lambda & 2 & 4 & 0 \\ 3 & -1 & 5-\lambda & 0 \end{bmatrix} \quad \text{Use row 1 to row reduce column 1.}$$

$$\begin{bmatrix} 1 & \lambda & -2 & 0 \\ 0 & 2-\lambda+\lambda^2 & 6-2\lambda & 0 \\ 0 & -1-3\lambda & 11-\lambda & 0 \end{bmatrix}$$ Interchange rows 2 and 3.

$$\begin{bmatrix} 1 & \lambda & -2 & 0 \\ 0 & -1-3\lambda & 11-\lambda & 0 \\ 0 & 2-\lambda+\lambda^2 & 6-2\lambda & 0 \end{bmatrix}$$ Change sign in row 2.
Use row 2 to cancel $2-\lambda+\lambda^2$ in row 3;
this requires that we have $1+3\lambda \neq 0$!

$$\begin{bmatrix} 1 & \lambda & -2 & 0 \\ 0 & 1+3\lambda & -11+\lambda & 0 \\ 0 & 0 & 6-2\lambda-\frac{2-\lambda+\lambda^2}{1+3\lambda}(-11+\lambda) & 0 \end{bmatrix}$$

$$\begin{bmatrix} 1 & \lambda & -2 & 0 \\ 0 & 1+3\lambda & -11+\lambda & 0 \\ 0 & 0 & \frac{28+3\lambda+6\lambda^2-\lambda^3}{1+3\lambda} & 0 \end{bmatrix}$$

Note that we are not allowed to have $1+3\lambda=0$ in this formula. If $1+3\lambda=0$, then we note that $2-\lambda+\lambda^2 \neq 0$ and $11-\lambda \neq 0$ so that the third display

$$\begin{bmatrix} 1 & \lambda & -2 & 0 \\ 0 & 2-\lambda+\lambda^2 & 6-2\lambda & 0 \\ 0 & -1-3\lambda & 11-\lambda & 0 \end{bmatrix}$$

guarantees that there are no nontrivial solutions in that case. This means that our analysis is valid and that multiplying the diagonal entries will get us the characteristic polynomial $28+3\lambda+6\lambda^2-\lambda^3$. First, observe that 7 is a root of this polynomial. We can then find the other two roots by dividing

$$\frac{28+3\lambda+6\lambda^2-\lambda^3}{\lambda-7} = -\lambda^2-\lambda-4$$

and using the quadratic formula: $-\frac{1}{2}+\frac{1}{2}i\sqrt{15}, -\frac{1}{2}-\frac{1}{2}i\sqrt{15}$.

These examples suggest a preliminary definition of the characteristic polynomial.

Definition 2.3.5. The *characteristic polynomial* of $A \in \text{Mat}_{n\times n}(\mathbb{F})$ is a polynomial $\chi_A(\lambda) \in \mathbb{F}[\lambda]$ of degree n such that all eigenvalues of A are roots of χ_A. In addition, we scale the polynomial so that the leading term is λ^n, i.e., the polynomial is monic.

To get a better understanding of the process that leads us to the characteristic polynomial, we study the 2×2 and 3×3 cases as well as a few specialized $n \times n$ situations.

Starting with $A \in \text{Mat}_{2\times 2}(\mathbb{F})$, we investigate

$$A - \lambda 1_{\mathbb{F}^2} = \begin{bmatrix} \alpha_{11}-\lambda & \alpha_{12} \\ \alpha_{21} & \alpha_{22}-\lambda \end{bmatrix}.$$

If $\alpha_{21} = 0$, the matrix is in upper triangular form and the characteristic polynomial is

$$\chi_A = (\alpha_{11} - \lambda)(\alpha_{22} - \lambda)$$
$$= \lambda^2 - (\alpha_{11} + \alpha_{22})\lambda + \alpha_{11}\alpha_{22}.$$

If $\alpha_{21} \neq 0$, then we switch the first and second row and then eliminate the bottom entry in the first column:

$$\begin{bmatrix} \alpha_{11} - \lambda & \alpha_{12} \\ \alpha_{21} & \alpha_{22} - \lambda \end{bmatrix}$$

$$\begin{bmatrix} \alpha_{21} & \alpha_{22} - \lambda \\ \alpha_{11} - \lambda & \alpha_{12} \end{bmatrix}$$

$$\begin{bmatrix} \alpha_{21} & \alpha_{22} - \lambda \\ 0 & \alpha_{12} - \frac{1}{\alpha_{21}}(\alpha_{11} - \lambda)(\alpha_{22} - \lambda) \end{bmatrix}.$$

Multiplying the diagonal entries gives

$$\alpha_{21}\alpha_{12} - (\alpha_{11} - \lambda)(\alpha_{22} - \lambda) = -\lambda^2 + (\alpha_{11} + \alpha_{22})\lambda$$
$$-\alpha_{11}\alpha_{22} + \alpha_{21}\alpha_{12}.$$

In both cases, the characteristic polynomial of a 2×2 matrix is given by

$$\chi_A = \lambda^2 - (\alpha_{11} + \alpha_{22})\lambda + (\alpha_{11}\alpha_{22} - \alpha_{21}\alpha_{12})$$
$$= \lambda^2 - \text{tr}(A)\lambda + \det(A).$$

We now make an attempt at the case where $A \in \text{Mat}_{3\times 3}(\mathbb{F})$. Thus, we consider

$$A - \lambda 1_{\mathbb{F}^3} = \begin{bmatrix} \alpha_{11} - \lambda & \alpha_{12} & \alpha_{13} \\ \alpha_{21} & \alpha_{22} - \lambda & \alpha_{23} \\ \alpha_{31} & \alpha_{32} & \alpha_{33} - \lambda \end{bmatrix}.$$

When $\alpha_{21} = \alpha_{31} = 0$, there is nothing to do in the first column, and we are left with the bottom right 2×2 matrix to consider. This is done as above.

If $\alpha_{21} = 0$ and $\alpha_{31} \neq 0$, then we switch the first and third rows and eliminate the last entry in the first row. This will look like

$$\begin{bmatrix} \alpha_{11} - \lambda & \alpha_{12} & \alpha_{13} \\ 0 & \alpha_{22} - \lambda & \alpha_{23} \\ \alpha_{31} & \alpha_{32} & \alpha_{33} - \lambda \end{bmatrix}$$

$$\begin{bmatrix} \alpha_{31} & \alpha_{32} & \alpha_{33} - \lambda \\ 0 & \alpha_{22} - \lambda & \alpha_{23} \\ \alpha_{11} - \lambda & \alpha_{12} & \alpha_{13} \end{bmatrix}$$

$$\begin{bmatrix} \alpha_{31} & \alpha_{32} & \alpha_{33} - \lambda \\ 0 & \alpha_{22} - \lambda & \alpha_{23} \\ 0 & \alpha\lambda + \beta & p\,(\lambda) \end{bmatrix}$$

where p has degree 2. If $\alpha\lambda + \beta$ is proportional to $\alpha_{22} - \lambda$, then we can eliminate it to get an upper triangular matrix. Otherwise, we can still eliminate $\alpha\lambda$ by multiplying the second row by α and adding it to the third row. This leads us to a matrix of the form

$$\begin{bmatrix} \alpha_{31} & \alpha_{32} & \alpha_{33} - \lambda \\ 0 & \alpha_{22} - \lambda & \alpha_{23} \\ 0 & \beta' & p'\,(\lambda) \end{bmatrix},$$

where β' is a scalar and p' a polynomial of degree 2. If $\beta' = 0$ we are finished. Otherwise, we switch the second and third rows and eliminate $\alpha_{22} - \lambda$ using β'.

If $\alpha_{21} \neq 0$, then we switch the first two rows and cancel below the diagonal in the first column. This gives us something like

$$\begin{bmatrix} \alpha_{11} - \lambda & \alpha_{12} & \alpha_{13} \\ \alpha_{21} & \alpha_{22} - \lambda & \alpha_{23} \\ \alpha_{31} & \alpha_{32} & \alpha_{33} - \lambda \end{bmatrix}$$

$$\begin{bmatrix} \alpha_{21} & \alpha_{22} - \lambda & \alpha_{23} \\ \alpha_{11} - \lambda & \alpha_{12} & \alpha_{13} \\ \alpha_{31} & \alpha_{32} & \alpha_{33} - \lambda \end{bmatrix}$$

$$\begin{bmatrix} \alpha_{21} & \alpha_{22} - \lambda & \alpha_{23} \\ 0 & p\,(\lambda) & \alpha'_{13} \\ 0 & q'\,(\lambda) & q\,(\lambda) \end{bmatrix},$$

where p has degree 2 and q, q' have degree 1. If $q' = 0$, we are finished. Otherwise, we switch the last two rows. If q' divides p, we can eliminate p to get an upper triangular matrix. If q' does not divide p, then we can still eliminate the degree 2 term in p to reduce it to a polynomial of degree 1. This lands us in a situation similar to what we ended up with when $\alpha_{21} = 0$. So we can finish using the same procedure.

Note that we avoided making any illegal moves in the above procedure. It is easy to formalize this procedure for $n \times n$ matrices. The idea is simply to treat λ as a variable and the entries as polynomials. To eliminate entries, we then use polynomial division to reduce the degrees of entries until they can be eliminated. Since we wish to treat λ as a variable, we shall rename it t when doing the Gauss elimination and only use λ for the eigenvalues and roots of the characteristic polynomial. More precisely, we claim the following:

Theorem 2.3.6. *Given* $A \in \mathrm{Mat}_{n \times n}(\mathbb{F})$, *there is a row reduction procedure that leads to a decomposition* $(t1_{\mathbb{F}^n} - A) = PU$, *where*

$$
U = \begin{bmatrix} p_1(t) & * & \cdots & * \\ 0 & p_2(t) & \cdots & * \\ \vdots & \vdots & \ddots & \vdots \\ 0 & 0 & \cdots & p_n(t) \end{bmatrix}
$$

with $p_1, \ldots, p_n \in \mathbb{F}[t]$ *being monic and unique, and* P *is the product of the elementary matrices:*

1. I_{kl} *interchanging rows* k *and* l.
2. $R_{kl}(r(t))$ *multiplies row* l *by* $r(t) \in \mathbb{F}[t]$ *and adds it to row* k.
3. $M_k(\alpha)$ *multiplies row* k *by* $\alpha \in \mathbb{F} - \{0\}$.

Proof. The procedure for obtaining the upper triangular form works as with row reduction with the twist that we think of all entries as being polynomials.

Starting with the first column, we look at all entries at or below the diagonal. We then select the nonzero entry with the lowest degree and make a row interchange to place this entry on the diagonal. Using that entry, we use polynomial division and the operation in (2) to reduce the degrees of all the entries below the diagonal. The degrees of all the entries below the diagonal are now strictly smaller than the degree of the diagonal entry. Moreover, if the diagonal entry actually divided a specific entry below the diagonal, then we get a 0 in that entry. We now repeat this process on the same column until we end up with a situation where all entries below the diagonal are 0.

Next, we must check that we actually get nonzero polynomials on the diagonal. This is clear for the first column as the first diagonal entry is non-zero. Should we end up with a situation where all entries on and below the diagonal vanish, then all values of $t \in \mathbb{F}$ must be eigenvalues. However, as we shall prove in Lemma 2.5.6, it is not possible for A to have more than n eigenvalues. So we certainly obtain a contradiction if we assume that $\mathbb{F}$ has characteristic zero as it will then have infinitely many elements. It is possible to use a more direct argument by carefully examining what happens in the process we described.

Next, we can multiply each row by a suitable nonzero scalar to ensure that the polynomials on the diagonal are monic.

Finally, to see that the polynomials on the diagonal are unique, we note that the matrix P is invertible and a multiple of elementary matrices. So if we have $PU = QV$ where U and V are both upper triangular, then $U = RV$ and $V = \tilde{R}U$ where both R and $\tilde{R}$ are matrices whose entries are polynomials. (In fact, they are products of the elementary matrices but we will not use that.) We claim that R and $\tilde{R}$ are also upper triangular and with all diagonal entries being 1. Clearly, this will show that the diagonal entries in P are unique. The proof goes by induction on n. For a general n, write $RU = V$ more explicitly as

$$\begin{bmatrix} r_{11} & r_{12} & \cdots & r_{1n} \\ r_{21} & r_{22} & & \\ \vdots & & \ddots & \\ r_{n1} & & & r_{nn} \end{bmatrix} \begin{bmatrix} p_1 & p_{12} & \cdots & p_{1n} \\ 0 & p_2 & & \\ \vdots & & \ddots & \\ 0 & 0 & & p_n \end{bmatrix} = \begin{bmatrix} q_1 & q_{12} & \cdots & q_{1n} \\ 0 & q_2 & & \\ \vdots & & \ddots & \\ 0 & 0 & & q_n \end{bmatrix}$$

and note that the entries in the first column on the right-hand side satisfy

$$r_{11} p_1 = q_1$$
$$r_{21} p_1 = 0$$
$$\vdots$$
$$r_{n1} p_1 = 0.$$

Since p_1 is nontrivial, it follows that $r_{21} = \cdots = r_{n1} = 0$. A similar argument
shows that the entries in the first column of $\tilde{R}$ satisfy $\tilde{r}_{21} = \cdots = \tilde{r}_{n1} = 0$. Next, we
note that $r_{11} p_1 = q_1$ and $\tilde{r}_{11} q_1 = p_1$ showing that $r_{11} = \tilde{r}_{11} = 1$. This shows that
our claim holds when $n = 1$. When, $n > 1$ we obtain

$$\begin{bmatrix} r_{22} & r_{23} & \cdots & r_{2n} \\ r_{32} & r_{33} & & \\ \vdots & & \ddots & \\ r_{n2} & & & r_{nn} \end{bmatrix} \begin{bmatrix} p_2 & p_{23} & \cdots & p_{2n} \\ 0 & p_3 & & \\ \vdots & & \ddots & \\ 0 & 0 & & p_n \end{bmatrix} = \begin{bmatrix} q_2 & q_{23} & \cdots & q_{2n} \\ 0 & q_3 & & \\ \vdots & & \ddots & \\ 0 & 0 & & q_n \end{bmatrix}$$

after deleting the first row and column in the matrices. This allows us to use
induction to finish the proof.   □

This gives us a solid definition of the characteristic polynomial although it is as
yet not completely clear why it has degree n. A very similar construction will be
given in Sect. 2.9. The main difference is that it also uses column operations. The
advantage of that more enhanced construction is that it calculates more invariants.
In addition, it shows that the characteristic polynomial has degree n and remains the
same for similar matrices.

Definition 2.3.7. The *characteristic polynomial* of $A \in \text{Mat}_{n \times n}(\mathbb{F})$ is the monic
polynomial $\chi_A(t) \in \mathbb{F}[t]$ we obtain by applying Gauss elimination to $A - t1_{\mathbb{F}^n}$
or $t1_{\mathbb{F}^n} - A$ until it is in upper triangular form and then multiplying the monic
polynomials in the diagonal entries, i.e., $\chi_A(t) = p_1(t) p_2(t) \cdots p_n(t)$.

The next example shows how the proof of Theorem 2.3.6 works in a specific
example.

Example 2.3.8. Let

$$A = \begin{bmatrix} 1 & 2 & 3 \\ 0 & 2 & 4 \\ 2 & 1 & -1 \end{bmatrix}.$$

The calculations go as follows:

$$A - t1_{\mathbb{F}^3} = \begin{bmatrix} 1-t & 2 & 3 \\ 0 & 2-t & 4 \\ 2 & 1 & -1-t \end{bmatrix}$$

$$\begin{bmatrix} 2 & 1 & -1-t \\ 0 & 2-t & 4 \\ 1-t & 2 & 3 \end{bmatrix}$$

$$\begin{bmatrix} 2 & 1 & -1-t \\ 0 & 2-t & 4 \\ 0 & 2-\frac{1-t}{2} & 3+\frac{(1-t)(1+t)}{2} \end{bmatrix}$$

$$\begin{bmatrix} 2 & 1 & -1-t \\ 0 & 2-t & 4 \\ 0 & \frac{3}{2}+\frac{t}{2} & 3+\frac{(1-t)(1+t)}{2} \end{bmatrix}$$

$$\begin{bmatrix} 2 & 1 & -1-t \\ 0 & 2-t & 4 \\ 0 & \frac{3}{2}+1 & 5+\frac{(1-t)(1+t)}{2} \end{bmatrix}$$

$$\begin{bmatrix} 2 & 1 & -1-t \\ 0 & \frac{5}{2} & 5+\frac{(1-t)(1+t)}{2} \\ 0 & 2-t & 4 \end{bmatrix}$$

$$\begin{bmatrix} 2 & 1 & -1-t \\ 0 & \frac{5}{2} & 5+\frac{(1-t)(1+t)}{2} \\ 0 & 0 & 4-2\frac{2-t}{5}\left(5+\frac{(1-t)(1+t)}{2}\right) \end{bmatrix}$$

$$\begin{bmatrix} 1 & \frac{1}{2} & \frac{-1-t}{2} \\ 0 & 1 & 2+\frac{(1-t)(1+t)}{5} \\ 0 & 0 & t^3-2t^2-11t+2 \end{bmatrix},$$

and the characteristic polynomial is

$$\chi_A(t) = t^3 - 2t^2 - 11t + 2$$

When the matrix A can be written in block triangular form, it becomes somewhat easier to calculate the characteristic polynomial.

Lemma 2.3.9. *Assume that $A \in \mathrm{Mat}_{n \times n}(\mathbb{F})$ has the form*

$$A = \begin{bmatrix} A_{11} & A_{12} \\ 0 & A_{22} \end{bmatrix},$$

where $A_{11} \in \text{Mat}_{k \times k} (\mathbb{F})$, $A_{22} \in \text{Mat} (\mathbb{F})$, and $A_{12} \in \text{Mat}_{k \times (n-k)} (\mathbb{F})$, then

$$\chi_A (t) = \chi_{A_{11}} (t) \, \chi_{A_{22}} (t) .$$

Proof. To compute $\chi_A (t)$, we do row operations on

$$t 1_{\mathbb{F}^n} - A = \begin{bmatrix} t 1_{\mathbb{F}^k} - A_{11} & -A_{12} \\ 0 & t 1_{\mathbb{F}^{n-k}} - A_{22} \end{bmatrix} .$$

This can be done by first doing row operations on the first k rows leading to a situation that looks like

$$\begin{bmatrix} q_1 (t) & & * & & \\ & \ddots & & & * \\ 0 & & q_k (t) & & \\ & & 0 & & t 1_{\mathbb{F}^{n-k}} - A_{22} \end{bmatrix} .$$

Having accomplished this, we then do row operations on the last $n - k$ rows. to get

$$\begin{bmatrix} p_1 (t) & & * & & & \\ & \ddots & & & * & \\ 0 & & p_k (t) & & & \\ & & & r_1 (t) & & * \\ & 0 & & & \ddots & \\ & & & 0 & & r_{n-k} (t) \end{bmatrix}$$

As these two sets of operations do not depend on each other, we see that

$$\chi_A (t) = q_1 (t) \cdots q_k (t) \, r_1 (t) \cdots r_{n-k} (t)$$
$$= \chi_{A_{11}} (t) \, \chi_{A_{22}} (t) . \qquad \square$$

Finally, we need to figure out how this matrix procedure generates eigenvalues for general linear maps $L : V \to V$. In case V is finite-dimensional, we can simply pick a basis and then study the matrix representation $[L]$. The diagram

$$\begin{array}{ccc} V & \xrightarrow{L} & V \\ \uparrow & & \uparrow \\ \mathbb{F}^n & \xrightarrow{[L]} & \mathbb{F}^n \end{array}$$

then quickly convinces us that eigenvectors in $\mathbb{F}^n$ for $[L]$ are mapped to eigenvectors in V for L without changing the eigenvalue, i.e.,

$$[L] \xi = \lambda \xi$$

is equivalent to

$$Lx = \lambda x$$

if $\xi \in \mathbb{F}^n$ is the coordinate vector for $x \in V$. Thus, we define the characteristic polynomial of L as $\chi_L(t) = \chi_{[L]}(t)$. While we do not have a problem with finding eigenvalues for L by finding them for $[L]$, it is less clear that $\chi_L(t)$ becomes well defined with this definition. To see that it is well defined, we would have to show that $\chi_{[L]}(t) = \chi_{B^{-1}[L]B}(t)$ where B the matrix transforming one basis into another basis. This is best done using determinants (see Sect. 5.7). Alternately, one would have to use the definition of the characteristic polynomial given in Sect. 2.7 which can be computed with a more elaborate procedure that uses both row and column operations (see Sect. 2.9). For now, we are going to take this on faith. Note, however, that computing $\chi_{[L]}(t)$ does give us a rigorous method for finding the eigenvalues as L. In particular, all of the matrix representations for L must have the same eigenvalues. Thus, there is nothing wrong with searching for eigenvalues using a fixed matrix representation.

In the case where $\mathbb{F} = \mathbb{Q}$ or $\mathbb{R}$, we can still think of $[L]$ as a complex matrix. As such, we might get complex eigenvalues that do not lie in the field $\mathbb{F}$. These roots of χ_L cannot be eigenvalues for L as we are not allowed to multiply elements in V by complex numbers.

Example 2.3.10. We now need to prove that our method for computing the characteristic polynomial of a matrix gives us the expected answer for the differential equation defined by the operator

$$L = D^n + \alpha_{n-1} D^{n-1} + \cdots + \alpha_1 D + \alpha_0.$$

The corresponding system is

$$L(x) = \dot{x} - Ax$$

$$= \dot{x} - \begin{bmatrix} 0 & 1 & \cdots & 0 \\ 0 & 0 & \ddots & \vdots \\ \vdots & \vdots & \ddots & 1 \\ -\alpha_0 & -\alpha_1 & \cdots & -\alpha_{n-1} \end{bmatrix} x$$

$$= 0.$$

So we consider the matrix

$$A = \begin{bmatrix} 0 & 1 & \cdots & 0 \\ 0 & 0 & \ddots & \vdots \\ \vdots & \vdots & \ddots & 1 \\ -\alpha_0 & -\alpha_1 & \cdots & -\alpha_{n-1} \end{bmatrix}$$

and with it

$$
t1_{\mathbb{F}^n} - A = \begin{bmatrix} t & -1 & \cdots & & 0 \\ 0 & t & \ddots & & \vdots \\ \vdots & \vdots & \ddots & & -1 \\ \alpha_0 & \alpha_1 & \cdots & t + \alpha_{n-1} \end{bmatrix}.
$$

We immediately run into a problem as we do not know if some or all of $\alpha_0, \ldots, \alpha_{n-1}$ are zero. Thus, we proceed without interchanging rows:

$$
\begin{bmatrix} -t & 1 & \cdots & & 0 \\ 0 & -t & \ddots & & \vdots \\ \vdots & \vdots & \ddots & & 1 \\ -\alpha_0 & -\alpha_1 & \cdots & -t - \alpha_{n-1} \end{bmatrix}
$$

$$
\begin{bmatrix} -t & 1 & & \cdots & & 0 \\ 0 & -t & & \ddots & & \vdots \\ \vdots & \vdots & & \ddots & & 1 \\ 0 & -\alpha_1 - \frac{\alpha_0}{t} & & \cdots & -\alpha_{n-1} - t \end{bmatrix}
$$

$$
\begin{bmatrix} -t & 1 & & & \cdots & & 0 \\ 0 & -t & 1 & & \ddots & & \vdots \\ \vdots & \vdots & & & & & 1 \\ 0 & 0 & -\alpha_2 - \frac{\alpha_1}{t} - \frac{\alpha_0}{t^2} & & \cdots & \alpha_{n-1} - t \end{bmatrix}
$$

$$
\vdots
$$

$$
\begin{bmatrix} t & -1 & & \cdots & & 0 \\ 0 & t & -1 & \ddots & & \vdots \\ \vdots & \vdots & & \ddots & & -1 \\ 0 & 0 & 0 & \cdots & t + \alpha_{n-1} + \frac{\alpha_{n-2}}{t} + \cdots + \frac{\alpha_1}{t^{n-2}} + \frac{\alpha_0}{t^{n-1}} \end{bmatrix}
$$

Note that $t = 0$ is the only value that might give us trouble. In case $t = 0$, we note that there cannot be a nontrivial kernel unless $\alpha_0 = 0$. Thus, $\lambda = 0$ is an eigenvalue if and only if $\alpha_0 = 0$. Fortunately, this gets build into our characteristic polynomial. After multiplying the diagonal entries together, we have

$$
\begin{aligned}
p(t) &= t^{n-1} \left(t + \alpha_{n-1} + \frac{\alpha_{n-2}}{t} + \cdots + \frac{\alpha_1}{t^{n-2}} + \frac{\alpha_0}{t^{n-1}} \right) \\
&= \left(t^n + \alpha_{n-1} t^{n-1} + \alpha_{n-2} t^{n-2} + \cdots + \alpha_1 t + \alpha_0 \right),
\end{aligned}
$$

where $\lambda = 0$ is a root precisely when $\alpha_0 = 0$ as hoped for. Finally, we see that $p(t) = 0$ is up to sign our old characteristic equation for $p(D) = 0$. In Proposition 2.6.3, we shall compute the characteristic polynomial for the transpose of A using only the techniques from Theorem 2.3.6.

There are a few useful facts that can help us find roots of polynomials.

Proposition 2.3.11. *Let $A \in \text{Mat}_{n \times n}(\mathbb{C})$ and*

$$\chi_A(t) = t^n + \alpha_{n-1}t^{n-1} + \cdots + \alpha_1 t + \alpha_0 = (t - \lambda_1) \cdots (t - \lambda_n).$$

1. $\text{tr}A = \lambda_1 + \cdots + \lambda_n = -\alpha_{n-1}$.
2. $\lambda_1 \cdots \lambda_n = (-1)^n \alpha_0$.
3. *If $\chi_A(t) \in \mathbb{R}[t]$ and $\lambda \in \mathbb{C}$ is a root, then $\bar{\lambda}$ is also a root. In particular, the number of real roots is even, respectively odd, if n is even, respectively odd.*
4. *If $\chi_A(t) \in \mathbb{R}[t]$, n is even, and $\alpha_0 < 0$, then there are at least two real roots one negative and one positive.*
5. *If $\chi_A(t) \in \mathbb{R}[t]$ and n is odd, then there is at least one real root, whose sign is the opposite of α_0.*
6. *If $\chi_A(t) \in \mathbb{Z}[t]$, then all rational roots are in fact integers that divide α_0.*

Proof. The proof of (3) follows from the fact that when the coefficients of χ_A are real, then $\overline{\chi_A(t)} = \chi_A(\bar{t})$. The proofs of (4) and (5) follow from the intermediate value theorem. Simply note that $\chi_A(0) = \alpha_0$ and that $\chi_A(t) \to \infty$ as $t \to \infty$ while $(-1)^n \chi_A(t) \to \infty$ as $t \to -\infty$.

For the first two facts, note that the relationship

$$t^n + \alpha_{n-1}t^{n-1} + \cdots + \alpha_1 t + \alpha_0 = (t - \lambda_1) \cdots (t - \lambda_n)$$

shows that

$$\lambda_1 + \cdots + \lambda_n = -\alpha_{n-1},$$
$$\lambda_1 \cdots \lambda_n = (-1)^n \alpha_0.$$

This establishes (2) and part of (1). Finally, the relation $\text{tr}A = \lambda_1 + \cdots + \lambda_n$ will be established when we can prove that complex matrices are similar to upper triangular matrices (see also Exercise 13 in Sect. 2.7). In other words, we will show that one can find $B \in Gl_n(\mathbb{C})$ such that $B^{-1}AB$ is upper triangular (see Sect. 4.8 or 2.8). We then observe that A and $B^{-1}AB$ have the same eigenvalues as $Ax = \lambda x$ if and only if $B^{-1}AB(B^{-1}x) = \lambda(B^{-1}x)$. However, as the eigenvalues for the upper triangular matrix $B^{-1}AB$ are precisely the diagonal entries, we see that

$$\lambda_1 + \cdots + \lambda_n = \text{tr}\left(B^{-1}AB\right)$$
$$= \text{tr}\left(ABB^{-1}\right)$$
$$= \text{tr}(A).$$

Another proof of $\operatorname{tr} A = -\alpha_{n-1}$ that works for all fields is presented below in the exercises to Sect. 2.7.

For (6), let p/q be a rational root in reduced form, then

$$\left(\frac{p}{q}\right)^n + \cdots + \alpha_1 \left(\frac{p}{q}\right) + \alpha_0 = 0,$$

and

$$
\begin{aligned}
0 &= p^n + \cdots + \alpha_1 p q^{n-1} + \alpha_0 q^n \\
&= p^n + q \left(\alpha_{n-1} p^{n-1} + \cdots + \alpha_1 p q^{n-2} + \alpha_0 q^{n-1}\right) \\
&= p \left(p^{n-1} + \cdots + \alpha_1 q^{n-1}\right) + \alpha_0 q^n.
\end{aligned}
$$

Thus, q divides p^n and p divides $a_0 q^n$. Since p and q have no divisors in common, the result follows. □

Exercises

1. Find the characteristic polynomial and if possible the eigenvalues and eigenvectors for each of the following matrices:

   (a)
   $$\begin{bmatrix} 1 & 0 & 1 \\ 0 & 1 & 0 \\ 1 & 0 & 1 \end{bmatrix}$$

   (b)
   $$\begin{bmatrix} 0 & 1 & 2 \\ 1 & 0 & 3 \\ 2 & 3 & 0 \end{bmatrix}$$

   (c)
   $$\begin{bmatrix} 0 & 1 & 2 \\ -1 & 0 & 3 \\ -2 & -3 & 0 \end{bmatrix}$$

2. Find the characteristic polynomial and if possible eigenvalues and eigenvectors for each of the following matrices:

   (a)
   $$\begin{bmatrix} 0 & i \\ i & 0 \end{bmatrix}$$

   (b)
   $$\begin{bmatrix} 0 & i \\ -i & 0 \end{bmatrix}$$

(c)
$$\begin{bmatrix} 1 & i & 0 \\ i & 1 & 0 \\ 0 & 2 & 1 \end{bmatrix}$$

3. Find the eigenvalues for the following matrices with a minimum of calculations (try not to compute the characteristic polynomial):

   (a)
   $$\begin{bmatrix} 1 & 0 & 1 \\ 0 & 0 & 0 \\ 1 & 0 & 1 \end{bmatrix}$$

   (b)
   $$\begin{bmatrix} 1 & 0 & 1 \\ 0 & 1 & 0 \\ 1 & 0 & 1 \end{bmatrix}$$

   (c)
   $$\begin{bmatrix} 0 & 0 & 1 \\ 0 & 1 & 0 \\ 1 & 0 & 0 \end{bmatrix}$$

4. Find the characteristic polynomial, eigenvalues, and eigenvectors for each of the following linear operators $L : P_3 \to P_3$:

   (a) $L = D$.
   (b) $L = tD = T \circ D$.
   (c) $L = D^2 + 2D + 1_{P_3}$.
   (d) $L = t^2 D^3 + D$.

5. Let $p \in \mathbb{C}[t]$ be a monic polynomial. Show that the characteristic polynomial for $D : \ker(p(D)) \to \ker(p(D))$ is $p(t)$. (To clarify the notation, see Exercise 12 in Sect. 2.2.)

6. Assume that $A \in \mathrm{Mat}_{n \times n}(\mathbb{F})$ is upper or lower triangular and let $p \in \mathbb{F}[t]$. Show that μ is an eigenvalue for $p(A)$ if and only if $\mu = p(\lambda)$ where λ is an eigenvalue for A. (Hint: See Exercise 7 in Sect. 1.6.)

7. Let $L : V \to V$ be a linear operator on a complex vector space. Assume that we have a polynomial $p \in \mathbb{C}[t]$ such that $p(L) = 0$. Show that all eigenvalues of L are roots of p.

8. Let $L : V \to V$ be a linear operator and $K : W \to V$ an isomorphism. Show that L and $K^{-1} \circ L \circ K$ have the same eigenvalues.

9. Let $K : V \to W$ and $L : W \to V$ be two linear maps.

   (a) Show that $K \circ L$ and $L \circ K$ have the same nonzero eigenvalues. Hint: If $x \in V$ is an eigenvector for $L \circ K$, then $K(x) \in W$ is an eigenvector for $K \circ L$.
   (b) Give an example where 0 is an eigenvalue for $L \circ K$ but not for $K \circ L$. Hint: Try to have different dimensions for V and W.

(c) If $\dim V = \dim W$, then (a) also holds for the zero eigenvalue. Hint: From Exercise 5 in Sect. 1.11, use that

$$\dim (\ker (K \circ L)) \geq \max \{\dim (\ker (L)) , \dim (\ker (K))\} ,$$

$$\dim (\ker (L \circ K)) \geq \max \{\dim (\ker (L)) , \dim (\ker (K))\} .$$

10. Let $A \in \mathrm{Mat}_{n \times n} (\mathbb{F})$.

   (a) Show that A and A^t have the same eigenvalues and that for each eigenvalue λ, we have

$$\dim (\ker (A - \lambda 1_{\mathbb{F}^n})) = \dim \left(\ker \left(A^t - \lambda 1_{\mathbb{F}^n}\right)\right) .$$

   (b) Show by example that A and A^t need not have the same eigenvectors.

11. Let $A \in \mathrm{Mat}_{n \times n} (\mathbb{F})$. Consider the following two linear operators on $\mathrm{Mat}_{n \times n} (\mathbb{F})$: $L_A (X) = AX$ and $R_A (X) = XA$ (see Example 1.7.6).

   (a) Show that λ is an eigenvalue for A if and only if λ is an eigenvalue for L_A.
   (b) Show that $\chi_{L_A} (t) = (\chi_A (t))^n$.
   (c) Show that λ is an eigenvalue for A^t if and only if λ is an eigenvalue for R_A.
   (d) Relate $\chi_{A^t} (t)$ and $\chi_{R_A} (t)$.

12. Let $A \in \mathrm{Mat}_{n \times n} (\mathbb{F})$ and $B \in \mathrm{Mat}_{m \times m} (\mathbb{F})$ and consider

$$L : \mathrm{Mat}_{n \times m} (\mathbb{F}) \to \mathrm{Mat}_{n \times m} (\mathbb{F}) ,$$

$$L (X) = AX - XB.$$

   (a) Show that if A and B have a common eigenvalue, then L has nontrivial kernel. Hint: Use that B and B^t have the same eigenvalues.
   (b) Show more generally that if λ is an eigenvalue of A and μ and eigenvalue for B, then $\lambda - \mu$ is an eigenvalue for L.

13. Find the characteristic polynomial, eigenvalues, and eigenvectors for

$$A = \begin{bmatrix} \alpha & -\beta \\ \beta & \alpha \end{bmatrix}, \alpha, \beta \in \mathbb{R}.$$

   as a map $A : \mathbb{C}^2 \to \mathbb{C}^2$.

14. Show directly, using the methods developed in this section, that the characteristic polynomial for a 3×3 matrix has degree 3.

15. Let

$$A = \begin{bmatrix} a & b \\ c & d \end{bmatrix}, a, b, c, d \in \mathbb{R}$$

   Show that the eigenvalues are either both real or are complex conjugates of each other.

16. Show that the eigenvalues of $\begin{bmatrix} a & b \\ \bar{b} & d \end{bmatrix}$, where $a, d \in \mathbb{R}$ and $b \in \mathbb{C}$, are real.

17. Show that the eigenvalues of $\begin{bmatrix} ia & -b \\ \bar{b} & id \end{bmatrix}$, where $a, d \in \mathbb{R}$ and $b \in \mathbb{C}$, are purely imaginary.

18. Show that the eigenvalues of $\begin{bmatrix} a & -\bar{b} \\ b & \bar{a} \end{bmatrix}$, where $a, b \in \mathbb{C}$ and $|a|^2 + |b|^2 = 1$, are complex numbers of unit length.

19. Let
$$A = \begin{bmatrix} 0 & 1 & \cdots & 0 \\ 0 & 0 & \ddots & \vdots \\ \vdots & \vdots & \ddots & 1 \\ -\alpha_0 & -\alpha_1 & \cdots & -\alpha_{n-1} \end{bmatrix}.$$

   (a) Show that all eigenspaces are one-dimensional.
   (b) Show that $\ker(A) \neq \{0\}$ if and only if $\alpha_0 = 0$.

20. Let
$$p(t) = (t - \lambda_1) \cdots (t - \lambda_n)$$
$$= t^n + \alpha_{n-1} t^{n-1} + \cdots + \alpha_1 t + \alpha_0,$$

where $\lambda_1, \ldots, \lambda_n \in \mathbb{F}$. Show that there is a change of basis such that
$$\begin{bmatrix} 0 & 1 & \cdots & 0 \\ 0 & 0 & \ddots & \vdots \\ \vdots & \vdots & \ddots & 1 \\ -\alpha_0 & -\alpha_1 & \cdots & -\alpha_{n-1} \end{bmatrix} = B \begin{bmatrix} \lambda_1 & 1 & & 0 \\ 0 & \lambda_2 & \ddots & \\ & & \ddots & 1 \\ 0 & & & \lambda_n \end{bmatrix} B^{-1}.$$

   Hint: Try $n = 2, 3$, assume that B is lower triangular with 1s on the diagonal, or alternately use Exercise 9 in Sect. 2.2.

21. Show that
   (a) The multiplication operator $T : C^\infty(\mathbb{R}, \mathbb{R}) \to C^\infty(\mathbb{R}, \mathbb{R})$ does not have any eigenvalues. Recall that $T(f)(t) = t \cdot f(t)$.
   (b) Show that the differential operator $D : \mathbb{C}[t] \to \mathbb{C}[t]$ only has 0 as an eigenvalue.
   (c) Show that $D : C^\infty(\mathbb{R}, \mathbb{R}) \to C^\infty(\mathbb{R}, \mathbb{R})$ has all real numbers as eigenvalues.
   (d) Show that $D : C^\infty(\mathbb{R}, \mathbb{C}) \to C^\infty(\mathbb{R}, \mathbb{C})$ has all complex numbers as eigenvalues.

2.4   The Minimal Polynomial

The minimal polynomial of a linear operator is, unlike the characteristic polynomial, fairly easy to define rigorously. It is, however, not as easy to calculate. The amazing properties contained in the minimal polynomial on the other hand seem to make it sufficiently desirable that it would be a shame to ignore it. See also Sect. 1.12 for a preliminary discussion of the minimal polynomial.

Recall that projections are characterized by a very simple polynomial relationship $L^2 - L = 0$. The purpose of this section is to find a polynomial $p(t)$ for a linear operator $L : V \to V$ such that $p(L) = 0$. This polynomial will, like the characteristic polynomial, also have the property that its roots are the eigenvalues of L. In subsequent sections, we shall study in more depth the properties of linear operators based on knowledge of the minimal polynomial.

Before passing on to the abstract constructions, let us consider two examples.

Example 2.4.1. An *involution* is a linear operator $L : V \to V$ such that $L^2 = 1_V$. This means that $p(L) = 0$ if $p(t) = t^2 - 1$. Our first observation is that this relationship implies that L is invertible and that $L^{-1} = L$. Next, note that any eigenvalue must satisfy $\lambda^2 = 1$ and hence be a root of p. It is possible to glean even more information out of this polynomial relationship. We claim that L is diagonalizable, i.e., V has a basis of eigenvectors for L; in fact

$$V = \ker(L - 1_V) \oplus \ker(L + 1_V).$$

First, we observe that these spaces have trivial intersection as they are eigenspaces for different eigenvalues. If $x \in \ker(L - 1_V) \cap \ker(L + 1_V)$, then

$$-x = L(x) = x$$

so $x = 0$. To show that

$$V = \ker(L - 1_V) + \ker(L + 1_V),$$

we observe that any $x \in V$ can be written as

$$x = \frac{1}{2}(x - L(x)) + \frac{1}{2}(x + L(x)).$$

Next, we see that

$$\begin{aligned} L(x \pm L(x)) &= L(x) \pm L^2(x) \\ &= L(x) \pm x \\ &= \pm(x \pm L(x)). \end{aligned}$$

Thus, $x + L(x) \in \ker(L - 1_V)$ and $x - L(x) \in \ker(L + 1_V)$. This proves the claim.

Example 2.4.2. Consider a linear operator $L : V \to V$ such that $(L - 1_V)^2 = 0$. This relationship implies that 1 is the only possible eigenvalue. Therefore, if L is diagonalizable, then $L = 1_V$ and hence also satisfies the simpler relationship $L - 1_V = 0$. Thus, L is not diagonalizable unless it is the identity map. By multiplying out the polynomial relationship, we obtain

$$L^2 - 2L + 1_V = 0.$$

This implies that

$$(2 \cdot 1_V - L) L = 1_V.$$

Hence, L is invertible with $L^{-1} = 2 \cdot 1_V - L$.

These two examples, together with our knowledge of projections, tell us that one can get a tremendous amount of information from knowing that an operator satisfies a polynomial relationship. To commence our more abstract developments we start with a very simple observation.

Proposition 2.4.3. *Let $L : V \to V$ be a linear operator and*

$$p(t) = t^k + \alpha_{k-1} t^{k-1} + \cdots + \alpha_1 t + \alpha_0 \in \mathbb{F}[t]$$

a polynomial such that

$$p(L) = L^k + \alpha_{k-1} L^{k-1} + \cdots + \alpha_1 L + \alpha_0 1_V = 0.$$

(1) All eigenvalues for L are roots of $p(t)$.
(2) If $p(0) = \alpha_0 \neq 0$, then L is invertible and

$$L^{-1} = \frac{-1}{\alpha_0} \left(L^{k-1} + \alpha_{k-1} L^{k-2} + \cdots + \alpha_1 1_V \right).$$

To begin with it would be nice to find a polynomial $p(t) \in \mathbb{F}[t]$ such that both of the above properties become bi-implications. In other words $\lambda \in \mathbb{F}$ is an eigenvalues for L if and only $p(\lambda) = 0$, and L is invertible if and only if $p(0) \neq 0$. It turns out that the characteristic polynomial does have this property, but there is a polynomial that has even more information as well as being much easier to define.

One defect of the characteristic polynomial can be seen by considering the two matrices

$$\begin{bmatrix} 1 & 0 \\ 0 & 1 \end{bmatrix}, \begin{bmatrix} 1 & 1 \\ 0 & 1 \end{bmatrix}$$

They clearly have the same characteristic polynomial $p(t) = (t - 1)^2$, but only the first matrix is diagonalizable.

Definition 2.4.4. We define the *minimal polynomial* $\mu_L(t)$ for a linear operator $L : V \to V$ on a finite-dimensional vector space in the following way. Consider $1_V, L, L^2, \ldots, L^k, \ldots \in \mathrm{Hom}(V, V)$. Since $\mathrm{Hom}(V, V)$ is finite-dimensional we can use Lemma 1.12.3 to find a smallest $k \geq 1$ such that L^k is a linear combination of $1_V, L, L^2, \ldots, L^{k-1}$:

$$L^k = -\left(\alpha_0 1_V + \alpha_1 L + \alpha_2 L^2 + \cdots + \alpha_{k-1} L^{k-1}\right), \text{ or}$$

$$0 = L^k + \alpha_{k-1} L^{k-1} + \cdots + \alpha_1 L + \alpha_0 1_V.$$

The minimal polynomial of L is defined as

$$\mu_L(t) = t^k + \alpha_{k-1} t^{k-1} + \cdots + \alpha_1 t + \alpha_0.$$

The first interesting thing to note is that the minimal polynomial for $L = 1_V$ is given by $\mu_{1_V}(t) = t - 1$. Hence, it is not the characteristic polynomial. The name "minimal" is justified by the next proposition.

Proposition 2.4.5. *Let $L : V \to V$ be a linear operator on a finite-dimensional space.*

(1) If $p(t) \in \mathbb{F}[t]$ satisfies $p(L) = 0$, then $\deg(p) \geq \deg(\mu_L)$.
(2) If $p(t) \in \mathbb{F}[t]$ satisfies $p(L) = 0$ and $\deg(p) = \deg(\mu_L)$, then $p(t) = \alpha \cdot \mu_L(t)$ for some $\alpha \in \mathbb{F}$.

Proof. (1) Assume that $p \neq 0$ and $p(L) = 0$, then

$$p(L) = \alpha_m L^m + \alpha_{m-1} L^{m-1} + \cdots + \alpha_1 L + \alpha_0 1_V$$

$$= 0.$$

If $\alpha_m \neq 0$, then L^m is a linear combination of lower order terms, and hence, $m \geq \deg(\mu_L)$.

(2) In case $m = \deg(\mu_L) = k$, we have that $1_V, L, \ldots, L^{k-1}$ are linearly independent. Thus, there is only one way in which to make L^k into a linear combination of $1_V, L, \ldots, L^{k-1}$. This implies the claim.   □

Before discussing further properties of the minimal polynomial, let us try to compute it for some simple matrices. See also Sect. 1.12 for similar examples.

Example 2.4.6. Let

$$A = \begin{bmatrix} \lambda & 1 \\ 0 & \lambda \end{bmatrix}$$

$$B = \begin{bmatrix} \lambda & 0 & 0 \\ 0 & \lambda & 1 \\ 0 & 0 & \lambda \end{bmatrix}$$

$$C = \begin{bmatrix} 0 & -1 & 0 \\ 1 & 0 & 0 \\ 0 & 0 & i \end{bmatrix}.$$

We note that A is not proportional to 1_V, while

$$A^2 = \begin{bmatrix} \lambda & 1 \\ 0 & \lambda \end{bmatrix}^2$$

$$= \begin{bmatrix} \lambda^2 & 2\lambda \\ 0 & \lambda^2 \end{bmatrix}$$

$$= 2\lambda \begin{bmatrix} \lambda & 1 \\ 0 & \lambda \end{bmatrix} - \lambda^2 \begin{bmatrix} 1 & 0 \\ 0 & 1 \end{bmatrix}.$$

Thus,
$$\mu_A (t) = t^2 - 2\lambda t + \lambda^2 = (t - \lambda)^2.$$

The calculation for B is similar and evidently yields the same minimal polynomial
$$\mu_B (t) = t^2 - 2\lambda t + \lambda^2 = (t - \lambda)^2.$$

Finally, for C, we note that

$$C^2 = \begin{bmatrix} -1 & 0 & 0 \\ 0 & -1 & 0 \\ 0 & 0 & -1 \end{bmatrix}$$

Thus,
$$\mu_C (t) = t^2 + 1.$$

The next proposition shows that the minimal polynomial contains much of the information that we usually get from the characteristic polynomial. In subsequent sections, we shall delve much deeper into the properties of the minimal polynomial and what it tells us about possible matrix representations for L.

Proposition 2.4.7. *Let $L : V \to V$ be a linear operator on a finite-dimensional vector space. Then, the following properties for the minimal polynomial hold:*

(1) If $p(L) = 0$ for some $p \in \mathbb{F}[t]$, then μ_L divides p, i.e., $p(t) = \mu_L(t) q(t)$ for some $q(t) \in \mathbb{F}[t]$.
(2) Let $\lambda \in \mathbb{F}$, then λ is an eigenvalue for L if and only if $\mu_L(\lambda) = 0$.
(3) L is invertible if and only if $\mu_L(0) \neq 0$.

Proof. (1) Assume that $p(L) = 0$. We know that $\deg(p) \geq \deg(\mu_L)$, so if we perform polynomial division (the Euclidean Algorithm 2.1.1), then $p(t) = q(t)\mu_L(t) + r(t)$, where $\deg(r) < \deg(\mu_L)$. Substituting L for t gives $p(L) = q(L)\mu_L(L) + r(L)$. Since both $p(L) = 0$, and $\mu_L(L) = 0$ we also have $r(L) = 0$. This will give us a contradiction with the definition of the minimal polynomial unless $r = 0$. Thus, μ_L divides p.

(2) We already know that eigenvalues are roots. Conversely, if $\mu_L(\lambda) = 0$, then we can write $\mu_L(t) = (t - \lambda)p(t)$. Thus,

$$0 = \mu_L(L) = (L - \lambda 1_V)p(L)$$

As $\deg(p) < \deg(\mu_L)$, we know that $p(L) \neq 0$, so the relationship $(L - \lambda 1_V)p(L) = 0$ shows that $L - \lambda 1_V$ is not invertible.

(3) If $\mu_L(0) \neq 0$, then we already know that L is invertible. Conversely, suppose that $\mu_L(0) = 0$. Then, 0 is an eigenvalue by (2) and hence L cannot be invertible.

$\square$

Example 2.4.8. The derivative map $D : P_n \to P_n$ has $\mu_D = t^{n+1}$. Certainly, D^{n+1} vanishes on P_n as all the polynomials in P_n have degree $\leq n$. This means that $\mu_D(t) = t^k$ for some $k \leq n + 1$. On the other hand, $D^n(t^n) = n! \neq 0$ forcing $k = n + 1$.

Example 2.4.9. Let $V = \text{span}\{\exp(\lambda_1 t), \ldots, \exp(\lambda_n t)\}$, with $\lambda_1, \ldots, \lambda_n$ being distinct, and consider again the derivative map $D : V \to V$. Then we have $D(\exp(\lambda_i t)) = \lambda_i \exp(\lambda_i t)$. In Example 1.12.15 (see also Sect. 1.13) it was shown that $\exp(\lambda_1 t), \ldots, \exp(\lambda_n t)$ form a basis for V. Now observe that

$$(D - \lambda_1 1_V) \cdots (D - \lambda_n 1_V)(\exp(\lambda_n)t) = 0.$$

By rearranging terms, it follows that also

$$(D - \lambda_1 1_V) \cdots (D - \lambda_n 1_V)(\exp(\lambda_i)t) = 0$$

and consequently

$$(D - \lambda_1 1_V) \cdots (D - \lambda_n 1_V) = 0 \text{ on } V.$$

On the other hand,

$$(D - \lambda_1 1_V) \cdots (D - \lambda_{n-1} 1_V)(\exp(\lambda_n)t) \neq 0.$$

This means that μ_D divides $(t - \lambda_1) \cdots (t - \lambda_n)$ and that it cannot divide $(t - \lambda_1) \cdots (t - \lambda_{n-1})$. Since the order of the λs is irrelevant, this shows that $\mu_D(t) = (t - \lambda_1) \cdots (t - \lambda_n)$ as μ_D cannot divide

$$\frac{(t - \lambda_1) \cdots (t - \lambda_n)}{t - \lambda_i}.$$

Finally, let us compute the minimal polynomials in two interesting and somewhat tricky situations.

Proposition 2.4.10. *The minimal polynomial for*

$$A = \begin{bmatrix} 0 & 1 & \cdots & 0 \\ 0 & 0 & \ddots & \vdots \\ \vdots & \vdots & \ddots & 1 \\ -\alpha_0 & -\alpha_1 & \cdots & -\alpha_{n-1} \end{bmatrix}$$

is given by

$$\mu_A(t) = t^n + \alpha_{n-1} t^{n-1} + \cdots + \alpha_1 t + \alpha_0.$$

Proof. It turns out to be easier to calculate the minimal polynomial for the transpose

$$B = A^t = \begin{bmatrix} 0 & 0 & \cdots & 0 & -\alpha_0 \\ 1 & 0 & \cdots & 0 & -\alpha_1 \\ 0 & 1 & \cdots & 0 & -\alpha_2 \\ \vdots & \vdots & \ddots & \vdots & \vdots \\ 0 & 0 & \cdots & 1 & -\alpha_{n-1} \end{bmatrix}$$

and it is not hard to show that a matrix and its transpose have the same minimal polynomials by noting that, if $p \in \mathbb{F}[t]$, then

$$(p(A))^t = p(A^t)$$

(see Exercise 3 in this section).
   Let

$$p(t) = t^n + \alpha_{n-1} t^{n-1} + \cdots + \alpha_1 t + \alpha_0.$$

We claim that $\mu_B(t) = p(t) = \chi_A(t)$. Recall from Example 2.3.10 that we already know that $\chi_A(t) = p(t)$. To prove the claim for μ_B, first note that $e_k = B(e_{k-1})$, for $k = 2, \ldots, n$ showing that $e_k = B^{k-1}(e_1)$, for $k = 2, \ldots, n$. Thus, the vectors $e_1, B(e_1), \ldots, B^{n-1}(e_1)$ are linearly independent. This shows that $1_{\mathbb{F}^n}, B, \ldots, B^{n-1}$ must also be linearly independent. Next, we can also show that $p(B) = 0$. This is because

$$p(B)(e_k) = p(B) \circ B^{k-1}(e_1)$$
$$= B^{k-1} \circ p(B)(e_1)$$

and $p(B)(e_1) = 0$ since

$$p(B)(e_1) = \left((B)^n + \alpha_{n-1}(B)^{n-1} + \cdots + \alpha_1 B + \alpha_0 1_{\mathbb{F}^n}\right) e_1$$

$$= (B)^n(e_1) + \alpha_{n-1}(B)^{n-1}(e_1) + \cdots + \alpha_1 B(e_1) + \alpha_0 1_{\mathbb{F}^n}(e_1)$$

$$= Be_n + \alpha_{n-1}e_n + \cdots + \alpha_1 e_2 + \alpha_0 e_1$$

$$= -\alpha_0 e_1 - \alpha_1 e_2 - \cdots - \alpha_{n-1}e_n$$

$$\quad + \alpha_{n-1}e_n + \cdots + \alpha_1 e_2 + \alpha_0 e_1$$

$$= 0. \qquad \qquad \square$$

Next, we show

Proposition 2.4.11. *The minimal polynomial for*

$$C = \begin{bmatrix} \lambda_1 & 1 & & 0 \\ 0 & \lambda_2 & \ddots & \\ & & \ddots & 1 \\ 0 & & & \lambda_n \end{bmatrix}$$

is given by

$$\mu_C(t) = (t - \lambda_1)\cdots(t - \lambda_n).$$

Proof. One strategy would be to show that C has the same minimal polynomial as A in the previous proposition (see also Exercise 20 in Sect. 2.3). But we can also prove the claim directly. Define $\alpha_0, \ldots, \alpha_{n-1}$ by

$$p(t) = t^n + \alpha_{n-1}t^{n-1} + \cdots + \alpha_1 t + \alpha_0 = (t - \lambda_1)\cdots(t - \lambda_n).$$

The claim is then established directly by first showing that $p(C) = 0$. This will imply that μ_C divides p. We then just need to show that $q_i(C) \neq 0$, where

$$q_i(t) = \frac{p(t)}{t - \lambda_i}.$$

The key observation for these facts follow from knowing how to multiply certain upper triangular matrices:

$$\begin{bmatrix} 0 & 1 & 0 & \\ 0 & \gamma_2 & 1 & \\ 0 & 0 & \gamma_3 & \ddots \\ & & & \ddots \end{bmatrix} \begin{bmatrix} \delta_1 & 1 & 0 & \\ 0 & 0 & 1 & \\ 0 & 0 & \delta_3 & \ddots \\ & & & \ddots \end{bmatrix} = \begin{bmatrix} 0 & 0 & 1 & 0 \\ 0 & 0 & * & \\ 0 & 0 & \gamma_3\delta_3 & \ddots \\ \vdots & \vdots & & \ddots \end{bmatrix}.$$

$$\begin{bmatrix} 0 & 0 & 1 & 0 \\ 0 & 0 & * & \\ 0 & 0 & \gamma_3\delta_3 & \end{bmatrix} \begin{bmatrix} \varepsilon_1 & 1 & 0 & \\ 0 & \varepsilon_2 & 1 & \\ 0 & 0 & 0 & 1 \\ & & & \ddots \end{bmatrix} = \begin{bmatrix} 0 & 0 & 0 & 1 \\ 0 & 0 & 0 & * \\ 0 & 0 & 0 & * \\ \vdots & \vdots & \vdots & \gamma_4\delta_4\varepsilon_4 \end{bmatrix}.$$

Therefore, when we do the multiplication

$$(C - \lambda_1 1_{\mathbb{F}^n})(C - \lambda_2 1_{\mathbb{F}^n}) \cdots (C - \lambda_n 1_{\mathbb{F}^n})$$

by starting from the left, we get that the first k columns are zero in

$$(C - \lambda_1 1_{\mathbb{F}^n})(C - \lambda_2 1_{\mathbb{F}^n}) \cdots (C - \lambda_k 1_{\mathbb{F}^n})$$

while the $(k+1)^{\text{th}}$ column has 1 as the first entry. Clearly, this shows that $p(C) = 0$ as well as $q_n(C) \neq 0$. By rearranging the λ_is, this also shows that $q_i(C) \neq 0$ for all $i = 1, \ldots, n$. □

Exercises

1. Find the minimal and characteristic polynomials for

$$A = \begin{bmatrix} 1 & 0 & 1 \\ 0 & 1 & 0 \\ 1 & 0 & 1 \end{bmatrix}.$$

2. Assume that $L : V \to V$ has an invariant subspace $M \subset V$, i.e., $L(M) \subset M$. Show that $\mu_{L|_M}$ divides μ_L.

3. Show that $\mu_A(t) = \mu_{A^t}(t)$, where A^t is the transpose of $A \in \text{Mat}_{n \times n}(\mathbb{F})$. More abstractly, one can show that a linear operator and its dual have the same minimal polynomials (see Sect. 1.14 for definitions related to dual spaces).

4. Let $L : V \to V$ be a linear operator on a finite-dimensional vector space over $\mathbb{F}$ and $p \in \mathbb{F}[t]$ a polynomial. Show that $\ker(p(L)) = \{0\}$ if and only if $\gcd\{p, \mu_L\} = 1$.

5. Let $L : V \to V$ be a linear operator on a finite-dimensional vector space over $\mathbb{F}$ and $p \in \mathbb{F}[t]$ a polynomial. Show that if p divides μ_L, then $\mu_{L|_{\ker(p(L))}}$ divides p.

6. Let $L : V \to V$ be a linear operator such that $L^2 + 1_V = 0$.

   (a) If V is real vector space show, that 1_V and L are linearly independent and that $\mu_L(t) = t^2 + 1$.

   (b) If V and L are complex, show that 1_V and L need not be linearly independent.

   (c) Find the possibilities for the minimal polynomial of $L^3 + 2L^2 + L + 3 \cdot 1_V$.

7. Assume that $L : V \to V$ has minimal polynomial $\mu_L (t) = t$. Find a matrix representation for L.
8. Assume that $L : V \to V$ has minimal polynomial $\mu_L (t) = t^3 + 2t + 1$. Find a polynomial $q (t)$ of degree ≤ 2 such that $L^4 = q (L)$.
9. Assume that $L : V \to V$ has minimal polynomial $\mu_L (t) = t^2 + 1$. Find a polynomial $p (t)$ such that $L^{-1} = p (L)$.
10. Show that if $l \geq \deg(\mu_L) = k$, then L^l is a linear combination of $1_V, L, \ldots, L^{k-1}$. If L is invertible, show the same for all $l < 0$.
11. Let $L : V \to V$ be a linear operator on a finite-dimensional vector space over $\mathbb{F}$ and $p \in \mathbb{F} [t]$ a polynomial. Show

$$\deg \mu_{p(L)} (t) \leq \deg \mu_L (t) .$$

12. Let $p \in \mathbb{C} [t]$. Show that the minimal polynomial for $D : \ker (p (D)) \to \ker (p (D))$ is $\mu_D = p$ (see also Exercise 5 in Sect. 2.3 and Example 2.3.10).
13. Let $A \in \mathrm{Mat}_{n \times n} (\mathbb{F})$ and consider the two linear operators $L_A, R_A : \mathrm{Mat}_{n \times n} (\mathbb{F}) \to \mathrm{Mat}_{n \times n} (\mathbb{F})$ defined by $L_A (X) = AX$ and $R_A (X) = XA$ (see also Exercise 11 in Sect. 2.3). Find the minimal polynomial of L_A, R_A given $\mu_A (t)$.
14. Consider two matrices A and B, show that the minimal polynomial for the block diagonal matrix

$$\begin{bmatrix} A & 0 \\ 0 & B \end{bmatrix}$$

is $\mathrm{lcm}\{\mu_A, \mu_B\}$ (see Proposition 2.1.5 for the definition of lcm). Generalize this to block diagonal matrices

$$\begin{bmatrix} A_1 & & \\ & \ddots & \\ & & A_k \end{bmatrix}.$$

2.5   Diagonalizability

In this section, we shall investigate how and when one can find a basis that puts a linear operator $L : V \to V$ into the simplest possible form. In Sect. 2.2, we saw that decoupling a system of differential equations by finding a basis of eigenvectors for a matrix considerably simplifies the problem of solving the differential equations. It is from that setup that we shall take our cue to the simplest form of a linear operator.

Definition 2.5.1. A linear operator $L : V \to V$ on a finite-dimensional vector space over a field $\mathbb{F}$ is said to be *diagonalizable* if we can find a basis for V that consists of eigenvectors for L, i.e., a basis $e_1, \ldots, e_n$ for V such that $L (e_i) = \lambda_i e_i$ and $\lambda_i \in \mathbb{F}$ for all $i = 1, \ldots, n$. This is the same as saying that

$$\left[L\left(e_1\right) \cdots L\left(e_n\right) \right] = \left[e_1 \cdots e_n \right] \begin{bmatrix} \lambda_1 & \cdots & 0 \\ \vdots & \ddots & \vdots \\ 0 & \cdots & \lambda_n \end{bmatrix}.$$

In other words, the matrix representation for L is a diagonal matrix.

One advantage of having a basis that diagonalizes a linear operator L is that it becomes much simpler to calculate the powers L^k since $L^k\left(e_i\right) = \lambda_i^k e_i$. More generally, if $p\left(t\right) \in \mathbb{F}\left[t\right]$, then we have $p\left(L\right)\left(e_i\right) = p\left(\lambda_i\right)e_i$. Thus, $p\left(L\right)$ is diagonalized with respect to the same basis and with eigenvalues $p\left(\lambda_i\right)$.

We are now ready for a few examples and then the promised application of diagonalizability.

Example 2.5.2. The derivative map $D : P_n \to P_n$ is not diagonalizable. We already know (see Example 1.7.3) that D has a matrix representation that is upper triangular and with zeros on the diagonal. Thus, the characteristic polynomial is t^{n+1}. So the only eigenvalue is 0. Therefore, had D been diagonalizable, it would have had to be the zero transformation 0_{P_n}. Since this is not true, we conclude that $D : P_n \to P_n$ is not diagonalizable.

Example 2.5.3. Let $V = \text{span}\{\exp\left(\lambda_1 t\right), \ldots, \exp\left(\lambda_n t\right)\}$ and consider again the derivative map $D : V \to V$. Then, we have $D\left(\exp\left(\lambda_i t\right)\right) = \lambda_i \exp\left(\lambda_i t\right)$. So if we extract a basis for V among the functions $\exp\left(\lambda_1 t\right), \ldots, \exp\left(\lambda_n t\right)$, then we have found a basis of eigenvectors for D.

These two examples show that diagonalizability is not just a property of the operator. It really matters what space the operator is restricted to live on. We can exemplify this with matrices as well.

Example 2.5.4. Consider

$$A = \begin{bmatrix} 0 & -1 \\ 1 & 0 \end{bmatrix}.$$

As a map $A : \mathbb{R}^2 \to \mathbb{R}^2$, this operator cannot be diagonalizable as it rotates vectors. However, as a map $A : \mathbb{C}^2 \to \mathbb{C}^2$, it has two eigenvalues $\pm i$ with eigenvectors

$$\begin{bmatrix} 1 \\ \mp i \end{bmatrix}.$$

As these eigenvectors form a basis for $\mathbb{C}^2$, we conclude that $A : \mathbb{C}^2 \to \mathbb{C}^2$ is diagonalizable.

We have already seen how decoupling systems of differential equations is related to being able to diagonalize a matrix (see Sect. 2.2). Below we give a related example showing that diagonalizability can be used to investigate a recurrence relation.

Example 2.5.5. Consider the Fibonacci sequence $1, 1, 2, 3, 5, 8, \ldots$ where each term is the sum of the previous two terms. Therefore, if ϕ_n is the nth term in the sequence, then $\phi_{n+2} = \phi_{n+1} + \phi_n$, with initial values $\phi_0 = 1, \phi_1 = 1$. If we record the elements in pairs

$$\Phi_n = \begin{bmatrix} \phi_n \\ \phi_{n+1} \end{bmatrix} \in \mathbb{R}^2,$$

then the relationship takes the form

$$\begin{bmatrix} \phi_{n+1} \\ \phi_{n+2} \end{bmatrix} = \begin{bmatrix} 0 & 1 \\ 1 & 1 \end{bmatrix} \begin{bmatrix} \phi_n \\ \phi_{n+1} \end{bmatrix},$$

$$\Phi_{n+1} = A\Phi_n.$$

The goal is to find a general formula for ϕ_n and to discover what happens as $n \to \infty$. The matrix relationship tells us that

$$\Phi_n = A^n \Phi_0,$$

$$\begin{bmatrix} \phi_n \\ \phi_{n+1} \end{bmatrix} = \begin{bmatrix} 0 & 1 \\ 1 & 1 \end{bmatrix}^n \begin{bmatrix} 1 \\ 1 \end{bmatrix}.$$

Thus, we must find a formula for

$$\begin{bmatrix} 0 & 1 \\ 1 & 1 \end{bmatrix}^n.$$

This is where diagonalization comes in handy. The matrix A has characteristic polynomial

$$t^2 - t - 1 = \left(t - \frac{1 + \sqrt{5}}{2} \right) \left(t - \frac{1 - \sqrt{5}}{2} \right).$$

The corresponding eigenvectors for $\frac{1 \pm \sqrt{5}}{2}$ are $\begin{bmatrix} 1 \\ \frac{1 \pm \sqrt{5}}{2} \end{bmatrix}$. So

$$\begin{bmatrix} 0 & 1 \\ 1 & 1 \end{bmatrix} \begin{bmatrix} 1 & 1 \\ \frac{1+\sqrt{5}}{2} & \frac{1-\sqrt{5}}{2} \end{bmatrix} = \begin{bmatrix} 1 & 1 \\ \frac{1+\sqrt{5}}{2} & \frac{1-\sqrt{5}}{2} \end{bmatrix} \begin{bmatrix} \frac{1+\sqrt{5}}{2} & 0 \\ 0 & \frac{1-\sqrt{5}}{2} \end{bmatrix}$$

or

$$\begin{bmatrix} 0 & 1 \\ 1 & 1 \end{bmatrix} = \begin{bmatrix} 1 & 1 \\ \frac{1+\sqrt{5}}{2} & \frac{1-\sqrt{5}}{2} \end{bmatrix} \begin{bmatrix} \frac{1+\sqrt{5}}{2} & 0 \\ 0 & \frac{1-\sqrt{5}}{2} \end{bmatrix} \begin{bmatrix} 1 & 1 \\ \frac{1+\sqrt{5}}{2} & \frac{1-\sqrt{5}}{2} \end{bmatrix}^{-1}$$

$$= \begin{bmatrix} 1 & 1 \\ \frac{1+\sqrt{5}}{2} & \frac{1-\sqrt{5}}{2} \end{bmatrix} \begin{bmatrix} \frac{1+\sqrt{5}}{2} & 0 \\ 0 & \frac{1-\sqrt{5}}{2} \end{bmatrix} \begin{bmatrix} \frac{1}{2} - \frac{1}{2\sqrt{5}} & \frac{1}{\sqrt{5}} \\ \frac{1}{2} + \frac{1}{2\sqrt{5}} & -\frac{1}{\sqrt{5}} \end{bmatrix}.$$

This means that

$$\begin{bmatrix} 0 & 1 \\ 1 & 1 \end{bmatrix}^n$$

$$= \begin{bmatrix} 1 & 1 \\ \frac{1+\sqrt{5}}{2} & \frac{1-\sqrt{5}}{2} \end{bmatrix} \begin{bmatrix} \frac{1+\sqrt{5}}{2} & 0 \\ 0 & \frac{1-\sqrt{5}}{2} \end{bmatrix}^n \begin{bmatrix} \frac{1}{2} - \frac{1}{2\sqrt{5}} & \frac{1}{\sqrt{5}} \\ \frac{1}{2} + \frac{1}{2\sqrt{5}} & -\frac{1}{\sqrt{5}} \end{bmatrix}$$

$$= \begin{bmatrix} 1 & 1 \\ \frac{1+\sqrt{5}}{2} & \frac{1-\sqrt{5}}{2} \end{bmatrix} \begin{bmatrix} \left(\frac{1+\sqrt{5}}{2}\right)^n & 0 \\ 0 & \left(\frac{1-\sqrt{5}}{2}\right)^n \end{bmatrix} \begin{bmatrix} \frac{1}{2} - \frac{1}{2\sqrt{5}} & \frac{1}{\sqrt{5}} \\ \frac{1}{2} + \frac{1}{2\sqrt{5}} & -\frac{1}{\sqrt{5}} \end{bmatrix}$$

$$= \begin{bmatrix} \left(\frac{1+\sqrt{5}}{2}\right)^n \left(\frac{1}{2} - \frac{1}{2\sqrt{5}}\right) + \left(\frac{1-\sqrt{5}}{2}\right)^n \left(\frac{1}{2} + \frac{1}{2\sqrt{5}}\right) \\ \left(\frac{1+\sqrt{5}}{2}\right)^{n+1} \left(\frac{1}{2} - \frac{1}{2\sqrt{5}}\right) + \left(\frac{1-\sqrt{5}}{2}\right)^{n+1} \left(\frac{1}{2} + \frac{1}{2\sqrt{5}}\right) \end{bmatrix}$$

$$\begin{bmatrix} \frac{1}{\sqrt{5}}\left(\frac{1+\sqrt{5}}{2}\right)^n - \frac{1}{\sqrt{5}}\left(\frac{1-\sqrt{5}}{2}\right)^n \\ \frac{1}{\sqrt{5}}\left(\frac{1+\sqrt{5}}{2}\right)^{n+1} - \frac{1}{\sqrt{5}}\left(\frac{1-\sqrt{5}}{2}\right)^{n+1} \end{bmatrix}$$

Hence

$$\phi_n = \left(\frac{1+\sqrt{5}}{2}\right)^n \left(\frac{1}{2} - \frac{1}{2\sqrt{5}}\right) + \left(\frac{1-\sqrt{5}}{2}\right)^n \left(\frac{1}{2} + \frac{1}{2\sqrt{5}}\right)$$

$$+ \frac{1}{\sqrt{5}} \left(\frac{1+\sqrt{5}}{2}\right)^n - \frac{1}{\sqrt{5}} \left(\frac{1-\sqrt{5}}{2}\right)^n$$

$$= \left(\frac{1}{2} + \frac{1}{2\sqrt{5}}\right) \left(\frac{1+\sqrt{5}}{2}\right)^n + \left(\frac{1-\sqrt{5}}{2}\right)^n \left(\frac{1}{2} - \frac{1}{2\sqrt{5}}\right)$$

$$= \left(\frac{1+\sqrt{5}}{2\sqrt{5}}\right) \left(\frac{1+\sqrt{5}}{2}\right)^n - \left(\frac{1-\sqrt{5}}{2}\right)^n \left(\frac{1-\sqrt{5}}{2\sqrt{5}}\right)$$

$$= \left(\frac{1}{\sqrt{5}}\right) \left(\frac{1+\sqrt{5}}{2}\right)^{n+1} - \left(\frac{1}{\sqrt{5}}\right) \left(\frac{1-\sqrt{5}}{2}\right)^{n+1}.$$

The ratio of successive Fibonacci numbers satisfies

$$\frac{\phi_{n+1}}{\phi_n} = \frac{\left(\frac{1}{\sqrt{5}}\right)\left(\frac{1+\sqrt{5}}{2}\right)^{n+2} - \left(\frac{1}{\sqrt{5}}\right)\left(\frac{1-\sqrt{5}}{2}\right)^{n+2}}{\left(\frac{1}{\sqrt{5}}\right)\left(\frac{1+\sqrt{5}}{2}\right)^{n+1} - \left(\frac{1}{\sqrt{5}}\right)\left(\frac{1-\sqrt{5}}{2}\right)^{n+1}}$$

$$= \frac{\left(\frac{1+\sqrt{5}}{2}\right)^{n+2} - \left(\frac{1-\sqrt{5}}{2}\right)^{n+2}}{\left(\frac{1+\sqrt{5}}{2}\right)^{n+1} - \left(\frac{1-\sqrt{5}}{2}\right)^{n+1}}$$

$$= \frac{\left(\frac{1+\sqrt{5}}{2}\right) - \left(\frac{1-\sqrt{5}}{2}\right)\left(\frac{1-\sqrt{5}}{1+\sqrt{5}}\right)^{n+1}}{1 - \left(\frac{1-\sqrt{5}}{1+\sqrt{5}}\right)^{n+1}},$$

where $\left(\frac{1-\sqrt{5}}{1+\sqrt{5}}\right)^{n+1} \to 0$ as $n \to \infty$. Thus,

$$\lim_{n\to\infty} \frac{\phi_{n+1}}{\phi_n} = \frac{1+\sqrt{5}}{2},$$

which is the golden ratio. This ratio is usually denoted by ϕ. The Fibonacci sequence is often observed in growth phenomena in nature and is also of fundamental importance in combinatorics.

It is not easy to come up with a criterion that guarantees that a matrix is diagonalizable and is also easy to use. It turns out that the minimal polynomial holds the key to diagonalizability of a general linear operator. In a different context, we shall show in Sect. 4.3 that symmetric matrices with real entries are diagonalizable.

The basic procedure for deciding diagonalizability of an operator $L : V \to V$ is to first compute the eigenvalues, then list them without multiplicities $\lambda_1, \dots, \lambda_k$, then calculate all the eigenspaces $\ker(L - \lambda_i 1_V)$, and, finally, check if one can find a basis of eigenvectors. To assist us in this process, there are some useful abstract results about the relationship between the eigenspaces.

Lemma 2.5.6. (Eigenspaces form Direct Sums) *If $\lambda_1, \dots, \lambda_k$ are distinct eigenvalues for a linear operator $L : V \to V$, then*

$$\ker(L - \lambda_1 1_V) + \cdots + \ker(L - \lambda_k 1_V) = \ker(L - \lambda_1 1_V) \oplus \cdots \oplus \ker(L - \lambda_k 1_V).$$

In particular, we have

$$k \le \dim(V).$$

Proof. The proof uses induction on k. When $k = 1$, there is nothing to prove. Assume that the result is true for any collection of k distinct eigenvalues for L

and suppose that we have $k + 1$ distinct eigenvalues $\lambda_1, \ldots, \lambda_{k+1}$ for L. Since we already know that

$$\ker(L - \lambda_1 1_V) + \cdots + \ker(L - \lambda_k 1_V) = \ker(L - \lambda_1 1_V) \oplus \cdots \oplus \ker(L - \lambda_k 1_V),$$

it will be enough to prove that

$$(\ker(L - \lambda_1 1_V) + \cdots + \ker(L - \lambda_k 1_V)) \cap \ker(L - \lambda_{k+1} 1_V) = \{0\}.$$

In other words, we claim that if $L(x) = \lambda_{k+1}x$ and $x = x_1 + \cdots + x_k$ where $x_i \in \ker(L - \lambda_i 1_V)$, then $x = 0$. We can prove this in two ways.

First, note that if $k = 1$, then $x = x_1$ implies that x is the eigenvector for two different eigenvalues. This is clearly not possible unless $x = 0$. Thus, we can assume that $k > 1$. In that case,

$$\begin{aligned} \lambda_{k+1}x &= L(x) \\ &= L(x_1 + \cdots + x_k) \\ &= \lambda_1 x_1 + \cdots + \lambda_k x_k. \end{aligned}$$

Subtracting yields

$$0 = (\lambda_1 - \lambda_{k+1})x_1 + \cdots + (\lambda_k - \lambda_{k+1})x_k.$$

Since we assumed that

$$\ker(L - \lambda_1 1_V) + \cdots + \ker(L - \lambda_k 1_V) = \ker(L - \lambda_1 1_V) \oplus \cdots \oplus \ker(L - \lambda_k 1_V),$$

it follows that $(\lambda_1 - \lambda_{k+1})x_1 = 0, \ldots, (\lambda_k - \lambda_{k+1})x_k = 0$. As $(\lambda_1 - \lambda_{k+1}) \neq 0$, $\ldots, (\lambda_k - \lambda_{k+1}) \neq 0$, we conclude that $x_1 = 0, \ldots, x_k = 0$, implying that $x = x_1 + \cdots + x_k = 0$.

The second way of doing the induction is slightly trickier and has the advantage that it is easy to generalize (see Exercise 20 in this section.) This proof will in addition give us an interesting criterion for when an operator is diagonalizable. Since $\lambda_1, \ldots, \lambda_{k+1}$ are different, the polynomials $t - \lambda_1, \ldots, t - \lambda_{k+1}$ have 1 as their greatest common divisor. Thus, also $(t - \lambda_1) \cdots (t - \lambda_k)$ and $(t - \lambda_{k+1})$ have 1 as their greatest common divisor. This means that we can find polynomials $p(t), q(t) \in \mathbb{F}[t]$ such that

$$1 = p(t)(t - \lambda_1) \cdots (t - \lambda_k) + q(t)(t - \lambda_{k+1})$$

(see Proposition 2.1.4). If we substitute the operator L into this formula in place of t, we obtain:

$$1_V = p(L)(L - \lambda_1 1_V) \cdots (L - \lambda_k 1_V) + q(L)(L - \lambda_{k+1} 1_V).$$

Applying this to x gives us

$$x = p(L)(L - \lambda_1 1_V) \cdots (L - \lambda_k 1_V)(x) + q(L)(L - \lambda_{k+1} 1_V)(x).$$

If

$$x \in (\ker(L - \lambda_1 1_V) + \cdots + \ker(L - \lambda_k 1_V)) \cap \ker(L - \lambda_{k+1} 1_V),$$

then

$$(L - \lambda_1 1_V) \cdots (L - \lambda_k 1_V)(x) = 0,$$
$$(L - \lambda_{k+1} 1_V)(x) = 0,$$

so also $x = 0$.   □

As applications of this lemma, we reexamine several examples.

Example 2.5.7. First, we wish to give a new proof (see Example 1.12.15) that $\exp(\lambda_1 t), \ldots, \exp(\lambda_n t)$ are linearly independent if $\lambda_1, \ldots, \lambda_n$ are distinct. For that, we consider $V = \operatorname{span}\{\exp(\lambda_1 t), \ldots, \exp(\lambda_n t)\}$ and $D : V \to V$. The result is now obvious as each of the functions $\exp(\lambda_i t)$ is an eigenvector with eigenvalue λ_i for $D : V \to V$. As $\lambda_1, \ldots, \lambda_n$ are distinct, we can conclude that the corresponding eigenfunctions are linearly independent. Thus, $\exp(\lambda_1 t), \ldots, \exp(\lambda_n t)$ form a basis for V which diagonalizes D.

Example 2.5.8. In order to solve the initial value problem for higher order differential equations, it was necessary to show that the *Vandermonde matrix*

$$\begin{bmatrix} 1 & \cdots & 1 \\ \lambda_1 & \cdots & \lambda_n \\ \vdots & \ddots & \vdots \\ \lambda_1^{n-1} & \cdots & \lambda_n^{n-1} \end{bmatrix}$$

is invertible, when $\lambda_1, \ldots, \lambda_n \in \mathbb{F}$ are distinct. This was done in Example 1.12.12 and will now be established using eigenvectors. Given the origins of this problem (in this book), it is not unnatural to consider a matrix

$$A = \begin{bmatrix} 0 & 1 & \cdots & 0 \\ 0 & 0 & \ddots & \vdots \\ \vdots & \vdots & \ddots & 1 \\ -\alpha_0 & -\alpha_1 & \cdots & -\alpha_{n-1} \end{bmatrix},$$

where

$$p(t) = t^n + \alpha_{n-1} t^{n-1} + \cdots + \alpha_1 t + \alpha_0$$
$$= (t - \lambda_1) \cdots (t - \lambda_n).$$

In Example 2.3.10, we saw that the characteristic polynomial for A is $p(t)$. In particular, $\lambda_1, \ldots, \lambda_n \in \mathbb{F}$ are the eigenvalues. When these eigenvalues are distinct, we consequently know that the corresponding eigenvectors are linearly independent. To find these eigenvectors, note that

$$
A \begin{bmatrix} 1 \\ \lambda_k \\ \vdots \\ \lambda_k^{n-1} \end{bmatrix} = \begin{bmatrix} 0 & 1 & \cdots & 0 \\ 0 & 0 & \ddots & \vdots \\ \vdots & \vdots & \ddots & 1 \\ -\alpha_0 & -\alpha_1 & \cdots & -\alpha_{n-1} \end{bmatrix} \begin{bmatrix} 1 \\ \lambda_k \\ \vdots \\ \lambda_k^{n-1} \end{bmatrix}
$$

$$
= \begin{bmatrix} \lambda_k \\ \lambda_k^2 \\ \vdots \\ -\alpha_0 - \alpha_1 \lambda_k - \cdots - \alpha_{n-1} \lambda_k^{n-1} \end{bmatrix}
$$

$$
= \begin{bmatrix} \lambda_k \\ \lambda_k^2 \\ \vdots \\ \lambda_k^n \end{bmatrix}, \text{ since } p(\lambda_k) = 0
$$

$$
= \lambda_k \begin{bmatrix} 1 \\ \lambda_k \\ \vdots \\ \lambda_k^{n-1} \end{bmatrix}.
$$

This implies that the columns in the Vandermonde matrix are the eigenvectors for a diagonalizable operator. Hence, the matrix must be invertible. Note that A is diagonalizable if and only if $\lambda_1, \ldots, \lambda_n$ are distinct as all eigenspaces for A are one-dimensional (we shall also prove and use this in the next Sect. 2.6).

Example 2.5.9. An interesting special case of the previous example occurs when $p(t) = t^n - 1$ and we assume that $\mathbb{F} = \mathbb{C}$. Then, the roots are the nth roots of unity, and the operator that has these numbers as eigenvalues looks like

$$
C = \begin{bmatrix} 0 & 1 & \cdots & 0 \\ 0 & 0 & \ddots & \vdots \\ \vdots & \vdots & \ddots & 1 \\ 1 & 0 & \cdots & 0 \end{bmatrix}.
$$

The powers of this matrix have the following interesting patterns:

$$C^2 = \begin{bmatrix} 0 & 0 & 1 & 0 & & 0 \\ & 0 & 0 & \ddots & \\ & & & 1 & 0 \\ 0 & & & & 0 & 1 \\ 1 & 0 & & & 0 & 0 \\ 0 & 1 & 0 & & & 0 \end{bmatrix},$$

$$\vdots$$

$$C^{n-1} = \begin{bmatrix} 0 & \cdots & \cdots & 1 \\ 1 & 0 & \ddots & \vdots \\ \vdots & \ddots & \ddots & 0 \\ 0 & \cdots & 1 & 0 \end{bmatrix},$$

$$C^n = \begin{bmatrix} 1 & 0 & \cdots & 0 \\ 0 & 1 & \ddots & \vdots \\ \vdots & \ddots & \ddots & 0 \\ 0 & \cdots & 0 & 1 \end{bmatrix} = 1_{\mathbb{F}^n}.$$

A linear combination of these powers looks like

$$C_{\alpha_0,\dots,\alpha_{n-1}} = \alpha_0 1_{\mathbb{F}^n} + \alpha_1 C + \cdots + \alpha_{n-1} C^{n-1}$$

$$= \begin{bmatrix} \alpha_0 & \alpha_1 & \alpha_2 & \alpha_3 & \cdots & \alpha_{n-1} \\ \alpha_{n-1} & \alpha_0 & \alpha_1 & \alpha_2 & \cdots & \alpha_{n-2} \\ \vdots & & \alpha_{n-1} & \alpha_0 & \ddots & \vdots \\ \alpha_3 & \vdots & & \alpha_{n-1} & \ddots & \\ \alpha_2 & \alpha_3 & \vdots & & \ddots & \alpha_0 & \alpha_1 \\ \alpha_1 & \alpha_2 & \alpha_3 & \cdots & \alpha_{n-1} & \alpha_0 \end{bmatrix}.$$

Since we have a basis that diagonalizes C and hence also all of its powers, we have also found a basis that diagonalizes $C_{\alpha_0,\dots,\alpha_{n-1}}$. This would probably not have been so easy to see if we had just been handed the matrix $C_{\alpha_0,\dots,\alpha_{n-1}}$.

The above lemma also helps us establish three criteria for diagonalizability.

Theorem 2.5.10. (First Criterion for Diagonalizability) *Let $L : V \to V$ be a linear operator on an n-dimensional vector space over $\mathbb{F}$. If $\lambda_1, \dots, \lambda_k \in \mathbb{F}$ are distinct eigenvalues for L such that*

$$n = \dim(\ker(L - \lambda_1 1_V)) + \cdots + \dim(\ker(L - \lambda_k 1_V)),$$

then L is diagonalizable. In particular, if L has n distinct eigenvalues in $\mathbb{F}$, then L is diagonalizable.

Proof. Our assumption together with Lemma 2.5.6 shows that

$$n = \dim(\ker(L - \lambda_1 1_V)) + \cdots + \dim(\ker(L - \lambda_k 1_V))$$
$$= \dim(\ker(L - \lambda_1 1_V) + \cdots + \ker(L - \lambda_k 1_V)).$$

Thus,

$$\ker(L - \lambda_1 1_V) \oplus \cdots \oplus \ker(L - \lambda_k 1_V) = V,$$

and we can find a basis of eigenvectors, by selecting a basis for each of the eigenspaces.

For the last statement, we only need to observe that $\dim(\ker(L - \lambda 1_V)) \geq 1$ for any eigenvalue $\lambda \in \mathbb{F}$.   □

The next characterization offers a particularly nice condition for diagonalizability and will also give us the minimal polynomial characterization of diagonalizability.

Theorem 2.5.11. (Second Criterion for Diagonalizability) *Let $L : V \to V$ be a linear operator on an n-dimensional vector space over $\mathbb{F}$. L is diagonalizable if and only if we can find $p \in \mathbb{F}[t]$ such that $p(L) = 0$ and*

$$p(t) = (t - \lambda_1) \cdots (t - \lambda_k),$$

where $\lambda_1, \ldots, \lambda_k \in \mathbb{F}$ are distinct.

Proof. Assuming that L is diagonalizable, we have

$$V = \ker(L - \lambda_1 1_V) \oplus \cdots \oplus \ker(L - \lambda_k 1_V).$$

So if we use

$$p(t) = (t - \lambda_1) \cdots (t - \lambda_k)$$

we see that $p(L) = 0$ as $p(L)$ vanishes on each of the eigenspaces (see also Exercise 16 in this section).

Conversely, assume that $p(L) = 0$ and

$$p(t) = (t - \lambda_1) \cdots (t - \lambda_k),$$

where $\lambda_1, \ldots, \lambda_k \in \mathbb{F}$ are distinct. If any of these λ_is are not eigenvalues for L, we can eliminate the factors $t - \lambda_i$ since $L - \lambda_i 1_V$ is an isomorphism unless λ_i is an eigenvalue. We then still have that L is a root of the new polynomial. The proof now goes by induction on the number of roots in p. If there is one root, the result is obvious. If $k \geq 2$, we can use Proposition 2.1.4 to write

$$1 = r(t)(t - \lambda_1) \cdots (t - \lambda_{k-1}) + s(t)(t - \lambda_k)$$
$$= r(t) q(t) + s(t)(t - \lambda_k).$$

We then claim that

$$V = \ker(q(L)) \oplus \ker(L - \lambda_k 1_V)$$

and that

$$L(\ker(q(L))) \subset \ker(q(L)).$$

This will finish the induction step as $L|_{\ker(q(L))}$ is a linear operator on the proper subspace $\ker(q(L))$ with the property that $q\left(L|_{\ker(q(L))}\right) = 0$. We can then use the induction hypothesis to conclude that the result holds for $L|_{\ker(q(L))}$. As it obviously holds for $(L - \lambda_k 1_V)|_{\ker(L - \lambda_k 1_V)}$, it follows that the result also holds for L.

To establish the decomposition observe that

$$x = q(L)(r(L)(x)) + (L - \lambda_k 1_V)(s(L)(x))$$
$$= y + z.$$

Here $y \in \ker(L - \lambda_k 1_V)$ since

$$(L - \lambda_k 1_V)(y) = (L - \lambda_k 1_V)(q(L)(r(L)(x)))$$
$$= p(L)(r(L)(x))$$
$$= 0,$$

and $z \in \ker(q(L))$ since

$$q(L)((L - \lambda_k 1_V)(s(L)(x))) = p(L)(s(L)(x)) = 0.$$

Thus,

$$V = \ker(q(L)) + \ker(L - \lambda_k 1_V).$$

If

$$x \in \ker(q(L)) \cap \ker(L - \lambda_k 1_V),$$

then we have

$$x = r(L)(q(L)(x)) + s(L)((L - \lambda_k 1_V)(x)) = 0.$$

This gives the direct sum decomposition.

Finally, if $x \in \ker(q(L))$, then we see that

$$q(L)(L(x)) = (q(L) \circ L)(x)$$
$$= (L \circ q(L))(x)$$

$$= L\left(q\left(L\right)\left(x\right)\right)$$

$$= 0.$$

Thus, showing that $L\left(x\right) \in \ker\left(q\left(L\right)\right)$.   □

Corollary 2.5.12. (The Minimal Polynomial Characterization of Diagonalizability) *Let* $L : V \to V$ *be a linear operator on an n-dimensional vector space over* $\mathbb{F}$. *L is diagonalizable if and only if the minimal polynomial factors*

$$\mu_L\left(t\right) = \left(t - \lambda_1\right) \cdots \left(t - \lambda_k\right),$$

and has no multiple roots, i.e., $\lambda_1, \ldots, \lambda_k \in \mathbb{F}$ *are distinct.*

Finally we can estimate how large $\dim\left(\ker\left(L - \lambda 1_V\right)\right)$ can be if we have factored the characteristic polynomial.

Lemma 2.5.13. *Let* $L : V \to V$ *be a linear operator on an n-dimensional vector space over* $\mathbb{F}$. *If* $\lambda \in \mathbb{F}$ *is an eigenvalue and* $\chi_L\left(t\right) = \left(t - \lambda\right)^m q\left(t\right)$, *where* $q\left(\lambda\right) \neq 0$, *then*

$$\dim\left(\ker\left(L - \lambda 1_V\right)\right) \leq m.$$

We call $\dim\left(\ker\left(L - \lambda 1_V\right)\right)$ the *geometric multiplicity* of λ and m the *algebraic multiplicity* of λ.

Proof. Select a complement N to $\ker\left(L - \lambda 1_V\right)$ in V. Then, choose a basis where $x_1, \ldots, x_k \in \ker\left(L - \lambda 1_V\right)$ and $x_{k+1}, \ldots, x_n \in N$. Since $L\left(x_i\right) = \lambda x_i$ for $i = 1, \ldots, k$, we see that the matrix representation has a block form that looks like

$$[L] = \begin{bmatrix} \lambda 1_{\mathbb{F}^k} & B \\ 0 & C \end{bmatrix}.$$

This implies that

$$\chi_L\left(t\right) = \chi_{[L]}\left(t\right)$$

$$= \chi_{\lambda 1_{\mathbb{F}^k}}\left(t\right) \chi_C\left(t\right)$$

$$= \left(t - \lambda\right)^k \chi_C\left(t\right)$$

and hence that λ has algebraic multiplicity $m \geq k$.   □

Clearly, the appearance of multiple roots in the characteristic polynomial is something that might prevent linear operators from becoming diagonalizable. The following criterion is often useful for deciding whether or not a polynomial has multiple roots.

Proposition 2.5.14. *A polynomial $p(t) \in \mathbb{F}[t]$ has $\lambda \in \mathbb{F}$ as a multiple root if and only if λ is a root of both p and Dp.*

Proof. If λ is a multiple root, then $p(t) = (t - \lambda)^m q(t)$, where $m \geq 2$. Thus,

$$Dp(t) = m(t - \lambda)^{m-1} q(t) + (t - \lambda)^m Dq(t)$$

also has λ as a root.

Conversely, if λ is a root of Dp and p, then we can write $p(t) = (t - \lambda) q(t)$ and

$$
\begin{aligned}
0 &= Dp(\lambda) \\
&= q(\lambda) + (\lambda - \lambda) Dq(\lambda) \\
&= q(\lambda).
\end{aligned}
$$

Thus, also $q(t)$ has λ as a root and hence λ is a multiple root of $p(t)$.   □

Example 2.5.15. If $p(t) = t^2 + \alpha t + \beta$, then $Dp(t) = 2t + \alpha$. Thus we have a double root only if the only root $t = -\frac{\alpha}{2}$ of Dp is a root of p. If we evaluate

$$
\begin{aligned}
p\left(-\frac{\alpha}{2}\right) &= \frac{\alpha^2}{4} - \frac{\alpha^2}{2} + \beta \\
&= -\frac{\alpha^2}{4} + \beta \\
&= -\frac{\alpha^2 - 4\beta}{4},
\end{aligned}
$$

we see that this occurs precisely when the discriminant vanishes. This conforms nicely with the quadratic formula.

Example 2.5.16. If $p(t) = t^3 + 12t^2 - 14$, then the roots are pretty nasty. We can, however, check for multiple roots by finding the roots of

$$Dp(t) = 3t^2 + 24t = 3t(t + 8)$$

and checking whether they are roots of p

$$
\begin{aligned}
p(0) &= -14 \neq 0, \\
p(8) &= 8^3 + 12 \cdot 8^2 - 14 \\
&= 8^2(8 + 12) - 14 > 0.
\end{aligned}
$$

Exercises

1. Decide whether or not the following matrices are diagonalizable:

   (a)
   $$\begin{bmatrix} 1 & 0 & 1 \\ 0 & 1 & 0 \\ 1 & 0 & 1 \end{bmatrix}$$

   (b)
   $$\begin{bmatrix} 0 & 1 & 2 \\ 1 & 0 & 3 \\ 2 & 3 & 0 \end{bmatrix}$$

   (c)
   $$\begin{bmatrix} 0 & 1 & 2 \\ -1 & 0 & 3 \\ -2 & -3 & 0 \end{bmatrix}$$

2. Decide whether or not the following matrices are diagonalizable:

   (a)
   $$\begin{bmatrix} 0 & i \\ i & 0 \end{bmatrix}$$

   (b)
   $$\begin{bmatrix} 0 & i \\ -i & 0 \end{bmatrix}$$

   (c)
   $$\begin{bmatrix} 1 & i & 0 \\ i & 1 & 0 \\ 0 & 2 & 1 \end{bmatrix}$$

3. Decide whether or not the following matrices are diagonalizable:

   (a)
   $$\begin{bmatrix} 1 & 0 & 1 \\ 0 & 0 & 0 \\ 1 & 0 & 1 \end{bmatrix}$$

   (b)
   $$\begin{bmatrix} 1 & 0 & 1 \\ 0 & 1 & 0 \\ 1 & 0 & 1 \end{bmatrix}$$

(c)
$$\begin{bmatrix} 0 & 0 & 1 \\ 0 & 1 & 0 \\ 1 & 0 & 0 \end{bmatrix}$$

4. Find the characteristic polynomial, eigenvalues, and eigenvectors for each of the following linear operators $L : P_3 \to P_3$. Then, decide whether they are diagonalizable by checking whether there is a basis of eigenvectors.

   (a) $L = D$.
   (b) $L = tD = T \circ D$.
   (c) $L = D^2 + 2D + 1_{P_3}$.
   (d) $L = t^2 D^3 + D$.

5. Consider the linear operator on $\mathrm{Mat}_{n \times n}(\mathbb{F})$ defined by $L(X) = X^t$. Show that L is diagonalizable. Compute the eigenvalues and eigenspaces.

6. For which $s, t \in \mathbb{C}$ is the matrix diagonalizable
$$\begin{bmatrix} 1 & 1 \\ s & t \end{bmatrix}?$$

7. For which $\alpha, \beta, \gamma \in \mathbb{C}$ is the matrix diagonalizable
$$\begin{bmatrix} 0 & 1 & 0 \\ 0 & 0 & 1 \\ -\alpha & -\beta & -\gamma \end{bmatrix}?$$

8. Assume $L : V \to V$ is diagonalizable. Show that $V = \ker(L) \oplus \mathrm{im}(L)$.

9. Assume that $L : V \to V$ is a diagonalizable real linear map on a finite-dimensional vector space. Show that $\mathrm{tr}(L^2) \geq 0$.

10. Assume that $A \in \mathrm{Mat}_{n \times n}(\mathbb{F})$ is diagonalizable.

    (a) Show that A^t is diagonalizable.
    (b) Show that $L_A(X) = AX$ defines a diagonalizable operator on $\mathrm{Mat}_{n \times n}(\mathbb{F})$ (see Example 1.7.6.)
    (c) Show that $R_A(X) = XA$ defines a diagonalizable operator on $\mathrm{Mat}_{n \times n}(\mathbb{F})$.

11. Show that if $E : V \to V$ is a projection on a finite-dimensional vector space, then $\mathrm{tr}(E) = \dim(\mathrm{im}(E))$.

12. Let $A \in \mathrm{Mat}_{n \times n}(\mathbb{F})$ and $B \in \mathrm{Mat}_{m \times m}(\mathbb{F})$ and consider
$$L : \mathrm{Mat}_{n \times m}(\mathbb{F}) \to \mathrm{Mat}_{n \times m}(\mathbb{F}),$$
$$L(X) = AX - XB.$$

Show that if B is diagonalizable, then all eigenvalues of L are of the form $\lambda - \mu$, where λ is an eigenvalue of A and μ an eigenvalue of B.

13. *(Restrictions of Diagonalizable Operators)* Let $L : V \to V$ be a linear operator on a finite-dimensional vector space and $M \subset V$ an invariant subspace, i.e., $L(M) \subset M$.

    (a) If $x+y \in M$, where $L(x) = \lambda x$, $L(y) = \mu y$, and $\lambda \neq \mu$, then $x, y \in M$.

    (b) If $x_1 + \cdots + x_k \in M$ and $L(x_i) = \lambda_i x_i$, where $\lambda_1, \ldots, \lambda_k$ are distinct, then $x_1, \ldots, x_k \in M$. Hint: Use induction on k.

    (c) If $L : V \to V$ is diagonalizable, use (a) and (b) to show that $L : M \to M$ is diagonalizable.

    (d) If $L : V \to V$ is diagonalizable, use Theorem 2.5.11 directly to show that $L : M \to M$ is diagonalizable.

14. Let $L : V \to V$ be a linear operator on a finite-dimensional vector space. Show that λ is a multiple root for $\mu_L(t)$ if and only if

$$\{0\} \subsetneq \ker(L - \lambda 1_V) \subsetneq \ker\left((L - \lambda 1_V)^2\right).$$

15. Assume that $L, K : V \to V$ are both diagonalizable, that $KL = LK$, and that V is finite-dimensional. Show that we can find a basis for V that diagonalizes both L and K. Hint: You can use Exercise 13 with M as an eigenspace for one of the operators as well as Exercise 3 in Sect. 1.11.

16. Let $L : V \to V$ be a linear operator on a vector space and $\lambda_1, \ldots, \lambda_k$ distinct eigenvalues. If $x = x_1 + \cdots + x_k$, where $x_i \in \ker(L - \lambda_i 1_V)$, then

$$(L - \lambda_1 1_V) \cdots (L - \lambda_k 1_V)(x) = 0.$$

17. Let $L : V \to V$ be a linear operator on a vector space and $\lambda \neq \mu$. Use the identity

$$\frac{1}{\mu - \lambda}(L - \lambda 1_V) - \frac{1}{\mu - \lambda}(L - \mu 1_V) = 1_V$$

to show that two eigenspaces associated to distinct eigenvalues for L have trivial intersection.

18. Consider an *involution* $L : V \to V$, i.e., $L^2 = 1_V$.

    (a) Show that $x \pm L(x)$ is an eigenvector for L with eigenvalue ± 1.

    (b) Show that $V = \ker(L + 1_V) \oplus \ker(L - 1_V)$.

    (c) Conclude that L is diagonalizable.

19. Assume $L : V \to V$ satisfies $L^2 + \alpha L + \beta 1_V = 0$ and that the roots λ_1, λ_2 of $\lambda^2 + \alpha \lambda + \beta$ are distinct and lie in $\mathbb{F}$.

    (a) Determine γ, δ so that

$$x = \gamma(L(x) - \lambda_1 x) + \delta(L(x) - \lambda_2 x).$$

    (b) Show that $L(x) - \lambda_1 x \in \ker(L - \lambda_2 1_V)$ and $L(x) - \lambda_2 x \in \ker(L - \lambda_1 1_V)$.

(c) Conclude that $V = \ker(L - \lambda_1 1_V) \oplus \ker(L - \lambda_2 1_V)$.

(d) Conclude that L is diagonalizable.

20. Let $L : V \to V$ be a linear operator on a finite-dimensional vector space. Show that

(a) If $p, q \in \mathbb{F}[t]$ and $\gcd\{p, q\} = 1$, then $V = \ker(p(L)) \oplus \ker(q(L))$. Hint: Look at the proof of Theorem 2.5.11.

(b) If $\mu_L(t) = p(t) q(t)$, where $\gcd\{p, q\} = 1$, then $\mu_L|_{\ker(p(L))} = p$ and $\mu_L|_{\ker(q(L))} = q$.

2.6   Cyclic Subspaces

The goal of this section is to find a relatively simple matrix representation for linear operators $L : V \to V$ on finite-dimensional vector spaces that are not necessarily diagonalizable. The way in which this is going to be achieved is by finding a decomposition $V = M_1 \oplus \cdots \oplus M_k$ into L-invariant subspaces M_i with the property that $L|_{M_i}$ has matrix representation that can be found by only knowing the characteristic or minimal polynomial for $L|_{M_i}$.

The invariant subspaces we are going to use are in fact a very natural generalization of eigenvectors. Observe that $x \in V$ is an eigenvector if $L(x) \in \mathrm{span}\{x\}$ or in other words $L(x)$ is a linear combination of x.

Definition 2.6.1. Let $L : V \to V$ be a linear operator on a finite-dimensional vector space. The *cyclic subspace* generated by $x \in V$ is the subspace spanned by the vectors $x, L(x), \ldots, L^k(x), \ldots$, i.e.,

$$C_x = \mathrm{span}\left\{x, L(x), L^2(x), \ldots, L^k(x), \ldots\right\}.$$

Assuming $x \neq 0$, we can use Lemma 1.12.3 to find a smallest $k \geq 1$ such that

$$L^k(x) \in \mathrm{span}\left\{x, L(x), L^2(x), \ldots, L^{k-1}(x)\right\}.$$

With this definition and construction behind us, we can now prove.

Lemma 2.6.2. *Let $L : V \to V$ be a linear operator on an finite-dimensional vector space. Then, C_x is L-invariant and we can find $k \leq \dim(V)$ so that $x, L(x)$, $L^2(x), \ldots, L^{k-1}(x)$ form a basis for C_x. The matrix representation for $L|_{C_x}$ with respect to this basis is*

$$\begin{bmatrix} 0 & 0 & \cdots & 0 & \alpha_0 \\ 1 & 0 & \cdots & 0 & \alpha_1 \\ 0 & 1 & \cdots & 0 & \alpha_2 \\ \vdots & \vdots & \ddots & \vdots & \vdots \\ 0 & 0 & \cdots & 1 & \alpha_{k-1} \end{bmatrix},$$

where

$$L^k (x) = \alpha_0 x + \alpha_1 L (x) + \cdots + \alpha_{k-1} L^{k-1} (x).$$

Proof. The vectors $x, L(x), L^2(x), \ldots, L^{k-1}(x)$ must be linearly independent if we pick k as the smallest k such that

$$L^k (x) = \alpha_0 x + \alpha_1 L (x) + \cdots + \alpha_{k-1} L^{k-1} (x).$$

To see that they span C_x, we need to show that

$$L^m (x) \in \text{span} \{x, L(x), L^2(x), \ldots, L^{k-1}(x)\}$$

for all $m \geq k$. We are going to use induction on m to prove this. If $m = 0, \ldots k - 1$, there is nothing to prove. Assuming that

$$L^{m-1} (x) = \beta_0 x + \beta_1 L (x) + \cdots + \beta_{k-1} L^{k-1} (x),$$

we get

$$L^m (x) = \beta_0 L (x) + \beta_1 L^2 (x) + \cdots + \beta_{k-1} L^k (x).$$

Since we already have that

$$L^k (x) \in \text{span} \{x, L(x), L^2(x), \ldots, L^{k-1}(x)\},$$

it follows that

$$L^m (x) \in \text{span} \{x, L(x), L^2(x), \ldots, L^{k-1}(x)\}.$$

This completes the induction step. This also explains why C_x is L-invariant, namely, if $z \in C_x$, then we have

$$z = \gamma_0 x + \gamma_1 L (x) + \cdots + \gamma_{k-1} L^{k-1} (x),$$

and

$$L (z) = \gamma_0 L (x) + \gamma_1 L^2 (x) + \cdots + \gamma_{k-1} L^k (x).$$

As $L^k (x) \in C_x$ we see that $L(z) \in C_x$ as well.

To find the matrix representation, we note that

$$\left[L(x) \ L(L(x)) \ \cdots \ L\left(L^{k-2}(x)\right) \ L\left(L^{k-1}(x)\right) \right]$$

$$= \left[L(x) \ L^2(x) \ \cdots \ L^{k-1}(x) \ L^k(x) \right]$$

$$= \left[x \ L(x) \ \cdots \ L^{k-2}(x) \ L^{k-1}(x) \right] \begin{bmatrix} 0 & 0 & \cdots & 0 & \alpha_0 \\ 1 & 0 & \cdots & 0 & \alpha_1 \\ 0 & 1 & \cdots & 0 & \alpha_2 \\ \vdots & \vdots & \ddots & \vdots & \vdots \\ 0 & 0 & \cdots & 1 & \alpha_{k-1} \end{bmatrix}.$$

This proves the lemma. $\qquad\square$

The matrix representation for $L|_{C_x}$ is apparently the transpose of the type of matrix coming from higher order differential equations that we studied in the previous sections. Therefore, we can expect our knowledge of those matrices to carry over without much effort. To be a little more precise, we define the *companion matrix* of a monic nonconstant polynomial $p(t) \in \mathbb{F}[t]$ as the matrix

$$C_p = \begin{bmatrix} 0 & 0 & \cdots & 0 & -\alpha_0 \\ 1 & 0 & \cdots & 0 & -\alpha_1 \\ 0 & 1 & \cdots & 0 & -\alpha_2 \\ \vdots & \vdots & \ddots & \vdots & \vdots \\ 0 & 0 & \cdots & 1 & -\alpha_{n-1} \end{bmatrix},$$

$$p(t) = t^n + \alpha_{n-1}t^{n-1} + \cdots + \alpha_1 t + \alpha_0.$$

It is worth mentioning that the companion matrix for $p = t + \alpha$ is simply the 1×1 matrix $[-\alpha]$.

Proposition 2.6.3. *The characteristic and minimal polynomials of C_p are both $p(t)$, and all eigenspaces are one-dimensional. In particular, C_p is diagonalizable if and only if all the roots of $p(t)$ are distinct and lie in $\mathbb{F}$.*

Proof. Even though we can prove these properties from our knowledge of the transpose of C_p, it is still worthwhile to give a complete proof. Recall that we computed the minimal polynomial in the proof of Proposition 2.4.10.

To compute the characteristic polynomial, we consider:

$$t1_{\mathbb{F}^n} - C_p = \begin{bmatrix} t & 0 & \cdots & 0 & \alpha_0 \\ -1 & t & \cdots & 0 & \alpha_1 \\ 0 & -1 & \cdots & 0 & \alpha_2 \\ \vdots & \vdots & \ddots & \vdots & \vdots \\ 0 & 0 & \cdots & -1 & t + \alpha_{n-1} \end{bmatrix}.$$

By switching rows 1 and 2, we see that this is row equivalent to

$$\begin{bmatrix} -1 & t & \cdots & 0 & \alpha_1 \\ t & 0 & \cdots & 0 & \alpha_0 \\ 0 & -1 & \cdots & 0 & \alpha_2 \\ \vdots & \vdots & \ddots & \vdots & \vdots \\ 0 & 0 & \cdots & -1 & t + \alpha_{n-1} \end{bmatrix}$$

eliminating t then gives us

$$\begin{bmatrix} -1 & t & \cdots & 0 & \alpha_1 \\ 0 & t^2 & \cdots & 0 & \alpha_0 + \alpha_1 t \\ 0 & -1 & \cdots & 0 & \alpha_2 \\ \vdots & \vdots & \ddots & \vdots & \vdots \\ 0 & 0 & \cdots & -1 & t + \alpha_{n-1} \end{bmatrix}.$$

Now, switch rows 2 and 3 to get

$$\begin{bmatrix} -1 & t & \cdots & 0 & \alpha_1 \\ 0 & -1 & \cdots & 0 & \alpha_2 \\ 0 & t^2 & \cdots & 0 & \alpha_0 + \alpha_1 t \\ \vdots & \vdots & \ddots & \vdots & \vdots \\ 0 & 0 & \cdots & -1 & t + \alpha_{n-1} \end{bmatrix}$$

and eliminate t^2

$$\begin{bmatrix} -1 & t & \cdots & 0 & \alpha_1 \\ 0 & -1 & \cdots & 0 & \alpha_2 \\ 0 & 0 & \cdots & 0 & \alpha_0 + \alpha_1 t + \alpha_2 t^2 \\ \vdots & \vdots & \ddots & \vdots & \vdots \\ 0 & 0 & \cdots & -1 & t + \alpha_{n-1} \end{bmatrix}.$$

Repeating this argument shows that $t1_{\mathbb{F}^n} - C_p$ is row equivalent to

$$\begin{bmatrix} -1 & t & \cdots & 0 & & \alpha_1 \\ 0 & -1 & \cdots & 0 & & \alpha_2 \\ 0 & 0 & \ddots & \vdots & & \vdots \\ \vdots & \vdots & & -1 & & t + \alpha_{n-1} \\ 0 & 0 & \cdots & 0 & t^n + \alpha_{n-1} t^{n-1} + \cdots + \alpha_1 t + \alpha_0 \end{bmatrix}.$$

This implies that the characteristic polynomial is $p(t)$.

To see that all eigenspaces are one-dimensional we note that if λ is a root of $p(t)$, then we have just shown that $\lambda 1_{\mathbb{F}^n} - C_p$ is row equivalent to the matrix

$$\begin{bmatrix} -1 & \lambda & \cdots & 0 & \alpha_1 \\ 0 & -1 & \cdots & 0 & \alpha_2 \\ 0 & 0 & \ddots & \vdots & \vdots \\ \vdots & \vdots & & -1 & \lambda + \alpha_{n-1} \\ 0 & 0 & \cdots & 0 & 0 \end{bmatrix}.$$

Since all but the last diagonal entry is nonzero we see that the kernel must be one-dimensional.   □

Cyclic subspaces lead us to a very elegant proof of the Cayley–Hamilton theorem.

Theorem 2.6.4. (The Cayley–Hamilton Theorem) *Let* $L : V \to V$ *be a linear operator on a finite-dimensional vector space. Then,* L *is a root of its own characteristic polynomial*

$$\chi_L(L) = 0.$$

In particular, the minimal polynomial divides the characteristic polynomial.

Proof. Select any $x \neq 0$ in V and a complement M to the cyclic subspace C_x generated by x. This gives us a nontrivial decomposition $V = C_x \oplus M$, where L maps C_x to itself and M into V. If we select a basis for V that starts with the cyclic basis for C_x, then L will have a matrix representation that looks like

$$[L] = \begin{bmatrix} C_p & B \\ 0 & D \end{bmatrix},$$

where C_p is the companion matrix representation for L restricted to C_x. This shows that

$$\chi_L(t) = \chi_{C_p}(t)\,\chi_D(t)$$
$$= p(t)\,\chi_D(t).$$

We know that $p(C_p) = 0$ from the previous result. This shows that $p(L|_{C_x}) = 0$ and in particular that $p(L)(x) = 0$. Thus,

$$\chi_L(L)(x) = \chi_D(L) \circ p(L)(x)$$
$$= 0.$$

Since x was arbitrary, this shows that $\chi_L(L) = 0$.   □

We now have quite a good understanding of the basic building blocks in the decomposition we are seeking.

Theorem 2.6.5. (The Cyclic Subspace Decomposition) *Let* $L : V \to V$ *be a linear operator on a finite-dimensional vector space. Then,* V *has a cyclic subspace decomposition*

$$V = C_{x_1} \oplus \cdots \oplus C_{x_k},$$

where each C_{x_i} *is a cyclic subspace. In particular,* L *has a block diagonal matrix representation where each block is a companion matrix*

$$[L] = \begin{bmatrix} C_{p_1} & 0 & & 0 \\ 0 & C_{p_2} & & \\ & & \ddots & \\ 0 & & & C_{p_k} \end{bmatrix}$$

and $\chi_L(t) = p_1(t) \cdots p_k(t)$. *Moreover, the geometric multiplicity satisfies*

$$\dim(\ker(L - \lambda 1_V)) = \text{ number of } p_i\text{s such that } p_i(\lambda) = 0.$$

Thus, L is diagonalizable if and only if all of the companion matrices C_{p_i} have distinct eigenvalues.

Proof. The proof uses induction on the dimension of the vector space. The theorem clearly holds if $\dim V = 1$, so assume that the theorem holds for all linear operators on vector spaces of dimension $< \dim V$. Our goal is to show that either $V = C_{x_1}$ for some $x_1 \in V$ or that $V = C_{x_1} \oplus M$ for some L-invariant subspace M.

Let $m \leq \dim V$ be the largest dimension of a cyclic subspace, i.e., $\dim C_x \leq m$ for all $x \in V$, and there is an $x_1 \in V$ such that $\dim C_{x_1} = m$. In other words, $L^m(x) \in \text{span}\{x, L(x), \ldots, L^{m-1}(x)\}$ for all $x \in V$, and we can find $x_1 \in V$ such that $x_1, L(x_1), \ldots, L^{m-1}(x_1)$ are linearly independent.

In case $m = \dim V$, it follows that $C_{x_1} = V$ and we are finished. Otherwise, we must show that there is an L-invariant complement to

$$C_{x_1} = \text{span}\{x_1, L(x_1), \ldots, L^{m-1}(x_1)\}$$

in V. To construct this complement, we consider the linear map $K : V \to \mathbb{F}^m$ defined by

$$K(x) = \begin{bmatrix} f(x) \\ f(L(x)) \\ \vdots \\ f(L^{m-1}(x)) \end{bmatrix},$$

where $f : V \to \mathbb{F}$ is a linear functional chosen so that

$$f(x_1) = 0,$$
$$f(L(x_1)) = 0,$$
$$\vdots \quad \vdots$$
$$f(L^{m-2}(x_1)) = 0,$$
$$f(L^{m-1}(x_1)) = 1.$$

Note that it is possible to choose such an f as $x_1, L(x_1), \ldots, L^{m-1}(x_1)$ are linearly independent and hence part of a basis for V.

We now claim that $K|_{C_{x_1}} : C_{x_1} \to \mathbb{F}^m$ is an isomorphism. To see this, we find the matrix representation for the restriction of K to C_{x_1}. Using the basis x_1, $L(x_1)$, $\ldots$, $L^{m-1}(x_1)$ for C_{x_1} and the canonical basis $e_1, \ldots, e_m$ for $\mathbb{F}^m$, we see that:

$$\left[K(x_1) \ K(L(x_1)) \ \cdots \ K\left(L^{m-1}(x_1)\right) \right]$$

$$= \left[e_1 \ e_2 \ \cdots \ e_m \right] \begin{bmatrix} 0 & 0 & \cdots & 1 \\ \vdots & & \ddots & * \\ 0 & 1 & & \vdots \\ 1 & * & \cdots & * \end{bmatrix},$$

where $*$ indicates that we do not know or care what the entry is. Since the matrix representation is clearly invertible, we have that $K|_{C_{x_1}} : C_{x_1} \to \mathbb{F}^m$ is an isomorphism.

Next, we need to show that $\ker(K)$ is L-invariant. Let $x \in \ker(K)$, i.e.,

$$K(x) = \begin{bmatrix} f(x) \\ f(L(x)) \\ \vdots \\ f\left(L^{m-1}(x)\right) \end{bmatrix} = \begin{bmatrix} 0 \\ 0 \\ \vdots \\ 0 \end{bmatrix}.$$

Then,

$$K(L(x)) = \begin{bmatrix} f(L(x)) \\ f\left(L^2(x)\right) \\ \vdots \\ f\left(L^{m-1}(x)\right) \\ f(L^m(x)) \end{bmatrix} = \begin{bmatrix} 0 \\ 0 \\ \vdots \\ 0 \\ f(L^m(x)) \end{bmatrix}.$$

Now, by the choice of m, $L^m(x)$ is a linear combination of x, $L(x)$, $\ldots$, $L^{m-1}(x)$ for all x. This shows that $f(L^m(x)) = 0$ and consequently $L(x) \in \ker(K)$.

Finally, we show that $V = C_{x_1} \oplus \ker(K)$. We have seen that $K|_{C_{x_1}} : C_{x_1} \to \mathbb{F}^m$ is an isomorphism. This implies that $C_{x_1} \cap \ker(K) = \{0\}$. From Theorem 1.11.7 and Corollary 1.10.14, we then get that

$$\dim(V) = \dim(\ker(K)) + \dim(\text{im}(K))$$

$$= \dim(\ker(K)) + m$$

$$= \dim(\ker(K)) + \dim(C_{x_1})$$

$$= \dim(\ker(K) + C_{x_1}).$$

Thus, $V = C_{x_1} + \ker(K) = C_{x_1} \oplus \ker(K)$.

To find the geometric multiplicity of λ, we need only observe that each of the blocks C_{p_i} has a one-dimensional eigenspace corresponding to λ if λ is an eigenvalue for C_{p_i}. We know in turn that λ is an eigenvalue for C_{p_i} precisely when $p_i(\lambda) = 0$.   □

It is important to understand that there can be several cyclic subspace decompositions. This fact, of course, makes our calculation of the geometric multiplicity of eigenvalues especially intriguing. A rather interesting example comes from companion matrices themselves. Clearly, they have the desired decomposition; however, if they are diagonalizable, then the space also has a different decomposition into cyclic subspaces given by the one-dimensional eigenspaces. The issue of obtaining a unique decomposition is discussed in the next section.

To see that this theorem really has something to say, we should give examples of linear maps that force the space to have a nontrivial cyclic subspace decomposition. Since a companion matrix always has one-dimensional eigenspaces, this is of course not hard at all.

Example 2.6.6. A very natural choice is the linear operator $L_A(X) = AX$ on $\text{Mat}_{n \times n}(\mathbb{C})$. In Example 1.7.6, we showed that it had a block diagonal form with As on the diagonal. This shows that any eigenvalue for A has geometric multiplicity at least n. We can also see this more directly. Assume that $Ax = \lambda x$, where $x \in \mathbb{C}^n$ and consider $X = \begin{bmatrix} \alpha_1 x \cdots \alpha_n x \end{bmatrix}$. Then,

$$
\begin{aligned}
L_A(X) &= A \begin{bmatrix} \alpha_1 x \cdots \alpha_n x \end{bmatrix} \\
&= \begin{bmatrix} \alpha_1 Ax \cdots \alpha_n Ax \end{bmatrix} \\
&= \lambda \begin{bmatrix} \alpha_1 x \cdots \alpha_n x \end{bmatrix} \\
&= \lambda X.
\end{aligned}
$$

Thus,

$$
M = \left\{ \begin{bmatrix} \alpha_1 x \cdots \alpha_n x \end{bmatrix} : \alpha_1, \ldots, \alpha_n \in \mathbb{C} \right\}
$$

forms an n-dimensional space of eigenvectors for L_A.

Example 2.6.7. Another interesting example of a cyclic subspace decomposition comes from permutation matrices. We first recall that a permutation matrix $A \in \text{Mat}_{n \times n}(\mathbb{F})$ is a matrix such that $Ae_i = e_{\sigma(i)}$, see also Example 1.7.7. We claim that we can find a cyclic subspace decomposition by simply rearranging the canonical basis $e_1, \ldots, e_n$ for $\mathbb{F}^n$. The proof works by induction on n. When $n = 1$, there is nothing to prove. For $n > 1$, we consider $C_{e_1} = \text{span}\{e_1, Ae_1, A^2 e_1, \ldots\}$. Since all of the powers $A^m e_1$ belong to the finite set $\{e_1, \ldots, e_n\}$, we can find integers $k > l > 0$ such that $A^k e_1 = A^l e_1$. Since A is invertible, this implies that $A^{k-l} e_1 = e_1$. Now, select the smallest integer $m > 0$ such that $A^m e_1 = e_1$. Then we have

$$
C_{e_1} = \text{span}\{e_1, Ae_1, A^2 e_1, \ldots, A^{m-1} e_1\}.
$$

Moreover, all of the vectors $e_1, Ae_1, A^2e_1, \ldots, A^{m-1}e_1$ must be distinct as we could otherwise find $l < k < m$ such that $A^{k-l}e_1 = e_1$. This contradicts minimality of m. Since all of $e_1, Ae_1, A^2e_1, \ldots, A^{m-1}e_1$ are also vectors from the basis $e_1, \ldots, e_n$, they must form a basis for C_{e_1}. In this basis, A is represented by the companion matrix to $p(t) = t^m - 1$ and hence takes the form

$$\begin{bmatrix} 0 & 0 & \cdots & 0 & 1 \\ 1 & 0 & \cdots & 0 & 0 \\ 0 & 1 & \cdots & 0 & 0 \\ \vdots & \vdots & \ddots & \vdots & \vdots \\ 0 & 0 & \cdots & 1 & 0 \end{bmatrix}.$$

The permutation that corresponds to $A : C_{e_1} \to C_{e_1}$ is also called a *cyclic permutation*. Evidently, it maps the elements $1, \sigma(1), \ldots, \sigma^{m-1}(1)$ to themselves in a cyclic manner. One often refers to such permutations by listing the elements as $(1, \sigma(1), \ldots, \sigma^{m-1}(1))$. This is not a unique representation as, e.g., $(\sigma^{m-1}(1), 1, \sigma(1), \ldots, \sigma^{m-2}(1))$ clearly describes the same permutation.

We used m of the basis vectors $e_1, \ldots, e_n$ to span C_{e_1}. Rename and reindex the complementary basis vectors $f_1, \ldots, f_{n-m}$. To get our induction to work we need to show that $Af_i = f_{\tau(i)}$ for each $i = 1, \ldots, n - m$. We know that $Af_i \in \{e_1, \ldots, e_n\}$. If $Af_i \in \{e_1, Ae_1, A^2e_1, \ldots, A^{m-1}e_1\}$, then either $f_i = e_1$ or $f_i = A^k e_1$. The former is impossible since $f_i \notin \{e_1, Ae_1, A^2e_1, \ldots, A^{m-1}e_1\}$. The latter is impossible as A leaves $\{e_1, Ae_1, A^2e_1, \ldots, A^{m-1}e_1\}$ invariant. Thus, it follows that $Af_i \in \{f_1, \ldots, f_{n-m}\}$ as desired. In this way, we see that it is possible to rearrange the basis $e_1, \ldots, e_n$ so as to get a cyclic subspace decomposition. Furthermore, on each cyclic subspace, A is represented by a companion matrix corresponding to $p(t) = t^k - 1$ for some $k \leq n$. Recall that if $\mathbb{F} = \mathbb{C}$, then all of these companion matrices are diagonalizable, in particular, A is itself diagonalizable.

Note that the cyclic subspace decomposition for a permutation matrix also decomposes the permutation σ into cyclic permutations that are disjoint. This is a basic construction in the theory of permutations.

The cyclic subspace decomposition qualifies as a central result in linear algebra for many reasons. While somewhat difficult and tricky to prove, it does not depend on several of our developments in this chapter. It could in fact be established without knowledge of eigenvalues, characteristic polynomials and minimal polynomials, etc. Second, it gives a matrix representation that is in block diagonal form and where we have a very good understanding of each of the blocks. Therefore, all of our developments in this chapter could be considered consequences of this decomposition. Finally, several important and difficult results such as the Frobenius and Jordan canonical forms become relatively easy to prove using this decomposition.

Exercises

1. Find all invariant subspaces for the following two matrices and show that they are not diagonalizable:

    (a)
    $$\begin{bmatrix} 0 & 1 \\ 0 & 0 \end{bmatrix}$$

    (b)
    $$\begin{bmatrix} \alpha & 1 \\ 0 & \alpha \end{bmatrix}$$

2. Show that the space of $n \times n$ companion matrices form an affine subspace isomorphic to the affine subspace of monic polynomials of degree n. Affine subspaces are defined in Exercise 8 in Sect. 1.10.

3. Given

    $$A = \begin{bmatrix} \lambda_1 & 1 & \cdots & 0 \\ 0 & \lambda_2 & \ddots & \vdots \\ \vdots & \vdots & \ddots & 1 \\ 0 & 0 & \cdots & \lambda_n \end{bmatrix}$$

    find $x \in \mathbb{F}^n$ such that $C_x = \mathbb{F}^n$. Hint: Try $n = 2, 3$ first.

4. Given a linear operator $L : V \to V$ on a finite-dimensional vector space and $x \in V$, show that

    $$C_x = \{p(L)(x) : p(t) \in \mathbb{F}[t]\}.$$

5. Let $p(t) = t^n + \alpha_{n-1}t^{n-1} + \cdots + \alpha_0 \in \mathbb{F}[t]$. Show that C_p and C_p^t are similar. Hint: Let

    $$B = \begin{bmatrix} \alpha_1 & \alpha_2 & \alpha_3 & \cdots & \alpha_{n-1} & 1 \\ \alpha_2 & \alpha_3 & \cdots & & 1 & 0 \\ \alpha_3 & \vdots & \ddots & & & \\ \vdots & \alpha_{n-1} & & & 0 & 0 \\ \alpha_{n-1} & 1 & & & & \vdots \\ 1 & 0 & \cdots & & 0 & 0 \end{bmatrix}$$

    and show

    $$C_p B = B C_p^t.$$

6. Use the previous exercise to show that $A \in \text{Mat}_{n \times n}(\mathbb{F})$ and its transpose are similar.

7. Show that if $V = C_x$ for some $x \in V$, then $\deg(\mu_L) = \dim(V)$.

8. For each $n \geq 2$, construct a matrix $A \in \text{Mat}_{n \times n}(\mathbb{F})$ such that $V \neq C_x$ for every $x \in V$.

9. For each $n \geq 2$, construct a matrix $A \in \text{Mat}_{n \times n}(\mathbb{F})$ such that $V = C_x$ for some $x \in V$.

10. Let $L : V \to V$ be a diagonalizable linear operator on a finite-dimensional vector space. Show that $V = C_x$ if and only if there are no multiple eigenvalues.

11. Let $L : V \to V$ be a linear operator on a finite-dimensional vector space. Assume that $V \neq C_{x_1}$, where C_{x_1} is the first cyclic subspace as constructed in the proof of the cyclic subspace decomposition. Show that it is possible to select another $y_1 \in V$ such that $\dim C_{y_1} = \dim C_{x_1} = m$, but $C_{x_1} \neq C_{y_1}$. This gives a different indication of why the cyclic subspace decomposition is not unique.

12. Let $L : V \to V$ be a linear operator on a finite-dimensional vector space such that $V = C_x$ for some $x \in V$.

    (a) Show that $K \circ L = L \circ K$ if and only if $K = p(L)$ for some $p \in \mathbb{F}[t]$. Hint: When $K \circ L = L \circ K$ define p by using that $K(x) = \alpha_0 + \cdots + \alpha_{n-1} L^{n-1}(x)$.

    (b) Show that all invariant subspaces for L are of the form $\ker(p(L))$ for some polynomial $p \in \mathbb{F}[t]$.

13. Let $L : V \to V$ be a linear operator on a finite-dimensional vector space. Define $\mathbb{F}[L] = \{p(L) : p(t) \in \mathbb{F}[t]\} \subset \text{Hom}(V, V)$ as the space of polynomials in L.

    (a) Show that $\mathbb{F}[L]$ is a subspace, that is also closed under composition of operators.

    (b) Show that $\dim(\mathbb{F}[L]) = \deg(\mu_L)$ and $\mathbb{F}[L] = \text{span}\{1_V, L, \ldots, L^{k-1}\}$, where $k = \deg(\mu_L)$.

    (c) Show that the map $\Phi : \mathbb{F}[t] \to \text{Hom}(V, V)$ defined by $\Phi(p(t)) = p(L)$ is linear and a ring homomorphism (preserves multiplication and sends $1 \in \mathbb{F}[t]$ to $1_V \in \text{Hom}(V, V)$) with image $\mathbb{F}[L]$.

    (d) Show that $\ker(\Phi) = \{p(t)\mu_L(t) : p(t) \in \mathbb{F}[t]\}$.

    (e) Show that for any $p(t) \in \mathbb{F}[t]$, we have $p(L) = r(L)$ for some $r(t) \in \mathbb{F}[t]$ with $\deg r(t) < \deg \mu_L(t)$.

    (f) Given an eigenvector $x \in V$ for L, show that x is an eigenvector for all $K \in \mathbb{F}[L]$ and that the map $\mathbb{F}[L] \to \mathbb{F}$ that sends K to the eigenvalue corresponding to x is a ring homomorphism.

    (g) Conversely, show that any ring nontrivial homomorphism $\phi : \mathbb{F}[L] \to \mathbb{F}$ is of the type described in (f).

2.7   The Frobenius Canonical Form

As we have already indicated, the above proof of the cyclic subspace decomposition actually proves quite a bit more than the result claims. It leads us to a unique matrix representation for the operator known as the Frobenius canonical form. This canonical form will be used in the next section to establish more refined canonical forms for complex operators.

Theorem 2.7.1. (The Frobenius Canonical Form) *Let* $L : V \rightarrow V$ *be a linear operator on a finite-dimensional vector space. Then,* V *has a cyclic subspace decomposition such that the block diagonal form of* L

$$[L] = \begin{bmatrix} C_{p_1} & 0 & & 0 \\ 0 & C_{p_2} & & \\ & & \ddots & \\ 0 & & & C_{p_k} \end{bmatrix}$$

has the property that p_i *divides* p_{i-1} *for each* $i = 2, \ldots, k$. *Moreover, the monic polynomials* $p_1, \ldots, p_k$ *are unique.*

Proof. We first establish that the polynomials constructed in the above version of the cyclic subspace decomposition have the desired divisibility properties.

Recall that $m \leq \dim V$ is the largest dimension of a cyclic subspace, i.e., $\dim C_x \leq m$ for all $x \in V$ and there is an $x_1 \in V$ such that $\dim C_{x_1} = m$. In other words, $L^m(x) \in \text{span}\{x, L(x), \ldots, L^{m-1}(x)\}$ for all $x \in V$ and we can find $x_1 \in V$ such that $x_1, L(x_1), \ldots, L^{m-1}(x_1)$ are linearly independent. With this choice of x_1, define

$$p_1(t) = t^m - \alpha_{m-1}t^{m-1} - \cdots - \alpha_0, \text{ where}$$

$$L^m(x_1) = \alpha_{m-1}L^{m-1}(x_1) + \cdots + \alpha_0 x_1,$$

and recall that in the proof of Theorem 2.6.5, we also found an L-invariant complementary subspace $M \subset V$.

With these choices we claim that $p_1(L)(z) = 0$ for all $z \in V$. Note that we already know this for $z = x_1$, and it is easy to also verify it for $z = L(x_1), \ldots,$ $L^{m-1}(x_1)$ by using that $p(L) \circ L^k = L^k \circ p(L)$. Thus, we only need to check the claim for $z \in M$. By construction of m we know that

$$L^m(x_1 + z) = \gamma_{m-1}L^{m-1}(x_1 + z) + \cdots + \gamma_0(x_1 + z).$$

Now, we rearrange the terms as follows:

$$L^m(x_1) + L^m(z) = L^m(x_1 + z)$$
$$= \gamma_{m-1}L^{m-1}(x_1) + \cdots + \gamma_0 x_1$$
$$+ \gamma_{m-1}L^{m-1}(z) + \cdots + \gamma_0 z.$$

Since

$$L^m(x_1), \ \gamma_{m-1}L^{m-1}(x_1) + \cdots + \gamma_0 x_1 \in C_{x_1}$$

and

$$L^m(z), \ \gamma_{m-1}L^{m-1}(z) + \cdots + \gamma_0 z \in M,$$

it follows that

$$\gamma_{m-1}L^{m-1}(x_1) + \cdots + \gamma_0 x_1 = L^m(x_1) = \alpha_{m-1}L^{m-1}(x_1) + \cdots + \alpha_0 x_1.$$

Since $x_1, L(x_1), \ldots, L^{m-1}(x_1)$ are linearly independent, this shows that $\gamma_i = \alpha_i$ for $i = 0, \ldots, m-1$. But then

$$
\begin{aligned}
0 &= p_1(L)(x_1 + z) \\
&= p_1(L)(x_1) + p_1(L)(z) \\
&= p_1(L)(z),
\end{aligned}
$$

which is what we wanted to prove.

Next, let $x_2 \in M$ and $p_2(t)$ be chosen in the same fashion as x_1 and p_1. We first note that $l = \deg p_2 \leq \deg p_1 = m$; this means that we can write $p_1 = q_1 p_2 + r$, where $\deg r < \deg p_2$. Thus,

$$
\begin{aligned}
0 &= p_1(L)(x_2) \\
&= q_1(L) \circ p_2(L)(x_2) + r(L)(x_2) \\
&= r(L)(x_2).
\end{aligned}
$$

Since $\deg r < l = \deg p_2$, the equation $r(L)(x_2) = 0$ takes the form

$$
\begin{aligned}
0 &= r(L)(x_2) \\
&= \beta_0 x_2 + \cdots + \beta_{l-1} L^{l-1}(x_2).
\end{aligned}
$$

However, p_2 was chosen to that $x_2, L(x_2), \ldots, L^{l-1}(x_2)$ are linearly independent, so

$$\beta_0 = \cdots = \beta_{l-1} = 0$$

and hence also $r = 0$. This shows that p_2 divides p_1.

We now show that p_1 and p_2 are unique, despite the fact that x_1 and x_2 need not be unique. To see that p_1 is unique, we simply check that it is the minimal polynomial of L. We have already seen that $p_1(L)(z) = 0$ for all $z \in V$. Thus, $p_1(L) = 0$ showing that $\deg \mu_L \leq \deg p_1$. On the other hand, we also know that $x_1, L(x_1), \ldots, L^{m-1}(x_1)$ are linearly independent; in particular, $1_V, L, \ldots, L^{m-1}$ must also be linearly independent. This shows that $\deg \mu_L \geq m = \deg p_1$. Hence, $\mu_L = p_1$ as they are both monic.

To see that p_2 is unique is a bit more tricky since the choice for C_{x_1} is not unique. We select two decompositions

$$C_{x'_1} \oplus M' = V = C_{x_1} \oplus M.$$

This yields two block diagonal matrix decompositions for L

$$\begin{bmatrix} C_{p_1} & 0 \\ 0 & [L|_{M'}] \end{bmatrix}$$

$$\begin{bmatrix} C_{p_1} & 0 \\ 0 & [L|_M] \end{bmatrix},$$

where the upper left-hand block is the same for both representations as p_1 is unique. Moreover, these two matrices are similar. Therefore, we only need to show that $\mu_{A_{22}} = \mu_{A'_{22}}$ if the two block diagonal matrices

$$\begin{bmatrix} A_{11} & 0 \\ 0 & A_{22} \end{bmatrix} \text{ and } \begin{bmatrix} A_{11} & 0 \\ 0 & A'_{22} \end{bmatrix}$$

are similar

$$\begin{bmatrix} A_{11} & 0 \\ 0 & A_{22} \end{bmatrix} = B^{-1} \begin{bmatrix} A_{11} & 0 \\ 0 & A'_{22} \end{bmatrix} B.$$

If p is any polynomial, then

$$\begin{bmatrix} p(A_{11}) & 0 \\ 0 & p(A_{22}) \end{bmatrix} = p\left(\begin{bmatrix} A_{11} & 0 \\ 0 & A_{22} \end{bmatrix} \right)$$

$$= p\left(B^{-1} \begin{bmatrix} A_{11} & 0 \\ 0 & A'_{22} \end{bmatrix} B \right)$$

$$= B^{-1} \left(p\left(\begin{bmatrix} A_{11} & 0 \\ 0 & A'_{22} \end{bmatrix} \right) \right) B$$

$$= B^{-1} \begin{bmatrix} p(A_{11}) & 0 \\ 0 & p(A'_{22}) \end{bmatrix} B.$$

In particular, the two matrices

$$\begin{bmatrix} p(A_{11}) & 0 \\ 0 & p(A_{22}) \end{bmatrix} \text{ and } \begin{bmatrix} p(A_{11}) & 0 \\ 0 & p(A'_{22}) \end{bmatrix}$$

always have the same rank. Since the upper left-hand corners are identical, this shows that $p(A_{22})$ and $p(A'_{22})$ have the same rank. As a special case, it follows that $p(A_{22}) = 0$ if and only if $p(A'_{22}) = 0$. This shows that A_{22} and A'_{22} have the same minimal polynomials and hence that p_2 is uniquely defined.   □

In some texts, the Frobenius canonical form is also known as the *rational canonical form*. The reason is that it will have rational entries if we start with an $n \times n$ matrix with rational entries. To see why this is, simply observe that the polynomials have rational coefficients starting with p_1, the minimal polynomial. In some other texts, the rational canonical form is refined by further factoring the characteristic or minimal polynomials into irreducible components over the rationals. One of the advantages of the Frobenius canonical form is that it does not depend on the scalar field. That is, if $A \in \mathrm{Mat}_{n \times n}(\mathbb{F}) \subset \mathrm{Mat}_{n \times n}(\mathbb{L})$, then the form does not depend on whether we compute it using $\mathbb{F}$ or $\mathbb{L}$.

Definition 2.7.2. The unique polynomials $p_1, \ldots, p_k$ are called the *similarity invariants, elementary divisors,* or *invariant factors* for L.

Clearly, two matrices are similar if they have the same similarity invariants as they have the same Frobenius canonical form. Conversely, similar matrices are both similar to the same Frobenius canonical form and hence have the same similarity invariants. It is possible to calculate the similarity invariants using only the elementary row and column operations (see Sect. 1.13.) The specific construction is covered in Sect. 2.9 and is related to the Smith normal form.

The following corollary shows that several of the matrices related to companion matrices are in fact similar. Various exercises have been devoted to establishing this fact, but using the Frobenius canonical form we get a very elegant characterization of when a linear map is similar to a companion matrix.

Corollary 2.7.3. *If two linear operators on an n-dimensional vector space have the same minimal polynomials of degree n, then they have the same Frobenius canonical form and are thus similar.*

Proof. If $\deg \mu_L = \dim V$, then the first block in the Frobenius canonical form is an $n \times n$ matrix. Thus, there is only one block in this decomposition. This proves the claim.   □

We can redefine the characteristic polynomial using similarity invariants. However, it is not immediately clear why it agrees with the definition given in Sect. 2.3 as we do not know that that definition gives the same answer for similar matrices (see, however, Sect. 5.7 for a proof that uses determinants).

Definition 2.7.4. The characteristic polynomial of a linear operator $L : V \to V$ on a finite-dimensional vector space is the product of its similarity invariants:

$$\chi_L(t) = p_1(t) \cdots p_k(t).$$

This gives us a way of defining the characteristic polynomial, but it does not tells us how to compute it. For that, the row reduction technique or determinants are the way to go. In this vein, we can also define the determinant as

$$\det L = (-1)^n \chi_L(0).$$

The problem is that one of the key properties of determinants

$$\det (K \circ L) = \det (K) \det (L)$$

does not follow easily from this definition. We do, however, get that similar matrices, and linear operators have the same determinant:

$$\det \left(K \circ L \circ K^{-1} \right) = \det (L).$$

Example 2.7.5. As a general sort of example, let us see what the Frobenius canonical form for

$$A = \begin{bmatrix} C_{q_1} & 0 \\ 0 & C_{q_2} \end{bmatrix}$$

is when q_1 and q_2 are relatively prime. Note that if

$$0 = p(A) = \begin{bmatrix} p\left(C_{q_1}\right) & 0 \\ 0 & p\left(C_{q_2}\right) \end{bmatrix},$$

then both q_1 and q_2 divide p. Conversely, if q_1 and q_2 both divide p, it also follows that $p(A) = 0$. Since the least common multiple of q_1 and q_2 is $q_1 \cdot q_2$, we see that $\mu_A = q_1 \cdot q_2 = \chi_A$. Thus, $p_1 = q_1 \cdot q_2$. This shows that the Frobenius canonical form is simply $C_{q_1 \cdot q_2}$. The general case where there might be a nontrivial greatest common divisor is relegated to the exercises.

Example 2.7.6. We now give a few examples showing that the characteristic and minimal polynomials alone do not yield sufficient information to determine all the similarity invariants when the dimension is ≥ 4 (see exercises for dimensions 2 and 3). We consider all possible canonical forms in dimension 4, where the characteristic polynomial is t^4. There are four nontrivial cases given by:

$$\begin{bmatrix} 0&0&0&0 \\ 1&0&0&0 \\ 0&1&0&0 \\ 0&0&1&0 \end{bmatrix}, \begin{bmatrix} 0&0&0&0 \\ 1&0&0&0 \\ 0&1&0&0 \\ 0&0&0&0 \end{bmatrix}, \begin{bmatrix} 0&0&0&0 \\ 1&0&0&0 \\ 0&0&0&0 \\ 0&0&0&0 \end{bmatrix}, \begin{bmatrix} 0&0&0&0 \\ 1&0&0&0 \\ 0&0&0&0 \\ 0&0&1&0 \end{bmatrix}$$

For the first, we know that $\mu = p_1 = t^4$. For the second, we have two blocks where $\mu = p_1 = t^3$ and $p_2 = t$. For the third, we have $\mu = p_1 = t^2$ while $p_2 = p_3 = t$. Finally, the fourth has $\mu = p_1 = p_2 = t^2$. The last two matrices clearly do not have the same canonical form, but they do have the same characteristic and minimal polynomials.

Example 2.7.7. Lastly, let us compute the Frobenius canonical form for a projection $E : V \to V$. As we shall see, this is clearly a situation where we should just stick to diagonalization as the Frobenius canonical form is far less informative. Apparently, we just need to find all possible Frobenius canonical forms that are also projections. The simplest are of course just 0_V and 1_V. In all other cases the minimal polynomial is $t^2 - t$. The companion matrix for that polynomial is

$$\begin{bmatrix} 0 & 0 \\ 1 & 1 \end{bmatrix}$$

so we expect to have one or several of those blocks, but note that we cannot have more than $\lfloor \frac{\dim V}{2} \rfloor$ of such blocks. The rest of the diagonal entries must now correspond to companion matrices for either t or $t - 1$. But we cannot use both as these two polynomials do not divide each other. This gives us two types of Frobenius canonical forms:

$$\begin{bmatrix} 0\,0 & & & & & \\ 1\,1 & & & & & \\ & \ddots & & & & \\ & & 0\,0 & & & \\ & & 1\,1 & & & \\ & & & 0 & & \\ & & & & \ddots & \\ & & & & & 0 \end{bmatrix}$$

or

$$\begin{bmatrix} 0\,0 & & & & & \\ 1\,1 & & & & & \\ & \ddots & & & & \\ & & 0\,0 & & & \\ & & 1\,1 & & & \\ & & & 1 & & \\ & & & & \ddots & \\ & & & & & 1 \end{bmatrix}$$

To find the correct canonical form for E, we just select the Frobenius canonical form that gives us the correct rank. If $\operatorname{rank} E \leq \lfloor \frac{\dim V}{2} \rfloor$ it will be of the first type and otherwise of the second.

Exercises

1. What are the similarity invariants for a companion matrix C_p?
2. Let $A \in \operatorname{Mat}_{n \times n}(\mathbb{R})$, and $n \geq 2$.

   (a) Show that when n is odd, then it is not possible to have $p_1(t) = t^2 + 1$.
   (b) Show by example that one can have $p_1(t) = t^2 + 1$ for all even n.
   (c) Show by example that one can have $p_1(t) = t^3 + t$ for all odd n.

3. If $L : V \to V$ is an operator on a 2-dimensional space, then either $p_1 = \mu_L = \chi_L$ or $L = \lambda 1_V$.

4. If $L : V \to V$ is an operator on a 3-dimensional space, then either $p_1 = \mu_L = \chi_L$, $p_1 = (t - \alpha)(t - \beta)$ and $p_2 = (t - \beta)$, or $L = \lambda 1_V$. Note that in the second case you know that p_1 has degree 2, the key is to show that it factors as described.

5. Let $L : V \to V$ be a linear operator on a finite-dimensional space. Show that $V = C_x$ for some $x \in V$ if and only if $\mu_L = \chi_L$.

6. Show that the matrix

$$\begin{bmatrix} \lambda_1 & 1 & \cdots & 0 \\ 0 & \lambda_2 & \ddots & \vdots \\ \vdots & \vdots & \ddots & 1 \\ 0 & 0 & \cdots & \lambda_n \end{bmatrix}$$

   is similar to a companion matrix.

7. Let $L : V \to V$ be a linear operator on a finite-dimensional vector space such that $V = C_x$ for some $x \in V$. Show that all invariant subspaces for L are of the form C_z for some $z \in V$. Hint: This relies on showing that if an invariant subspace is not cyclic, then $\deg \mu_L < \dim V$.

8. Consider two companion matrices C_p and C_q; show that the similarity invariants for the block diagonal matrix

$$\begin{bmatrix} C_p & 0 \\ 0 & C_q \end{bmatrix}$$

   are $p_1 = \operatorname{lcm}\{p,q\}$ and $p_2 = \gcd\{p,q\}$. Hint: Use Propositions 2.1.4 and 2.1.5 to show that $p_1 \cdot p_2 = p \cdot q$.

9. Is it possible to find the similarity invariants for

$$\begin{bmatrix} C_p & 0 & 0 \\ 0 & C_q & 0 \\ 0 & 0 & C_r \end{bmatrix}?$$

   Note that you can easily find $p_1 = \operatorname{lcm}\{p,q,r\}$, so the issue is whether it is possible to decide what p_2 should be.

10. Show that $A, B \in \operatorname{Mat}_{n \times n}(\mathbb{F})$ are similar if and only if $\operatorname{rank}(p(A)) = \operatorname{rank}(p(B))$ for all $p \in \mathbb{F}[t]$. (Recall that two matrices have the same rank if and only if they are equivalent and that equivalent matrices certainly need not be similar. This is what makes the exercise interesting.)

11. The previous exercise can be made into a checkable condition: Show that $A, B \in \operatorname{Mat}_{n \times n}(\mathbb{F})$ are similar if and only if $\chi_A = \chi_B$ and $\operatorname{rank}(p(A)) = \operatorname{rank}(p(B))$ for all p that divide χ_A. (Note that as χ_A has a unique prime factorization (see Theorem 2.1.7), this means that we only have to check a finite number of conditions.)

12. Show that any linear map with the property that

$$\chi_L (t) = (t - \lambda_1) \cdots (t - \lambda_n) \in \mathbb{F}[t]$$

   for $\lambda_1, \ldots, \lambda_n \in \mathbb{F}$ has an upper triangular matrix representation.
13. Let $L : V \to V$ be a linear operator on a finite-dimensional vector space. Use the Frobenius canonical form to show that $\operatorname{tr}(L) = -\alpha_{n-1}$, where $\chi_L (t) = t^n + \alpha_{n-1} t^{n-1} + \cdots + \alpha_0$. This is the result mentioned in Proposition 2.3.11.
14. Assume that $L : V \to V$ satisfies $(L - \lambda_0 1_V)^k = 0$, for some $k > 1$, but $(L - \lambda_0 1_V)^{k-1} \neq 0$. Show that $\ker (L - \lambda_0 1_V)$ is neither $\{0\}$ nor V. Show that $\ker (L - \lambda_0 1_V)$ does not have a complement in V that is L-invariant.
15. *(The Cayley–Hamilton Theorem)* Show the Cayley–Hamilton theorem using the Frobenius canonical form.

2.8   The Jordan Canonical Form*

In this section, we present a proof of the Jordan canonical form. We start with a somewhat more general point of view that in the end is probably the most important feature of this special canonical form.

Theorem 2.8.1. (The Jordan–Chevalley Decomposition) *Let $L : V \to V$ be a linear operator on an n-dimensional complex vector space. Then, $L = S + N$, where S is diagonalizable, $N^n = 0$, and $SN = NS$.*

Proof. First, use the Fundamental Theorem of Algebra 2.1.8 to factor the minimal polynomial

$$\mu_L (t) = (t - \lambda_1)^{m_1} \cdots (t - \lambda_k)^{m_k} ,$$

where $\lambda_1, \ldots, \lambda_k$ are distinct. If we define

$$M_i = \ker (L - \lambda_i)^{m_i} ,$$

then the proof of Lemma 2.5.6 together with Exercise 20 in Sect. 2.5 shows that

$$V = M_1 \oplus \cdots \oplus M_k.$$

We can now define

$$S|_{M_i} = \lambda_i 1_V|_{M_i} = \lambda_i 1_{M_i}$$
$$N|_{M_i} = (L - \lambda_i 1_V)|_{M_i} = L|_{M_i} - \lambda_i 1_{M_i}$$

Clearly, $L = S + N$, S is diagonalizable and $SN = NS$. Finally, since $\mu_L (L) = 0$, it follows that $N^n = 0$. $\square$

It is in fact possible to show that the Jordan–Chevalley decomposition is unique, i.e., the operators S and N are uniquely determined by L (see exercises to this chapter).

As a corollary of the above proof, we obtain:

Corollary 2.8.2. *Let C_p be a companion matrix with $p(t) = (t - \lambda)^n$. Then, C_p is similar to a* Jordan block*:*

$$[J] = \begin{bmatrix} \lambda & 1 & 0 & \cdots & 0 & 0 \\ 0 & \lambda & 1 & \cdots & 0 & 0 \\ 0 & 0 & \lambda & \ddots & \vdots & \vdots \\ 0 & 0 & 0 & \ddots & 1 & 0 \\ \vdots & \vdots & \vdots & \cdots & \lambda & 1 \\ 0 & 0 & 0 & \cdots & 0 & \lambda \end{bmatrix}.$$

Moreover, the eigenspace for λ is one-dimensional and is generated by the first basis vector.

We can now give a simple proof of the so-called Jordan canonical form. Interestingly, the famous analyst Weierstrass deserves equal credit as he too proved the result at about the same time.

Theorem 2.8.3. (The Jordan–Weierstrass Canonical form) *Let $L : V \to V$ be a linear operator on a finite-dimensional complex vector space. Then, we can find L-invariant subspaces $M_1, \ldots, M_s$ such that*

$$V = M_1 \oplus \cdots \oplus M_s$$

and each $L|_{M_i}$ has a matrix representation of the form

$$\begin{bmatrix} \lambda_i & 1 & 0 & \cdots & 0 \\ 0 & \lambda_i & 1 & \cdots & 0 \\ 0 & 0 & \lambda_i & \ddots & \vdots \\ \vdots & \vdots & \vdots & \ddots & 1 \\ 0 & 0 & 0 & \cdots & \lambda_i \end{bmatrix},$$

where λ_i is an eigenvalue for L.

Proof. First, we invoke the Jordan–Chevalley decomposition $L = S + N$. Then, we decompose V into eigenspaces for S:

$$V = \ker(S - \lambda_1 1_V) \oplus \cdots \oplus \ker(S - \lambda_k 1_V).$$

Each of these eigenspaces is invariant for N since S and N commute. Specifically, if $S(x) = \lambda x$, then

$$S(N(x)) = N(S(x)) = N(\lambda x) = \lambda N(x),$$

showing that $N(x)$ is also an eigenvector for the eigenvalue λ.

This reduces the problem to showing that operators of the form $\lambda 1_W + N$, where $N^n = 0$ have the desired decomposition. Since the homothety $\lambda \cdot 1_W$ is always diagonal in any basis, it then suffices to show the theorem holds for operators N such that $N^n = 0$. The similarity invariants for such an operator all have to look like t^k so the blocks in the Frobenius canonical form must look like

$$\begin{bmatrix} 0 & 0 & \cdots & 0 \\ 1 & 0 & & \ddots \\ \vdots & \ddots & \ddots & 0 \\ 0 & & 1 & 0 \end{bmatrix}.$$

If $e_1, \ldots, e_k$ is the basis yielding this matrix representation, then

$$N \begin{bmatrix} e_1 & \cdots & e_k \end{bmatrix} = \begin{bmatrix} e_2 & \cdots & e_k & 0 \end{bmatrix}$$

$$= \begin{bmatrix} e_1 & \cdots & e_k \end{bmatrix} \begin{bmatrix} 0 & 0 & \cdots & 0 \\ 1 & 0 & & \ddots \\ \vdots & \ddots & \ddots & 0 \\ 0 & & 1 & 0 \end{bmatrix}.$$

Reversing the basis to $e_k, \ldots, e_1$ then gives us the desired block

$$N \begin{bmatrix} e_k & \cdots & e_1 \end{bmatrix} = \begin{bmatrix} 0 & e_k & \cdots & e_2 \end{bmatrix}$$

$$= \begin{bmatrix} e_k & \cdots & e_1 \end{bmatrix} \begin{bmatrix} 0 & 1 & \cdots & 0 \\ 0 & 0 & & \ddots \\ \vdots & & \ddots & 1 \\ 0 & & & 0 \end{bmatrix}.$$

$\square$

In this decomposition, it is possible for several of the subspaces M_i to correspond to the same eigenvalue. Given that the eigenspace for each Jordan block is one-dimensional we see that each eigenvalue corresponds to as many blocks as the geometric multiplicity of the eigenvalue. The job of calculating the Jordan canonical form is in general quite hard. Here we confine ourselves to the 2- and 3-dimensional situations.

Corollary 2.8.4. *Let $L : V \to V$ be a complex linear operator where $\dim(V) = 2$. Either L is diagonalizable and there is a basis where*

$$[L] = \begin{bmatrix} \lambda_1 & 0 \\ 0 & \lambda_2 \end{bmatrix},$$

or L is not diagonalizable and there is a basis where

$$[L] = \begin{bmatrix} \lambda & 1 \\ 0 & \lambda \end{bmatrix}.$$

Note that in case L is diagonalizable, we either have that $L = \lambda 1_V$ or that the eigenvalues are distinct. In the nondiagonalizable case, there is only one eigenvalue.

Corollary 2.8.5. *Let $L : V \to V$ be a complex linear operator where* $\dim(V) = 3$. *Either L is diagonalizable and there is a basis where*

$$[L] = \begin{bmatrix} \lambda_1 & 0 & 0 \\ 0 & \lambda_2 & 0 \\ 0 & 0 & \lambda_3 \end{bmatrix},$$

or L is not diagonalizable and there is a basis where one of the following two situations occur:

$$[L] = \begin{bmatrix} \lambda_1 & 0 & 0 \\ 0 & \lambda_2 & 1 \\ 0 & 0 & \lambda_2 \end{bmatrix}$$

or

$$[L] = \begin{bmatrix} \lambda & 1 & 0 \\ 0 & \lambda & 1 \\ 0 & 0 & \lambda \end{bmatrix}.$$

Remark 2.8.6. It is possible to check which of these situations occur by knowing the minimal and characteristic polynomials. We note that the last case happens precisely when there is only one eigenvalue with geometric multiplicity 1. The second case happens if either L has two eigenvalues each with geometric multiplicity 1 or if L has one eigenvalue with geometric multiplicity 2.

Exercises

1. Find the Jordan canonical forms for the matrices

$$\begin{bmatrix} 1 & 0 \\ 1 & 1 \end{bmatrix}, \begin{bmatrix} 1 & 1 \\ 0 & 2 \end{bmatrix}, \begin{bmatrix} 2 & -1 \\ 4 & -2 \end{bmatrix}.$$

2. Find the basis that yields the Jordan canonical form for

$$\begin{bmatrix} \lambda & -1 \\ \lambda^2 & -\lambda \end{bmatrix}.$$

3. Find the Jordan canonical form for the matrix

$$\begin{bmatrix} \lambda_1 & 1 \\ 0 & \lambda_2 \end{bmatrix}.$$

   Hint: The answer depends on λ_1 and λ_2.

4. Find the Jordan canonical forms for the matrix

$$\begin{bmatrix} 0 & 1 \\ -\lambda_1\lambda_2 & \lambda_1 + \lambda_2 \end{bmatrix}.$$

5. Find the Jordan canonical forms for the matrix

$$\begin{bmatrix} \lambda^2 & -2\lambda & 1 \\ \lambda^3 & -2\lambda^2 & \lambda \\ \lambda^4 & -2\lambda^3 & \lambda^2 \end{bmatrix}.$$

6. Find the Jordan canonical forms for the matrix

$$\begin{bmatrix} \lambda_1 & 1 & 0 \\ 0 & \lambda_2 & 1 \\ 0 & 0 & \lambda_3 \end{bmatrix}.$$

7. Find the Jordan canonical forms for the matrix

$$\begin{bmatrix} 0 & 1 & 0 \\ 0 & 0 & 1 \\ \lambda_1\lambda_2\lambda_3 & -(\lambda_1\lambda_2 + \lambda_2\lambda_3 + \lambda_1\lambda_3) & \lambda_1 + \lambda_2 + \lambda_3 \end{bmatrix}.$$

8. Find the Jordan canonical forms for the matrices

$$\begin{bmatrix} 0 & 1 & 0 \\ 0 & 0 & 1 \\ 2 & -5 & 4 \end{bmatrix}, \begin{bmatrix} 0 & 1 & 0 \\ 0 & 0 & 1 \\ 1 & -3 & 3 \end{bmatrix}, \begin{bmatrix} 0 & 1 & 0 \\ 0 & 0 & 1 \\ 6 & -11 & 6 \end{bmatrix}.$$

9. An operator $L : V \to V$ on an n-dimensional vector space over any field is said to be *nilpotent* if $L^k = 0$ for some k.

   (a) Show that $\chi_L(t) = t^n$.
   (b) Show that L can be put in triangular form.
   (c) Show that L is diagonalizable if and only if $L = 0$.
   (d) Find a real matrix such that its real eigenvalues are 0 but which is not nilpotent.

10. Let $L : V \to V$ be a linear operator on an n-dimensional complex vector space. Show that for $p \in \mathbb{C}\,[t]$, the operator $p\,(L)$ is nilpotent if and only if the eigenvalues of L are roots of p. What goes wrong with this statement in the real case when $p\,(t) = t^2 + 1$ and dim V is odd?

11. Show that if

$$\ker\left((L - \lambda 1_V)^k\right) \neq \ker\left((L - \lambda 1_V)^{k-1}\right),$$

then the algebraic multiplicity of λ is $\geq k$. Give an example where the algebraic multiplicity $> k$ and

$$\ker\left((L - \lambda 1_V)^{k+1}\right) = \ker\left((L - \lambda 1_V)^k\right) \neq \ker\left((L - \lambda 1_V)^{k-1}\right).$$

12. Show that if $L : V \to V$ is a linear operator such that

$$\chi_L\,(t) = (t - \lambda_1)^{n_1} \cdots (t - \lambda_k)^{n_k}\,,$$
$$\mu_L\,(t) = (t - \lambda_1)^{m_1} \cdots (t - \lambda_k)^{m_k}\,,$$

then m_i corresponds to the largest Jordan block that has λ_i on the diagonal. Using that show that m_i is the first integer such that

$$\ker((L - \lambda_i 1_V)^{m_i}) = \ker\left((L - \lambda_i 1_V)^{m_i+1}\right).$$

13. Show that if $L : V \to V$ is a linear operator on an n-dimensional complex vector space with distinct eigenvalues $\lambda_1, \ldots, \lambda_k$, then $p\,(L) = 0$, where

$$p\,(t) = (t - \lambda_1)^{n-k+1} \cdots (t - \lambda_k)^{n-k+1}.$$

Hint: Try $k = 2$.

14. Assume that $L = S + N$ is a Jordan–Chevalley decompositions, i.e., $SN = NS$, S is diagonalizable, and $N^n = (N')^n = 0$, where n is the dimension of the vector space.

(a) Show that S and N commute with L.
(b) Show that L and S have the same eigenvalues.
(c) Show that if λ is an eigenvalue for L, then

$$\ker\left((L - \lambda 1_V)^n\right) = \ker\left((S - \lambda 1_V)\right).$$

(d) Show that the Jordan–Chevalley decomposition is unique.
(e) Find polynomials p, q such that $S = p\,(L)$ and $N = q\,(L)$.

2.9 The Smith Normal Form*

In this section, we show that the row reduction method we developed in Sect. 2.3 to compute the characteristic polynomial can be enhanced to give a direct method for computing similarity invariants provided we also use column operations in addition to row operations.

Let $\mathrm{Mat}_{n \times n} (\mathbb{F}[t])$ be the set of all $n \times n$ matrices that have entries in $\mathbb{F}[t]$. The operations we allow are those coming from multiplying by the elementary matrices: I_{kl}, $R_{kl}(r(t))$, and $M_k(\alpha)$. Recall that we used these operations to compute the characteristic polynomial in Sect. 2.3. When multiplied on the left, these matrices have the effect of:

- I_{kl} interchanges rows k and l.
- $R_{kl}(r(t))$ multiplies row l by $r(t) \in \mathbb{F}[t]$ and adds it to row k.
- $M_k(\alpha)$ multiplies row k by $\alpha \in \mathbb{F} - \{0\}$.

While when multiplied on right:

- I_{kl} interchanges columns k and l.
- $R_{kl}(r(t))$ multiplies column k by $r(t) \in \mathbb{F}[t]$ and adds it to column l.
- $M_k(\alpha)$ multiplies column k by $\alpha \in \mathbb{F} - \{0\}$.

Define

$$Gl_n(\mathbb{F}[t]) \subset \mathrm{Mat}_{n \times n}(\mathbb{F}[t])$$

as the set of all matrices P such that we can find an inverse $Q \in \mathrm{Mat}_{n \times n}(\mathbb{F}[t])$, i.e., $PQ = QP = 1_{\mathbb{F}^n}$. As for regular matrices (see Theorem 1.13.14), we obtain

Proposition 2.9.1. *The elementary matrices generate* $Gl_n(\mathbb{F}[t])$.

Proof. The elementary matrices I_{kl}, $R_{kl}(r(t))$, and $M_k(\alpha)$ all lie in $Gl_n(\mathbb{F}[t])$ as they have inverses given by I_{kl}, $R_{kl}(-r(t))$, and $M_k(\alpha^{-1})$, respectively. Let $P \in Gl_n(\mathbb{F}[t])$, then $P^{-1} \in Gl_n(\mathbb{F}[t])$. Now perform row operations on P^{-1} as in Theorem 2.3.6 until we obtain an upper triangular matrix:

$$U = \begin{bmatrix} p_1(t) & * & \cdots & * \\ 0 & p_2(t) & \cdots & * \\ \vdots & \vdots & \ddots & \vdots \\ 0 & 0 & \cdots & p_n(t) \end{bmatrix} \in Gl_n(\mathbb{F}[t])$$

However, the upper triangular matrix U cannot be invertible unless its diagonal entries are nonzero scalars. Thus, we can assume that $p_i = 1$. We can then perform row operations to eliminate all entries above the diagonal as well. This shows that there is a matrix Q that is a product of elementary matrices and such that

$$QP^{-1} = 1_{\mathbb{F}^n}.$$

Multiplying by P on the right on both sides then shows that

$$P = Q$$

which in turn proves our claim.   □

We can now explain how far it is possible to reduce a matrix with polynomials as entries using row and column operations. As with regular matrices, we obtain a diagonal form (see Corollary 1.13.19), but the diagonal entries have a more complicated relationship between each other.

Theorem 2.9.2. (The Smith Normal Form) *Let $C \in \mathrm{Mat}_{n \times n}\,(\mathbb{F}\,[t])$, then we can find $P, Q \in Gl_n\,(\mathbb{F}\,[t])$ such that*

$$PCQ = \begin{bmatrix} q_1\,(t) & 0 & \cdots & 0 \\ 0 & q_2\,(t) & \cdots & 0 \\ \vdots & \vdots & \ddots & \vdots \\ 0 & 0 & \cdots & q_n\,(t) \end{bmatrix},$$

where $q_i\,(t) \in \mathbb{F}\,[t]$ divides $q_{i+1}\,(t)$ and $q_1\,(t), \ldots, q_n\,(t)$ are monic if they are nonzero. Moreover, with these conditions, $q_1\,(t), \ldots, q_n\,(t)$ are unique.

Proof. Note that having found the diagonal form, we can always make the nonzero polynomials monic so we are not going to worry about that issue.
    We start by giving a construction for finding $P, Q \in Gl_n\,(\mathbb{F}\,[t])$ such that

$$PCQ = \begin{bmatrix} q_1\,(t) & 0 \\ 0 & D \end{bmatrix},$$

where $D \in \mathrm{Mat}_{(n-1) \times (n-1)}\,(\mathbb{F}\,[t])$ and $q_1\,(t)$ divides all of the entries in D.
    If $C = 0$, there is nothing to prove so assume $C \neq 0$.

Step 1: Use row and column interchanges until the $(1, 1)$ entry is the entry with the lowest degree among all nonvanishing entries.
Step 2: For each entry p_{1j}, $j > 1$ in the first row, write it as $p_{1j} = s_{1j} p_{11} + r_{1j}$ where $\deg r_{1j} < \deg p_{11}$ and apply the column operation $CR_{1j}\,(-s_{1j})$ so that the $(1, j)$ entry becomes r_{1j}.
Step 3: For each entry p_{i1}, $i > 1$ in the first column, write it as $p_{i1} = s_{i1} p_{11} + r_{i1}$ where $\deg r_{i1} < \deg p_{11}$ and apply the row operation $R_{i1}\,(-s_{i1})\,C$ so that the $(i, 1)$ entry becomes r_{i1}.
Step 4: If some nonzero entry has degree $< \deg p_{11}$, go back to Step 1. Otherwise, go to Step 5.
Step 5: We know that p_{11} is the only nonzero entry in the first row and column and all other nonzero entries have degree $\geq \deg p_{11}$. If p_{11} divides all entries, we have the desired form. Otherwise, use the column operation $CR_{i1}\,(1)$ for some $i > 1$ to obtain a matrix where the first column has non-zero entries of degree $\geq \deg p_{11}$ and go back to Step 3.

This process will terminate in a finite number of steps and yields

$$PCQ = \begin{bmatrix} q_1(t) & 0 \\ 0 & D \end{bmatrix},$$

where $D \in \mathrm{Mat}_{(n-1) \times (n-1)}(\mathbb{F}[t])$ and $q_1(t)$ divides all of the entries in D.

To obtain the desired diagonal form, we can repeat this process with D or use induction on n.

Next, we have to check uniqueness of the diagonal entries. We concentrate on showing uniqueness of $q_1(t)$ and $q_2(t)$.

Note that if

$$C = P^{-1} \begin{bmatrix} q_1(t) & 0 \\ 0 & D \end{bmatrix} Q^{-1}$$

and $q_1(t)$ divides the entries in D, then it also divides all entries in C. Conversely, the relationship

$$PCQ = \begin{bmatrix} q_1(t) & 0 \\ 0 & D \end{bmatrix}$$

shows that any polynomial that divides all of the entries in C must in particular divide $q_1(t)$. This implies that $q_1(t)$ is the greatest common divisor of the entries in C, i.e., the monic polynomial of highest degree which divides all of the entries in C. Thus, $q_1(t)$ is uniquely defined.

To see that $q_2(t)$ is also uniquely defined, we need to show that if C is equivalent to both

$$\begin{bmatrix} q_1(t) & 0 \\ 0 & D \end{bmatrix} \quad \text{and} \quad \begin{bmatrix} q_1(t) & 0 \\ 0 & D' \end{bmatrix},$$

then D and D' have the same greatest common divisors for their entries. It suffices to show that the greatest common divisor q for all entries in D divides all entries in D'. To show this, first, observe that

$$\begin{bmatrix} q_1(t) & 0 \\ 0 & D' \end{bmatrix} = P \begin{bmatrix} q_1(t) & 0 \\ 0 & D \end{bmatrix} Q$$

$$= \begin{bmatrix} P_{11} & P_{12} \\ P_{21} & P_{22} \end{bmatrix} \begin{bmatrix} q_1(t) & 0 \\ 0 & D \end{bmatrix} \begin{bmatrix} Q_{11} & Q_{12} \\ Q_{21} & Q_{22} \end{bmatrix}$$

$$= \begin{bmatrix} P_{11}q_1(t) & P_{12}D \\ P_{21}q_1(t) & P_{22}D \end{bmatrix} \begin{bmatrix} Q_{11} & Q_{12} \\ Q_{21} & Q_{22} \end{bmatrix}$$

$$= \begin{bmatrix} P_{11}q_1(t)Q_{11} + P_{12}DQ_{21} & P_{11}q_1(t)Q_{12} + P_{12}DQ_{22} \\ P_{21}q_1(t)Q_{11} + P_{22}DQ_{21} & P_{21}q_1(t)Q_{12} + P_{22}DQ_{22} \end{bmatrix}.$$

As q_1 divides all entries of D, we have that $q = pq_1$. Looking at the relationship

$$D' = P_{21}q_1(t)Q_{12} + P_{22}DQ_{22},$$

we observe that if p divides all of the entries in, say, P_{21}, then q will divide all entries in $P_{21}q_1(t)Q_{12}$ and consequently also in D'. To show that p divides the entries in P_{21}, we use the relationship

$$q_1 = P_{11}q_1(t)Q_{11} + P_{12}DQ_{21}.$$

As every element in D is a multiple of q_1, we can write it as $D = q_1E$, where p divides every entry in E. This gives us

$$1 = P_{11}Q_{11} + P_{12}EQ_{21}.$$

Thus, p divides $1 - P_{11}Q_{11}$, which implies that p and Q_{11} are relatively prime. The relationship

$$0 = P_{21}q_1(t)Q_{11} + P_{22}DQ_{21}.$$

in turn shows that p divides the entries in $P_{21}Q_{11}$. As p and Q_{11} are relatively prime, this shows that p divides the entries in P_{21}. $\square$

Example 2.9.3. If $A = \lambda 1_{\mathbb{F}^n}$, then $t1_{\mathbb{F}^n} - A = (t - \lambda)1_{\mathbb{F}^n}$ is already in diagonal form. Thus, we see that $q_1(t) = \cdots = q_n(t) = (t - \lambda)$.

The Smith normal form can be used to give a very effective way of solving fairly complicated systems of higher order linear differential equations. If we start with $C \in \mathrm{Mat}_{n \times n}(\mathbb{F}[t])$, then we can create a system of n differential equations for the functions $x_1, \ldots, x_n$ by interpreting the variable t in the polynomials as the derivative D:

$$
Cx = \begin{bmatrix} p_{11}(D) & p_{12}(D) & \cdots & p_{1n}(D) \\ p_{21}(D) & p_{22}(D) & \cdots & p_{2n}(D) \\ \vdots & \vdots & \ddots & \vdots \\ p_{n1}(D) & p_{n2}(D) & \cdots & p_{nn}(D) \end{bmatrix} \begin{bmatrix} x_1 \\ x_2 \\ \vdots \\ x_n \end{bmatrix}
$$

$$
= \begin{bmatrix} p_{11}(D)x_1 + p_{12}(D)x_2 + \cdots + p_{1n}(D)x_n \\ p_{21}(D)x_1 + p_{22}(D)x_2 + \cdots + p_{2n}(D)x_n \\ \vdots \\ p_{n1}(D)x_1 + p_{n2}(D)x_2 + \cdots + p_{nn}(D)x_n \end{bmatrix}
$$

$$
= \begin{bmatrix} b_1 \\ b_2 \\ \vdots \\ b_n \end{bmatrix},
$$

where $b_1, \ldots, b_n$ are given functions. To solve such a system, we use the Smith normal form

$$PCQ = \begin{bmatrix} q_1(t) & 0 & \cdots & 0 \\ 0 & q_2(t) & \cdots & 0 \\ \vdots & \vdots & \ddots & \vdots \\ 0 & 0 & \cdots & q_n(t) \end{bmatrix}$$

and define

$$\begin{bmatrix} y_1 \\ y_2 \\ \vdots \\ y_n \end{bmatrix} = Q^{-1} \begin{bmatrix} x_1 \\ x_2 \\ \vdots \\ x_n \end{bmatrix}$$

$$\begin{bmatrix} c_1 \\ c_2 \\ \vdots \\ c_n \end{bmatrix} = P \begin{bmatrix} b_1 \\ b_2 \\ \vdots \\ b_n \end{bmatrix}.$$

We then start by solving the decoupled system

$$\begin{bmatrix} q_1(D) & 0 & \cdots & 0 \\ 0 & q_2(D) & \cdots & 0 \\ \vdots & \vdots & \ddots & \vdots \\ 0 & 0 & \cdots & q_n(D) \end{bmatrix} \begin{bmatrix} y_1 \\ y_2 \\ \vdots \\ y_n \end{bmatrix} = \begin{bmatrix} c_1 \\ c_2 \\ \vdots \\ c_n \end{bmatrix}$$

which is really just n-independent higher order equations

$$q_1(D) y_1 = c_1$$
$$q_2(D) y_2 = c_2$$
$$\vdots$$
$$q_n(D) y_n = c_n$$

and then we find the original functions by

$$\begin{bmatrix} x_1 \\ x_2 \\ \vdots \\ x_n \end{bmatrix} = Q \begin{bmatrix} y_1 \\ y_2 \\ \vdots \\ y_n \end{bmatrix}.$$

This use of the Smith normal form is similarly very effective in solving systems of linear recurrences (recurrences were discussed at the end of Sect. 2.2 in relation to solving higher order differential equations).

Example 2.9.4. Consider the 2×2 system of differential equations that comes from

$$C = \begin{bmatrix} (t - \lambda_1)(t - \lambda_2)(t - \lambda_3) & (t - \lambda_1)(t - \lambda_2) \\ (t - \lambda_1)(t - \lambda_3) & (t - \lambda_1) \end{bmatrix},$$

i.e.,

$$\begin{bmatrix} (D - \lambda_1)(D - \lambda_2)(D - \lambda_3) & (D - \lambda_1)(D - \lambda_2) \\ (D - \lambda_1)(D - \lambda_3) & (D - \lambda_1) \end{bmatrix} \begin{bmatrix} x_1 \\ x_2 \end{bmatrix} = \begin{bmatrix} b_1 \\ b_2 \end{bmatrix}.$$

Here

$$R_{21}\left(-(t - \lambda_2)\right) I_{12} C I_{12} R_{12}\left(-(t - \lambda_3)\right) = \begin{bmatrix} t - \lambda_1 & 0 \\ 0 & 0 \end{bmatrix}.$$

So we have to start by solving

$$\begin{bmatrix} t - \lambda_1 & 0 \\ 0 & 0 \end{bmatrix} \begin{bmatrix} y_1 \\ y_2 \end{bmatrix} = R_{21}\left(-(t - \lambda_2)\right) I_{12} \begin{bmatrix} b_1 \\ b_2 \end{bmatrix}$$

$$= \begin{bmatrix} b_2 \\ b_1 - (D - \lambda_2) b_2 \end{bmatrix}$$

In order to solve that system, we have to require that b_1, b_2 are related by

$$(D - \lambda_2) b_2 = Db_2 - \lambda_2 b_2 = b_1.$$

If that is the case, then y_2 can be any function and y_1 is found by solving

$$Dy_1 - \lambda_1 y_1 = b_2.$$

We then find the original solutions from

$$\begin{bmatrix} x_1 \\ x_2 \end{bmatrix} = I_{12} R_{12}\left(-(t - \lambda_3)\right) \begin{bmatrix} y_1 \\ y_2 \end{bmatrix} = \begin{bmatrix} y_2 \\ y_1 - (Dy_2 - \lambda_3 y_2) \end{bmatrix}.$$

Definition 2.9.5. The monic polynomials $q_1(t), \ldots, q_n(t)$ are called the *invariant factors* of $C \in \mathrm{Mat}_{n \times n}(\mathbb{F}[t])$. Note that some of these polynomials can vanish: $q_{k+1}(t) = \cdots = q_n(t) = 0$.

In case $C = t 1_{\mathbb{F}^n} - A$, where $A \in \mathrm{Mat}_{n \times n}(\mathbb{F})$, the invariant factors are related to the *similarity invariants* of A that we defined in Sect. 2.7. Before proving this, we need to gain a better understanding of the invariant factors.

Proposition 2.9.6. *The invariant factors of* $t1_{\mathbb{F}^n} - A$ *and* $t1_{\mathbb{F}^n} - A'$ *are the same if* A *and* A' *are similar.*

Proof. Assume that $A = BA'B^{-1}$, then

$$t1_{\mathbb{F}^n} - A = B\left(t1_{\mathbb{F}^n} - A'\right)B^{-1}.$$

In particular, $t1_{\mathbb{F}^n} - A$ and $t1_{\mathbb{F}^n} - A'$ are equivalent. Since the Smith normal form is unique this shows that they have the same Smith Normal Form. $\square$

This proposition allows to define the invariant factors related to a linear operator $L : V \to V$ on a finite-dimensional vector space by computing the invariant factors of $t1_{\mathbb{F}^n} - [L]$ for any matrix representation $[L]$ of L.

Next, we check what happens for companion matrices.

Proposition 2.9.7. *The invariant factors of* $t1_{\mathbb{F}^n} - C_p$, *where* C_p *is the companion matrix for a monic polynomial* p *of degree* n *are given by* $q_1 = \cdots = q_{n-1} = 1$ *and* $q_n = p$.

Proof. Let C_p be the companion matrix for $p(t) = t^n + \alpha_{n-1}t^{n-1} + \cdots + \alpha_1 t + \alpha_0 \in \mathbb{F}[t]$, i.e.,

$$C_p = \begin{bmatrix} 0 & 0 & \cdots & 0 & -\alpha_0 \\ 1 & 0 & \cdots & 0 & -\alpha_1 \\ 0 & 1 & \cdots & 0 & -\alpha_2 \\ \vdots & \vdots & \ddots & \vdots & \vdots \\ 0 & 0 & \cdots & 1 & -\alpha_{n-1} \end{bmatrix}.$$

Then,

$$t1_{\mathbb{F}^n} - C_p = \begin{bmatrix} t & 0 & \cdots & 0 & \alpha_0 \\ -1 & t & \cdots & 0 & \alpha_1 \\ 0 & -1 & \cdots & 0 & \alpha_2 \\ \vdots & \vdots & \ddots & \vdots & \vdots \\ 0 & 0 & \cdots & -1 & t+\alpha_{n-1} \end{bmatrix}.$$

We know from Proposition 2.6.3 that $t1_{\mathbb{F}^n} - C_p$ is row equivalent to a matrix of the form

$$\begin{bmatrix} -1 & t & \cdots & 0 & & \alpha_1 \\ 0 & -1 & \cdots & 0 & & \alpha_2 \\ 0 & 0 & \ddots & \vdots & & \vdots \\ \vdots & \vdots & & -1 & & t+\alpha_{n-1} \\ 0 & 0 & \cdots & 0 & & t^n + \alpha_{n-1}t^{n-1} + \cdots + \alpha_1 t + \alpha_0 \end{bmatrix}.$$

We can change the -1 diagonal entries to 1. We can then use column operations to eliminate the t entries to the right of the diagonal entries in columns $2, \ldots, n-1$ as well as the entries in the last column that are in the rows $1, \ldots, n-1$. This results in the equivalent diagonal matrix

$$\begin{bmatrix} 1 & 0 & \cdots & 0 \\ 0 & \ddots & & \vdots \\ \vdots & & 1 & 0 \\ 0 & \cdots & 0 & p \end{bmatrix}.$$

This must be the Smith normal form. $\qquad\square$

We can now show how the similarity invariant of a linear operator can be computed using the Smith normal form.

Theorem 2.9.8. *Let $L : V \to V$ be a linear operator on a finite-dimensional vector space, $p_1, \ldots, p_k \in \mathbb{F}[t]$ the similarity invariants, and $q_1, \ldots, q_n \in \mathbb{F}[t]$ the invariant factors of $t1_{\mathbb{F}^n} - [L]$. Then, $q_{n-i} = p_{i+1}$ for $i = 0, \ldots, k-1$ and $q_j = 1$ for $j = 1, \ldots, n-k$.*

Proof. We start by selecting the Frobenius canonical form

$$[L] = \begin{bmatrix} C_{p_1} & 0 & & 0 \\ 0 & C_{p_2} & & \\ & & \ddots & \\ 0 & & & C_{p_k} \end{bmatrix}$$

as the matrix representation for L. The previous proposition gives the invariant factors of the blocks $t1_{\mathbb{F}^{\deg p_i}} - C_{p_i}$. This tells us that if we only do row and column operations that respect the block diagonal form, then $t1_{\mathbb{F}^n} - [L]$ is equivalent to a block diagonal matrix

$$C = \begin{bmatrix} C_1 & 0 & & 0 \\ 0 & C_2 & & \\ & & \ddots & \\ 0 & & & C_k \end{bmatrix},$$

where C_i is a diagonal matrix

$$C_i = \begin{bmatrix} 1 & 0 & & 0 \\ 0 & 1 & & \\ & & \ddots & \\ 0 & & & p_i \end{bmatrix}.$$

We can now perform row and column interchanges on C to obtain the diagonal matrix

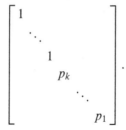

Since the Smith normal form is unique and the Frobenius normal form has the property that p_{i+1} divides p_i for $i = 1, \ldots k - 1$ we have obtained the Smith normal form for $t 1_{\mathbb{F}^n} - [L]$ and proven the claim.   □

Exercises

1. Find the Smith normal form for the matrices

   (a)
   $$\begin{bmatrix} (t - \lambda_1)(t - \lambda_2)(t - \lambda_3) & (t - \lambda_1)(t - \lambda_2) \\ (t - \lambda_1)(t - \lambda_2) & (t - \lambda_1) \end{bmatrix}$$

   (b)
   $$\begin{bmatrix} t - 1 & t^2 + t - 2 & t - 1 \\ t^3 + t^2 - 4t + 2 & 0 & t^2 - t \\ t - 1 & 0 & t - 1 \end{bmatrix}$$

   (c)
   $$\begin{bmatrix} -t & 0 & 0 & 0 \\ 1 & -t & 0 & 0 \\ 0 & 0 & -t & 0 \\ 0 & 0 & 0 & -t \end{bmatrix}$$

   (d)
   $$\begin{bmatrix} -t & 0 & 0 & 0 \\ 1 & -t & 0 & 0 \\ 0 & 0 & -t & 0 \\ 0 & 0 & 1 & -t \end{bmatrix}.$$

2. Show that if $C \in \mathrm{Mat}_{n \times n} (\mathbb{F}[t])$ has a $k \times k$ minor that belongs to $Gl_k (\mathbb{F}[t])$, then $q_1 (t) = \cdots = q_k (t) = 1$. A $k \times k$ *minor* is a $k \times k$ matrix that is obtained form C by deleting all but k columns and k rows.

3. Let $A \in \mathrm{Mat}_{n \times n}(\mathbb{F})$ and consider the two linear operators $L_A, R_A :$ $\mathrm{Mat}_{n \times n}(\mathbb{F}) \to \mathrm{Mat}_{n \times n}(\mathbb{F})$ defined by $L_A(X) = AX$ and $R_A(X) = XA$. Are L_A and R_A similar?

4. Let C_p and C_q be companion matrices. Show that

$$\begin{bmatrix} C_p & 0 \\ 0 & C_q \end{bmatrix}$$

has

$$q_1(t) = \cdots = q_{n-2}(t) = 1,$$

$$q_{n-1}(t) = \gcd(p, q),$$

$$q_n(t) = \mathrm{lcm}(p, q).$$

5. Find the similarity invariants for

$$\begin{bmatrix} C_p & 0 & 0 \\ 0 & C_q & 0 \\ 0 & 0 & C_r \end{bmatrix}$$

6. Find the Smith normal form of a diagonal matrix

$$\begin{bmatrix} p_1 & & & \\ & p_2 & & \\ & & \ddots & \\ & & & p_n \end{bmatrix}$$

where $p_1, \ldots, p_n \in \mathbb{F}[t]$. Hint: Start with $n = 2$.

7. Show that $A, B \in \mathrm{Mat}_{n \times n}(\mathbb{F})$ are similar if $(t1_{\mathbb{F}^n} - A), (t1_{\mathbb{F}^n} - B) \in \mathrm{Mat}_{n \times n}(\mathbb{F}[t])$ are equivalent. Hint: Use the Smith normal form. It is interesting to note that there is a proof which does not use the Smith normal form (see [Serre, Theorem 6.3.2]).

Chapter 3
Inner Product Spaces

So far, we have only discussed vector spaces without adding any further structure to the space. In this chapter, we shall study so-called inner product spaces. These are vector spaces where in addition we know the length of each vector and the angle between two vectors. Since this is what we are used to from the plane and space, it would seem like a reasonable extra layer of information.

We shall cover some of the basic constructions such as Gram–Schmidt orthogonalization, orthogonal projections, and orthogonal complements. We also prove the Cauchy–Schwarz and Bessel inequalities. In the last sections, we introduce the adjoint of linear maps. The adjoint helps us understand the connections between image and kernel and leads to a very interesting characterization of orthogonal projections. Finally, we also explain matrix exponentials and how they can be used to solve systems of linear differential equations.

In this and the following chapter, vector spaces always have either real or complex scalars.

3.1 Examples of Inner Products

3.1.1 Real Inner Products

We start by considering the (real) plane $\mathbb{R}^2 = \{(\alpha_1, \alpha_2) : \alpha_1, \alpha_2 \in \mathbb{R}\}$. The length of a vector is calculated via the Pythagorean theorem:

$$\|(\alpha_1, \alpha_2)\| = \sqrt{\alpha_1^2 + \alpha_2^2}.$$

The angle between two vectors $x = (\alpha_1, \alpha_2)$ and $y = (\beta_1, \beta_2)$ is a little trickier to compute. First, we normalize the vectors

P. Petersen, *Linear Algebra*, Undergraduate Texts in Mathematics,
DOI 10.1007/978-1-4614-3612-6_3, © Springer Science+Business Media New York 2012

Fig. 3.1 Definition of angle

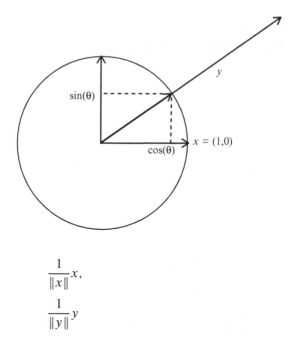

$$\frac{1}{\|x\|}x,$$

$$\frac{1}{\|y\|}y$$

so that they lie on the unit circle. We then trace the arc on the unit circle between the vectors in order to find the angle θ. If $x = (1,0)$, the definitions of cosine and sine (see Fig. 3.1) tell us that this angle can be computed via

$$\cos\theta = \frac{\beta_1}{\|y\|},$$

$$\sin\theta = \frac{\beta_2}{\|y\|}.$$

This suggests that if we define

$$\cos\theta_1 = \frac{\alpha_1}{\|x\|}, \quad \sin\theta_1 = \frac{\alpha_2}{\|x\|},$$

$$\cos\theta_2 = \frac{\beta_1}{\|y\|}, \quad \sin\theta_2 = \frac{\beta_2}{\|y\|},$$

then

$$\cos\theta = \cos(\theta_2 - \theta_1)$$

$$= \cos\theta_1\cos\theta_2 + \sin\theta_1\sin\theta_2$$

$$= \frac{\alpha_1\beta_1 + \alpha_2\beta_2}{\|x\| \cdot \|y\|}.$$

So if the *inner or dot product* of x and y is defined by

$$(x|y) = \alpha_1 \beta_1 + \alpha_2 \beta_2,$$

then we obtain the relationship

$$(x|y) = \|x\| \, \|y\| \cos \theta.$$

The length of vectors can also be calculated via

$$(x|x) = \|x\|^2.$$

The $(x|y)$ notation is used so as not to confuse the expression with pairs of vectors (x, y). One also often sees $\langle x, y \rangle$ or $\langle x|y \rangle$ used for inner products.

The key properties that we shall use to generalize the idea of an inner product are:

1. $(x|x) = \|x\|^2 > 0$ unless $x = 0$.
2. $(x|y) = (y|x)$.
3. $x \to (x|y)$ is linear.

One can immediately generalize this algebraically defined inner product to $\mathbb{R}^3$ and even $\mathbb{R}^n$ by

$$(x|y) = \left(\begin{bmatrix} \alpha_1 \\ \vdots \\ \alpha_n \end{bmatrix} \middle| \begin{bmatrix} \beta_1 \\ \vdots \\ \beta_n \end{bmatrix} \right)$$

$$= x^t y$$

$$= \begin{bmatrix} \alpha_1 & \cdots & \alpha_n \end{bmatrix} \begin{bmatrix} \beta_1 \\ \vdots \\ \beta_n \end{bmatrix}$$

$$= \alpha_1 \beta_1 + \cdots + \alpha_n \beta_n.$$

The three above-mentioned properties still remain true, but we seem to have lost the connection with the angle. This is settled by observing that Cauchy's inequality holds:

$$(x|y)^2 \le (x|x)\,(y|y), \text{ or}$$

$$(\alpha_1 \beta_1 + \cdots + \alpha_n \beta_n)^2 \le (\alpha_1^2 + \cdots + \alpha_n^2)(\beta_1^2 + \cdots + \beta_n^2).$$

In other words,

$$-1 \le \frac{(x|y)}{\|x\| \, \|y\|} \le 1.$$

Fig. 3.2 Projection

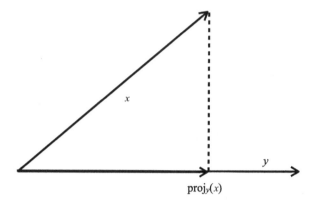

$$\text{proj}_y(x)$$

This implies that the angle can be redefined up to sign through the equation

$$\cos \theta = \frac{(x|y)}{\|x\| \, \|y\|}.$$

In addition, as we shall see, the three properties can be used as axioms for inner products.

Two vectors are said to be *orthogonal* or *perpendicular* if their inner product vanishes. With this definition, the proof of the Pythagorean theorem becomes completely algebraic:

$$\|x\|^2 + \|y\|^2 = \|x + y\|^2,$$

if x and y are orthogonal. To see why this is true, note that the properties of the inner product imply:

$$\|x + y\|^2 = (x + y|x + y)$$
$$= (x|x) + (y|y) + (x|y) + (y|x)$$
$$= (x|x) + (y|y) + 2\,(x|y)$$
$$= \|x\|^2 + \|y\|^2 + 2\,(x|y).$$

Thus, the relation $\|x\|^2 + \|y\|^2 = \|x + y\|^2$ holds precisely when $(x|y) = 0$.

The inner product also comes in handy in expressing several other geometric constructions.

The *projection* of a vector x onto the line in the direction of y (see Fig. 3.2) is given by

$$\text{proj}_y(x) = \left(x \left| \frac{y}{\|y\|} \right. \right) \frac{y}{\|y\|}$$
$$= \frac{(x|y)\,y}{(y|y)}.$$

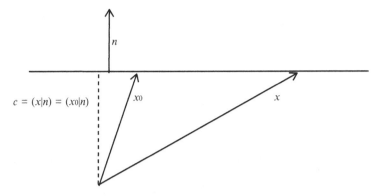

Fig. 3.3 Plane

All *planes* that have normal n, i.e., are perpendicular to n, are defined by the equation

$$(x|n) = c,$$

where c is determined by any point x_0 that lies in the plane: $c = (x_0|n)$ (see also Fig. 3.3).

3.1.2 Complex Inner Products

Let us now see what happens if we try to use complex scalars. Our geometric picture seems to disappear, but we shall insist that the real part of a complex inner product must have the (geometric) properties we have already discussed. Let us start with the complex plane $\mathbb{C}$. Recall that if $z = \alpha_1 + \alpha_2 i$, then the complex conjugate is the reflection of z in the first coordinate axis and is defined by $\bar{z} = \alpha_1 - \alpha_2 i$. Note that $z \to \bar{z}$ is not complex linear but only linear with respect to real scalar multiplication. Conjugation has some further important properties:

$$\|z\| = \sqrt{z \cdot \bar{z}},$$

$$\overline{z \cdot w} = \bar{z} \cdot \bar{w},$$

$$z^{-1} = \frac{\bar{z}}{\|z\|^2}$$

$$\mathrm{Re}\,(z) = \frac{z + \bar{z}}{2}$$

$$\mathrm{Im}\,(z) = \frac{z - \bar{z}}{2i}.$$

Given that $\|z\|^2 = z\bar{z}$, it seems natural to define the *complex inner product* by $(z|w) = z\bar{w}$. Thus, it is not just complex multiplication. If we take the real part, we also note that we retrieve the real inner product defined above:

$$\text{Re}\,(z|w) = \text{Re}\,(z\bar{w})$$
$$= \text{Re}\,((\alpha_1 + \alpha_2 i)\,(\beta_1 - \beta_2 i))$$
$$= \alpha_1\beta_1 + \alpha_2\beta_2.$$

Having established this, we should be happy and just accept the fact that complex inner products include conjugations.

The three important properties for complex inner products are

1. $(x|x) = \|x\|^2 > 0$ unless $x = 0$.
2. $(x|y) = \overline{(y|x)}$.
3. $x \to (x|y)$ is complex linear.

The *inner product* on $\mathbb{C}^n$ is defined by

$$(x|y) = \left(\begin{bmatrix} \alpha_1 \\ \vdots \\ \alpha_n \end{bmatrix} \middle| \begin{bmatrix} \beta_1 \\ \vdots \\ \beta_n \end{bmatrix}\right)$$
$$= x^t \bar{y}$$
$$= \begin{bmatrix} \alpha_1 \cdots \alpha_n \end{bmatrix} \begin{bmatrix} \bar{\beta}_1 \\ \vdots \\ \bar{\beta}_n \end{bmatrix}$$
$$= \alpha_1\bar{\beta}_1 + \cdots + \alpha_n\bar{\beta}_n.$$

If we take the real part of this inner product, we get the inner product on $\mathbb{R}^{2n} \simeq \mathbb{C}^n$.

We say that two complex vectors are orthogonal if their inner product vanishes. This is not quite the same as in the real case, as the two vectors 1 and i in $\mathbb{C}$ are not complex orthogonal even though they are orthogonal as real vectors. To spell this out a little further, let us consider the Pythagorean theorem for complex vectors. Note that

$$\|x + y\|^2 = (x + y|x + y)$$
$$= (x|x) + (y|y) + (x|y) + (y|x)$$
$$= (x|x) + (y|y) + (x|y) + \overline{(x|y)}$$
$$= \|x\|^2 + \|y\|^2 + 2\text{Re}\,(x|y).$$

Thus, only the real part of the inner product needs to vanish for this theorem to hold. This should not come as a surprise as we already knew the result to be true in this case.

3.1.3   A Digression on Quaternions*

Another very interesting space that contains some new algebra as well as geometry is $\mathbb{C}^2 \simeq \mathbb{R}^4$. This is the space-time of special relativity. In this short section, we mention some of the important features of this space.

In analogy with writing $\mathbb{C} = \text{span}_{\mathbb{R}} \{1, i\}$, let us define

$$\mathbb{H} = \text{span}_{\mathbb{C}} \{1, j\}$$
$$= \text{span}_{\mathbb{R}} \{1, i, 1 \cdot j, i \cdot j\}$$
$$= \text{span}_{\mathbb{R}} \{1, i, j, k\} .$$

The three vectors i, j, k form the usual basis for the three-dimensional space $\mathbb{R}^3$. The remaining coordinate in $\mathbb{H}$ is the time coordinate. In $\mathbb{H}$, we also have a conjugation that changes the sign in front of the imaginary numbers i, j, k

$$\bar{q} = \overline{\alpha_0 + \alpha_1 i + \alpha_2 j + \alpha_3 k}$$
$$= \alpha_0 - \alpha_1 i - \alpha_2 j - \alpha_3 k.$$

To make perfect sense of things, we need to figure out how to multiply i, j, k. In line with $i^2 = -1$, we also define $j^2 = -1$ and $k^2 = -1$. As for the mixed products, we have already defined $ij = k$. More generally, we can decide how to compute these products by using the cross product in $\mathbb{R}^3$. Thus,

$$ij = k = -ji,$$
$$jk = i = -kj,$$
$$ki = j = -ik.$$

This enables us to multiply $q_1, q_2 \in \mathbb{H}$. The multiplication is not commutative, but it is associative (unlike the cross product), and nonzero elements have inverses. The fact that the imaginary numbers i, j, k anti-commute shows that conjugation must reverse the order of multiplication (like taking inverses of matrices and quaternions)

$$\overline{pq} = \bar{q}\,\bar{p}.$$

As with real and complex numbers, we have that

$$q\bar{q} = |q|^2 = \alpha_0^2 + \alpha_1^2 + \alpha_2^2 + \alpha_3^2.$$

This shows that every nonzero quaternion has an inverse given by

$$q^{-1} = \frac{\bar{q}}{|q|^2}.$$

The space $\mathbb{H}$ with usual vector addition and this multiplication is called the space of *quaternions*. The name was chosen by Hamilton who invented these numbers and wrote voluminous material on their uses.

As with complex numbers, we have a real part, namely, the part without i, j, k, that can be calculated by

$$\mathrm{Re}q = \frac{q + \bar{q}}{2}.$$

The usual real inner product on $\mathbb{R}^4$ can now be defined by

$$(p|q) = \mathrm{Re}\,(p \cdot \bar{q}).$$

If we ignore the conjugation but still take the real part, we obtain something else entirely

$$\begin{aligned}
(p|q)_{1,3} &= \mathrm{Re}\,(pq) \\
&= \mathrm{Re}\,(\alpha_0 + \alpha_1 i + \alpha_2 j + \alpha_3 k)\,(\beta_0 + \beta_1 i + \beta_2 j + \beta_3 k) \\
&= \alpha_0 \beta_0 - \alpha_1 \beta_1 - \alpha_2 \beta_2 - \alpha_3 \beta_3.
\end{aligned}$$

We note that restricted to the time axis this is the usual inner product while if restricted to the space part it is the negative of the usual inner product. This *pseudo-inner* product is what is used in special relativity. The subscript 1,3 refers to the signs that appear in the formula, 1 plus and 3 minuses.

Note that one can have $(q|q)_{1,3} = 0$ without $q = 0$. The geometry of such an inner product is thus quite different from the usual ones we introduced above.

The purpose of this very brief encounter with quaternions and space-times is to show that they appear quite naturally in the context of linear algebra. While we will not use them here, they are used quite a bit in more advanced mathematics and physics.

Exercises

1. Using the algebraic properties of inner products, show the law of cosines

$$c^2 = a^2 + b^2 - 2ab \cos \theta,$$

where a and b are adjacent sides in a triangle forming an angle θ and c is the opposite side.

2. Here are some matrix constructions of both complex and quaternion numbers.

   (a) Show that $\mathbb{C}$ is isomorphic (same addition and multiplication) to the set of real 2×2 matrices of the form

   $$\begin{bmatrix} \alpha & -\beta \\ \beta & \alpha \end{bmatrix}.$$

   (b) Show that $\mathbb{H}$ is isomorphic to the set of complex 2×2 matrices of the form

   $$\begin{bmatrix} z & -\bar{w} \\ w & \bar{z} \end{bmatrix}.$$

   (c) Show that $\mathbb{H}$ is isomorphic to the set of real 4×4 matrices

   $$\begin{bmatrix} A & -B^t \\ B & A^t \end{bmatrix}$$

   that consists of 2×2 blocks

   $$A = \begin{bmatrix} \alpha & -\beta \\ \beta & \alpha \end{bmatrix}, B = \begin{bmatrix} \gamma & -\delta \\ \delta & \gamma \end{bmatrix}.$$

   (d) Show that the quaternionic 2×2 matrices of the form

   $$\begin{bmatrix} p & -\bar{q} \\ q & \bar{p} \end{bmatrix}$$

   form a real vector space isomorphic to $\mathbb{R}^8$ but that matrix multiplication does not necessarily give us a matrix of this type. What goes wrong in this case?

3. If $q \in \mathbb{H} - \{0\}$, consider the map $\mathrm{Ad}_q : \mathbb{H} \to \mathbb{H}$ defined by $\mathrm{Ad}_q (x) = qxq^{-1}$.

   (a) Show that $x = 1$ is an eigenvector with eigenvalue 1.
   (b) Show that Ad_q maps $\mathrm{span}_{\mathbb{R}} \{i, j, k\}$ to itself and defines an isometry on $\mathbb{R}^3$.
   (c) Show that $\mathrm{Ad}_{q_1} = \mathrm{Ad}_{q_2}$ if and only if $q_1 = \lambda q_2$, where $\lambda \in \mathbb{R}$.

3.2 Inner Products

Recall that we only use real or complex vector spaces. Thus, the field $\mathbb{F}$ of scalars is always $\mathbb{R}$ or $\mathbb{C}$.

Definition 3.2.1. An *inner product* on a vector space V over $\mathbb{F}$ is an $\mathbb{F}$-valued pairing $(x|y)$ for $x, y \in V$, i.e., a map $(x|y) : V \times V \to \mathbb{F}$, that satisfies:

(1) $(x|x) \geq 0$ and vanishes only when $x = 0$.
(2) $(x|y) = \overline{(y|x)}$.
(3) For each $y \in V$, the map $x \to (x|y)$ is linear.

A vector space with an inner product is called an inner product space. In the real case, the inner product is also called a *Euclidean structure*, while in the complex situation, the inner product is known as an *Hermitian structure*. Observe that a complex inner product $(x|y)$ always defines a real inner product $\mathrm{Re}\,(x|y)$ that is symmetric and linear with respect to real scalar multiplication. One also uses the term *dot product* for the standard inner products in $\mathbb{R}^n$ and $\mathbb{C}^n$. The term *scalar product* is also used quite often as a substitute for inner product. In fact, this terminology seems better as it indicates that the product of two vectors becomes a scalar.

We note that the second property really only makes sense when the inner product is complex valued. If V is a real vector space, then the inner product is real valued and hence symmetric in x and y, i.e., $(x|y) = (y|x)$. In the complex case, property 2 implies that $(x|x)$ is real, thus showing that the condition in property 1 makes sense. If we combine the second and third conditions, we get the sesqui-linearity properties:

$$(\alpha_1 x_1 + \alpha_2 x_2 | y) = \alpha_1\,(x_1|y) + \alpha_2\,(x_2|y)\,,$$
$$(x|\beta_1 y_1 + \beta_2 y_2) = \bar{\beta}_1\,(x|y_1) + \bar{\beta}_2\,(x|y_2)\,.$$

In particular, we have the scaling property

$$(\alpha x|\alpha x) = \alpha\bar{\alpha}\,(x|x)$$
$$= |\alpha|^2\,(x|x)\,.$$

We define the *length* or *norm* of a vector by

$$\|x\| = \sqrt{(x|x)}.$$

In case $(x|y)$ is complex, we see that $(x|y)$ and $\mathrm{Re}\,(x|y)$ define the same norm. Note that $\|x\|$ is nonnegative and only vanishes when $x = 0$. We also have the scaling property $\|\alpha x\| = |\alpha|\,\|x\|$. The triangle inequality: $\|x + y\| \le \|x\| + \|y\|$ will be established later in this section after some important preparatory work (see Corollary 3.2.11). Before studying the properties of inner products further, let us list some important examples. In Sect. 3.1, we already introduced what we shall refer to as the standard inner product structures on $\mathbb{R}^n$ and $\mathbb{C}^n$.

Example 3.2.2. If we have an inner product on V, then we also get an inner product on all of the subspaces of V.

Example 3.2.3. If we have inner products on V and W, both with respect to $\mathbb{F}$, then we get an inner product on $V \times W$ defined by

$$((x_1, y_1)\,|(x_2, y_2)) = (x_1|x_2) + (y_1|y_2)\,.$$

Note that $(x, 0)$ and $(0, y)$ always have zero inner product.

Example 3.2.4. Given that $\mathrm{Mat}_{n \times m}(\mathbb{C}) = \mathbb{C}^{n \cdot m}$, we have an inner product on this space. As we shall see, it has an interesting alternate construction. Let $A, B \in \mathrm{Mat}_{n \times m}(\mathbb{C})$ the adjoint B^* is the transpose combined with conjugating each entry

$$B^* = \begin{bmatrix} \bar{\beta}_{11} & \cdots & \bar{\beta}_{n1} \\ \vdots & \ddots & \vdots \\ \bar{\beta}_{1m} & \cdots & \bar{\beta}_{nm} \end{bmatrix}.$$

The inner product $(A|B)$ can now be defined as

$$(A|B) = \mathrm{tr}\left(AB^*\right)$$
$$= \mathrm{tr}\left(B^*A\right).$$

In case $m = 1$, we have $\mathrm{Mat}_{n \times 1}(\mathbb{C}) = \mathbb{C}^n$, and we recover the standard inner product from the entry in the 1×1 matrix B^*A. In the general case, we note that it also defines the usual inner product as

$$(A|B) = \mathrm{tr}\left(AB^*\right)$$
$$= \sum_{i,j} \alpha_{ij} \bar{\beta}_{ij}.$$

Example 3.2.5. Let $V = C^0\left([a,b],\mathbb{C}\right)$ and define

$$(f|g) = \int_a^b f(t)\overline{g(t)}\mathrm{d}t.$$

Then,

$$\|f\|_2 = \sqrt{(f,f)}.$$

If $V = C^0\left([a,b],\mathbb{R}\right)$, then we have the real inner product

$$(f|g) = \int_a^b f(t)g(t)\,\mathrm{d}t.$$

In the above example, it is often convenient to normalize the inner product so that the function $f = 1$ is of unit length. This normalized inner product is defined as

$$(f|g) = \frac{1}{b-a}\int_a^b f(t)\overline{g(t)}\mathrm{d}t.$$

Example 3.2.6. Another important infinite-dimensional inner product space is the space ℓ^2 first investigated by Hilbert. It is the collection of all real or complex sequences (α_n) such that $\sum_n |\alpha_n|^2 < \infty$. We have not specified the index set for

n, but we always think of it as being $\mathbb{N}$, $\mathbb{N}_0$, or $\mathbb{Z}$. Because these index sets are all bijectively equivalent, they all the define the same space but with different indices for the coordinates α_n. Addition and scalar multiplication are defined by

$$(\alpha_n) + (\beta_n) = (\alpha_n + \beta_n),$$
$$\beta(\alpha_n) = (\beta\alpha_n).$$

Since

$$\sum_n |\beta\alpha_n|^2 = |\beta|^2 \sum_n |\alpha_n|^2,$$

$$\sum_n |\alpha_n + \beta_n|^2 \le \sum_n \left(2|\alpha_n|^2 + 2|\beta_n|^2\right)$$

$$= 2\sum_n |\alpha_n|^2 + 2\sum_n |\beta_n|^2,$$

it follows that ℓ^2 is a subspace of the vector space of all sequences. The inner product $((\alpha_n) \mid (\beta_n))$ is defined by

$$((\alpha_n) \mid (\beta_n)) = \sum_n \alpha_n \bar{\beta}_n.$$

For that to make sense, we need to know that

$$\sum_n |\alpha_n \bar{\beta}_n| < \infty.$$

This follows from

$$|\alpha_n \bar{\beta}_n| = |\alpha_n| \, |\bar{\beta}_n|$$
$$= |\alpha_n| \, |\beta_n|$$
$$\le |\alpha_n|^2 + |\beta_n|^2$$

and the fact that

$$\sum_n \left(|\alpha_n|^2 + |\beta_n|^2\right) < \infty.$$

Definition 3.2.7. We say that two vectors x and y are *orthogonal* or *perpendicular* if $(x|y) = 0$, and we denote this by $x \perp y$.

The proof of the Pythagorean theorem for both $\mathbb{R}^n$ and $\mathbb{C}^n$ clearly carries over to this more abstract situation. So if $(x|y) = 0$, then $\|x + y\|^2 = \|x\|^2 + \|y\|^2$.

Definition 3.2.8. The *orthogonal projection* of a vector x onto a nonzero vector y is defined by

Fig. 3.4 Orthogonal
projection

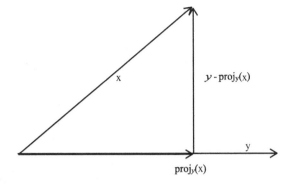

$$\mathrm{proj}_y\,(x) = \left(x\,\bigg|\,\frac{y}{\|y\|}\right)\frac{y}{\|y\|}$$

$$= \frac{(x|y)}{(y|y)}y.$$

This projection creates a vector in the subspace spanned by y. The fact that it makes sense to call it the orthogonal projection is explained in the next proposition (Fig. 3.4).

Proposition 3.2.9. *Given a nonzero y, the map $x \to \mathrm{proj}_y\,(x)$ is linear and a projection with the further property that $x - \mathrm{proj}_y\,(x)$ and $\mathrm{proj}_y\,(x)$ are orthogonal. In particular,*

$$\|x\|^2 = \left\|x - \mathrm{proj}_y\,(x)\right\|^2 + \left\|\mathrm{proj}_y\,(x)\right\|^2,$$

and

$$\left\|\mathrm{proj}_y\,(x)\right\| \le \|x\|.$$

Proof. The definition of $\mathrm{proj}_y\,(x)$ immediately implies that it is linear from the linearity of the inner product and that it is a projection that follows from

$$\mathrm{proj}_y\,\big(\mathrm{proj}_y\,(x)\big) = \mathrm{proj}_y\left(\frac{(x|y)}{(y|y)}y\right)$$

$$= \frac{(x|y)}{(y|y)}\mathrm{proj}_y\,(y)$$

$$= \frac{(x|y)}{(y|y)}\frac{(y|y)}{(y|y)}y$$

$$= \frac{(x|y)}{(y|y)}y$$

$$= \mathrm{proj}_y\,(x)\,.$$

3 Inner Product Spaces

To check orthogonality, simply compute

$$\left(x - \text{proj}_y\left(x\right) | \text{proj}_y\left(x\right)\right) = \left(x - \frac{(x|y)}{(y|y)}y \left| \frac{(x|y)}{(y|y)}y\right.\right)$$

$$= \left(x \left| \frac{(x|y)}{(y|y)}y\right.\right) - \left(\frac{(x|y)}{(y|y)}y \left| \frac{(x|y)}{(y|y)}y\right.\right)$$

$$= \frac{\overline{(x|y)}}{(y|y)}(x|y) - \frac{|(x|y)|^2}{|(y|y)|^2}(y|y)$$

$$= \frac{|(x|y)|^2}{(y|y)} - \frac{|(x|y)|^2}{(y|y)}$$

$$= 0.$$

The Pythagorean theorem now implies the relationship

$$\|x\|^2 = \left\|x - \text{proj}_y\left(x\right)\right\|^2 + \left\|\text{proj}_y\left(x\right)\right\|^2.$$

Using $\left\|x - \text{proj}_y\left(x\right)\right\|^2 \geq 0$, we then obtain the inequality $\left\|\text{proj}_y\left(x\right)\right\| \leq \|x\|$. $\quad\square$

Two important corollaries follow almost directly from this result.

Corollary 3.2.10. (The Cauchy–Schwarz Inequality)

$$|(x|y)| \leq \|x\| \, \|y\| \, .$$

Proof. If $y = 0$, the inequality is trivial. Otherwise, use

$$\|x\| \geq \left\|\text{proj}_y\left(x\right)\right\|$$

$$= \left|\frac{(x|y)}{(y|y)}\right| \|y\|$$

$$= \frac{|(x|y)|}{\|y\|}. \qquad\qquad\qquad \square$$

Corollary 3.2.11. (The Triangle Inequality)

$$\|x + y\| \leq \|x\| + \|y\| \, .$$

Proof. We expand $\|x + y\|$ and use the Cauchy–Schwarz inequality

$$\|x + y\|^2 = (x + y | x + y)$$

$$= \|x\|^2 + 2\text{Re}\,(x|y) + \|y\|^2$$

$$\leq \|x\|^2 + 2\,|(x|y)| + \|y\|^2$$

$$\leq \|x\|^2 + 2\,\|x\|\,\|y\| + \|y\|^2$$
$$= (\|x\| + \|y\|)^2 .$$
$\square$

Exercises

1. Show that a hyperplane $H = \{x \in V : (a|x) = \alpha\}$ in a real n-dimensional inner product space V can be represented as an affine subspace

$$H = \{t_1 x_1 + \cdots + t_n x_n : t_1 + \cdots + t_n = 1\},$$

   where $x_1, \ldots, x_n \in H$. Find conditions on $x_1, \ldots, x_n$ so that they generate a hyperplane (see Exercise 8 in Sect. 1.10 for the definition of an affine subspace).
2. Let $x = (2, 1)$ and $y = (3, 1)$ in $\mathbb{R}^2$. If $z \in \mathbb{R}^2$ satisfies $(z|x) = 1$ and $(z|y) = 2$, then find the coordinates for z.
3. Show that it is possible to find k vectors $x_1, \ldots, x_k \in \mathbb{R}^n$ such that $\|x_i\| = 1$ and $(x_i|x_j) < 0$, $i \neq j$ only when $k \leq n + 1$. Show that for any such choice of k vectors, we get a linearly independent set by deleting any one of the k vectors.
4. In a real inner product space V select $y \neq 0$. For fixed $\alpha \in \mathbb{R}$, show that $H = \{x \in V : \text{proj}_y (x) = \alpha y\}$ describes a hyperplane with normal y.
5. Let V be an inner product space and let $y, z \in V$. Show that $y = z$ if and only if $(x|y) = (x|z)$ for all $x \in V$.
6. Prove the Cauchy–Schwarz inequality by expanding the right-hand side of the inequality

$$0 \leq \left\| x - \frac{(x|y)}{\|y\|^2} y \right\|^2 .$$

7. Let V be an inner product space and $x_1, \ldots, x_n, y_1, \ldots, y_n \in V$. Show the following generalized Cauchy–Schwarz inequality:

$$\left(\sum_{i=1}^{n} |(x_i|y_i)| \right)^2 \leq \left(\sum_{i=1}^{n} \|x_i\|^2 \right) \left(\sum_{i=1}^{n} \|y_i\|^2 \right).$$

8. Let $S^{n-1} = \{x \in \mathbb{R}^n : \|x\| = 1\}$ be the unit sphere. When $n = 1$, it consists of two points, When $n = 2$, it is a circle, and when $n = 3$ a sphere. A finite subset $\{x_1, \ldots, x_k\} \in S^{n-1}$ is said to consist of equidistant points if $\angle(x_i, x_j) = \theta$ for all $i \neq j$.

   (a) Show that this is equivalent to assuming that $(x_i|x_j) = \cos\theta$ for all $i \neq j$.
   (b) Show that S^0 contains a set of two equidistant points, S^1 a set of three equidistant points, and S^2 a set of four equidistant points.
   (c) Using induction on n, show that a set of equidistant points in S^{n-1} contains no more than $n + 1$ elements.

9. In an inner product space, show the *parallelogram rule*

$$\|x - y\|^2 + \|x + y\|^2 = 2 \|x\|^2 + 2 \|y\|^2 .$$

Here x and y describe the sides in a parallelogram and $x + y$ and $x - y$ the diagonals.

10. In a complex inner product space, show that

$$4 (x|y) = \sum_{k=0}^{3} i^k \|x + i^k y\|^2 .$$

3.3 Orthonormal Bases

Let us fix an inner product space V.

Definition 3.3.1. A possibly infinite collection $e_1, \ldots, e_n, \ldots$ of vectors in V is said to be *orthogonal* if $(e_i|e_j) = 0$ for $i \neq j$. If in addition these vectors are of unit length, i.e., $(e_i|e_j) = \delta_{ij}$, then we call the collection *orthonormal*.

The usual bases for $\mathbb{R}^n$ and $\mathbb{C}^n$ are evidently orthonormal collections. Since they are also bases, we call them *orthonormal bases*.

Lemma 3.3.2. *Let $e_1, \ldots, e_n$ be orthonormal. Then, $e_1, \ldots, e_n$ are linearly independent and any element $x \in \text{span} \{e_1, \ldots, e_n\}$ has the expansion*

$$x = (x|e_1) e_1 + \cdots (x|e_n) e_n.$$

Proof. Note that if $x = \alpha_1 e_1 + \cdots + \alpha_n e_n$, then

$$\begin{aligned}
(x|e_i) &= (\alpha_1 e_1 + \cdots + \alpha_n e_n|e_i) \\
&= \alpha_1 (e_1|e_i) + \cdots + \alpha_n (e_n|e_i) \\
&= \alpha_1 \delta_{1i} + \cdots + \alpha_n \delta_{ni} \\
&= \alpha_i .
\end{aligned}$$

In case $x = 0$, this gives us linear independence, and in case $x \in \text{span} \{e_1, \ldots, e_n\}$, we have computed the ith coordinate using the inner product. □

We shall use the equation

$$(\alpha_1 e_1 + \cdots + \alpha_n e_n|e_i) = \alpha_i$$

from the above proof repeatedly throughout the next two chapters.

This allows us to construct a special isomorphism between $\text{span} \{e_1, \ldots, e_n\}$ and $\mathbb{F}^n$.

Definition 3.3.3. We say that two inner product spaces V and W over $\mathbb{F}$ are *isometric*, if we can find an *isometry* $L : V \to W$, i.e., an isomorphism such that $(L(x)|L(y)) = (x|y)$.

Lemma 3.3.4. *If V admits a basis that is orthonormal, then V is isometric to $\mathbb{F}^n$.*

Proof. Choose an orthonormal basis $e_1, \ldots, e_n$ for V and define the usual isomorphism $L : \mathbb{F}^n \to V$ by

$$L\left(\begin{bmatrix} \alpha_1 \\ \vdots \\ \alpha_n \end{bmatrix}\right) = [e_1 \cdots e_n] \begin{bmatrix} \alpha_1 \\ \vdots \\ \alpha_n \end{bmatrix}$$

$$= \alpha_1 e_1 + \cdots + \alpha_n e_n.$$

Let

$$a = \begin{bmatrix} \alpha_1 \\ \vdots \\ \alpha_n \end{bmatrix} \text{ and } b = \begin{bmatrix} \beta_1 \\ \vdots \\ \beta_n \end{bmatrix}$$

then

$$(L(a)|L(b)) = (L(a)|\beta_1 e_1 + \cdots + \beta_n e_n)$$
$$= \bar{\beta}_1 (L(a)|e_1) + \cdots + \bar{\beta}_n (L(a)|e_n)$$
$$= \bar{\beta}_1 (\alpha_1 e_1 + \cdots + \alpha_n e_n|e_1) + \cdots + \bar{\beta}_n (\alpha_1 e_1 + \cdots + \alpha_n e_n|e_n)$$
$$= \bar{\beta}_1 \alpha_1 + \cdots + \bar{\beta}_n \alpha_n$$
$$= (a|b).$$

which is what we wanted to prove. □

Remark 3.3.5. Note that the inverse map that computes the coordinates of a vector is explicitly given by

$$L^{-1}(x) = \begin{bmatrix} (x|e_1) \\ \vdots \\ (x|e_n) \end{bmatrix}.$$

We are now left with the nagging possibility that orthonormal bases might be very special and possibly not exist.

The procedure for constructing orthonormal collections is known as the *Gram–Schmidt procedure*. It is not clear who invented the process, but these two people definitely promoted and used it to great effect.

Given a linearly independent set $x_1, \ldots, x_m$ in an inner product space V, it is possible to construct an orthonormal collection $e_1, \ldots, e_m$ such that

$$\text{span}\{x_1, \ldots, x_m\} = \text{span}\{e_1, \ldots, e_m\}.$$

The procedure is actually iterative and creates $e_1, \ldots, e_m$ in such a way that

$$\text{span}\{x_1\} = \text{span}\{e_1\},$$
$$\text{span}\{x_1, x_2\} = \text{span}\{e_1, e_2\},$$
$$\vdots \qquad \vdots$$
$$\text{span}\{x_1, \ldots, x_m\} = \text{span}\{e_1, \ldots, e_m\}.$$

This basically forces us to define e_1 as

$$e_1 = \frac{1}{\|x_1\|} x_1.$$

Then, e_2 is constructed by considering

$$\begin{aligned}
z_2 &= x_2 - \text{proj}_{x_1}(x_2) \\
&= x_2 - \text{proj}_{e_1}(x_2) \\
&= x_2 - (x_2|e_1)\, e_1,
\end{aligned}$$

and defining

$$e_2 = \frac{1}{\|z_2\|} z_2.$$

Having constructed an orthonormal set $e_1, \ldots, e_k$, we can then define

$$z_{k+1} = x_{k+1} - (x_{k+1}|e_1)\, e_1 - \cdots - (x_{k+1}|e_k)\, e_k.$$

As

$$\text{span}\{x_1, \ldots, x_k\} = \text{span}\{e_1, \ldots, e_k\},$$
$$x_{k+1} \notin \text{span}\{x_1, \ldots, x_k\},$$

we have that $z_{k+1} \neq 0$. Thus, we can define

$$e_{k+1} = \frac{1}{\|z_{k+1}\|} z_{k+1}.$$

To see that e_{k+1} is perpendicular to $e_1, \ldots, e_k$, we note that

$$\begin{aligned}
(e_{k+1}|e_i) &= \frac{1}{\|z_{k+1}\|}\, (z_{k+1}|e_i) \\
&= \frac{1}{\|z_{k+1}\|}\, (x_{k+1}|e_i) - \frac{1}{\|z_{k+1}\|}\, ((x_{k+1}|e_1)\, e_1 + \cdots + (x_{k+1}|e_k)\, e_k \,|\, e_i)
\end{aligned}$$

$$= \frac{1}{\|z_{k+1}\|} (x_{k+1}|e_i) - \frac{1}{\|z_{k+1}\|} (x_{k+1}|e_i)$$

$$= 0.$$

Since

$$\text{span}\{x_1\} = \text{span}\{e_1\},$$

$$\text{span}\{x_1, x_2\} = \text{span}\{e_1, e_2\},$$

$$\vdots \qquad \vdots$$

$$\text{span}\{x_1, \ldots, x_m\} = \text{span}\{e_1, \ldots, e_m\},$$

we have constructed $e_1, \ldots, e_m$ in such a way that

$$[\, e_1 \cdots e_m \,] = [\, x_1 \cdots x_m \,] B,$$

where B is an upper triangular $m \times m$ matrix with positive diagonal entries. Conversely, we have

$$[\, x_1 \cdots x_m \,] = [\, e_1 \cdots e_m \,] R,$$

where $R = B^{-1}$ is also upper triangular with positive diagonal entries. Given that we have a formula for the expansion of each x_k in terms of $e_1, \ldots, e_k$, we see that

$$R = \begin{bmatrix} (x_1|e_1) & (x_2|e_1) & (x_3|e_1) & \cdots & (x_m|e_1) \\ 0 & (x_2|e_2) & (x_3|e_2) & \cdots & (x_m|e_2) \\ 0 & 0 & (x_3|e_3) & \cdots & (x_m|e_3) \\ \vdots & \vdots & \vdots & \ddots & \vdots \\ 0 & 0 & 0 & \cdots & (x_m|e_m) \end{bmatrix}.$$

We often abbreviate

$$A = [\, x_1 \cdots x_m \,],$$

$$Q = [\, e_1 \cdots e_m \,]$$

and obtain the *QR-factorization* $A = QR$. In case V is $\mathbb{R}^n$ or $\mathbb{C}^n$ A is a general $n \times m$ matrix of rank m, Q is also an $n \times m$ matrix of rank m with the added feature that its columns are orthonormal, and R is an upper triangular $m \times m$ matrix. Note that in this interpretation, the QR-factorization is an improved Gauss elimination: $A = PU$, $P \in Gl_n$ and U upper triangular (see Sect. 1.13).

With that in mind, it is not surprising that the QR-factorization gives us a way of inverting the linear map

$$[\, x_1 \cdots x_n \,] : \mathbb{F}^n \to V$$

when $x_1, \ldots, x_n$ is a basis. First, recall that the isometry

$$\begin{bmatrix} e_1 \cdots e_n \end{bmatrix} : \mathbb{F}^n \to V$$

is easily inverted and the inverse can be symbolically represented as

$$\begin{bmatrix} e_1 \cdots e_n \end{bmatrix}^{-1} = \begin{bmatrix} \overline{(e_1|\cdot)} \\ \vdots \\ \overline{(e_n|\cdot)} \end{bmatrix},$$

or more precisely

$$\begin{bmatrix} e_1 \cdots e_n \end{bmatrix}^{-1}(x) = \begin{bmatrix} \overline{(e_1|x)} \\ \vdots \\ \overline{(e_n|x)} \end{bmatrix}$$

$$= \begin{bmatrix} (x|e_1) \\ \vdots \\ (x|e_n) \end{bmatrix}.$$

This is the great feature of orthonormal bases, namely, that one has an explicit formula for the coordinates in such a basis. Next on the agenda is the construction of R^{-1}. Given that it is upper triangular, this is a reasonably easy problem in the theory of solving linear systems. However, having found the orthonormal basis through Gram–Schmidt, we have already found this inverse since

$$\begin{bmatrix} x_1 \cdots x_n \end{bmatrix} = \begin{bmatrix} e_1 \cdots e_n \end{bmatrix} R$$

implies that

$$\begin{bmatrix} e_1 \cdots e_n \end{bmatrix} = \begin{bmatrix} x_1 \cdots x_n \end{bmatrix} R^{-1}$$

and the goal of the process was to find $e_1, \ldots, e_n$ as a linear combination of $x_1, \ldots, x_n$. Thus, we obtain the formula

$$\begin{bmatrix} x_1 \cdots x_n \end{bmatrix}^{-1} = R^{-1} \begin{bmatrix} e_1 \cdots e_n \end{bmatrix}^{-1}$$

$$= R^{-1} \begin{bmatrix} \overline{(e_1|\cdot)} \\ \vdots \\ \overline{(e_n|\cdot)} \end{bmatrix}.$$

The Gram–Schmidt process, therefore, not only gives us an orthonormal basis but it also gives us a formula for the coordinates of a vector with respect to the original basis.

It should also be noted that if we start out with a set $x_1, \ldots, x_m$ that is not linearly independent, then this fact will be revealed in the process of constructing $e_1, \ldots, e_m$.

We know from Lemma 1.12.3 that either $x_1 = 0$ or there is a smallest k such that x_{k+1} is a linear combination of $x_1, \ldots, x_k$. In the latter case, we get to construct $e_1, \ldots, e_k$ since $x_1, \ldots, x_k$ were linearly independent. As $x_{k+1} \in \text{span}\{e_1, \ldots, e_k\}$, we must have that

$$z_{k+1} = x_{k+1} - (x_{k+1}|e_1)\, e_1 - \cdots - (x_{k+1}|e_k)\, e_k = 0$$

since the way in which x_{k+1} is expanded in terms of $e_1, \ldots, e_k$ is given by

$$x_{k+1} = (x_{k+1}|e_1)\, e_1 + \cdots + (x_{k+1}|e_k)\, e_k.$$

Thus, we fail to construct the unit vector e_{k+1}.

With all this behind us, we have proved the important result.

Theorem 3.3.6. (Uniqueness of Inner Product Spaces) *An n-dimensional inner product space over $\mathbb{R}$, respectively $\mathbb{C}$, is isometric to $\mathbb{R}^n$, respectively $\mathbb{C}^n$.*

Definition 3.3.7. The *operator norm*, for a linear map $L : V \to W$ between inner product spaces is defined as

$$\|L\| = \sup_{\|x\|=1} \|L(x)\|.$$

The operator norm is finite provided V is finite-dimensional.

Theorem 3.3.8. *Let $L : V \to W$ be a linear map. Then,*

$$\|L(x)\| \le \|L\|\, \|x\|$$

for all $x \in V$. And if V is a finite-dimensional inner product space, then

$$\|L\| = \sup_{\|x\|=1} \|L(x)\| < \infty.$$

Proof. To establish the first claim, we only need to consider $x \in V - \{0\}$. Then,

$$\left\| L\left(\frac{x}{\|x\|}\right) \right\| \le \|L\|,$$

and the claim follows by using linearity of L and the scaling property of the norm.

When V is finite-dimensional select an orthonormal basis $e_1, \ldots, e_n$ for V. Then, by using the Cauchy–Schwarz inequality (Corollary 3.2.10) and the triangle inequality (Corollary 3.2.11), we obtain

$$\|L(x)\| = \left\| L\left(\sum_{i=1}^{n} (x|e_i)\, e_i\right) \right\|$$

$$= \left\| \sum_{i=1}^{n} (x|e_i)\, L(e_i) \right\|$$

$$\leq \sum_{i=1}^{n} |(x|e_i)| \, \|L(e_i)\|$$

$$\leq \sum_{i=1}^{n} \|x\| \, \|L(e_i)\|$$

$$= \left(\sum_{i=1}^{n} \|L(e_i)\| \right) \|x\| \, .$$

Thus,

$$\|L\| \leq \sum_{i=1}^{n} \|L(e_i)\| \, . \qquad \qquad \square$$

To finish the section, let us try to do a few concrete examples.

Example 3.3.9. Consider the vectors $x_1 = (1,1,0)$, $x_2 = (1,0,1)$, and $x_3 = (0,1,1,)$ in $\mathbb{R}^3$. If we perform Gram–Schmidt, then the QR factorization is

$$\begin{bmatrix} 1 & 1 & 0 \\ 1 & 0 & 1 \\ 0 & 1 & 1 \end{bmatrix} = \begin{bmatrix} \frac{1}{\sqrt{2}} & \frac{1}{\sqrt{6}} & -\frac{1}{\sqrt{3}} \\ \frac{1}{\sqrt{2}} & -\frac{1}{\sqrt{6}} & \frac{1}{\sqrt{3}} \\ 0 & \frac{2}{\sqrt{6}} & \frac{1}{\sqrt{3}} \end{bmatrix} \begin{bmatrix} \sqrt{2} & \frac{1}{\sqrt{2}} & \frac{1}{\sqrt{2}} \\ 0 & \frac{3}{\sqrt{6}} & \frac{1}{\sqrt{6}} \\ 0 & 0 & \frac{2}{\sqrt{3}} \end{bmatrix} \, .$$

Example 3.3.10. The Legendre polynomials of degrees 0, 1, and 2 on $[-1, 1]$ are by definition the polynomials obtained via Gram–Schmidt from $1, t, t^2$ with respect to the inner product

$$(f|g) = \int_{-1}^{1} f(t) \, \overline{g(t)} dt.$$

We see that $\|1\| = \sqrt{2}$, so the first polynomial is

$$p_0(t) = \frac{1}{\sqrt{2}}.$$

To find $p_1(t)$, we first find

$$z_1 = t - (t|p_0) \, p_0$$

$$= t - \left(\int_{-1}^{1} t \frac{1}{\sqrt{2}} dt \right) \frac{1}{\sqrt{2}}$$

$$= t.$$

Then,

$$p_1(t) = \frac{t}{\|t\|} = \sqrt{\frac{3}{2}} t.$$

Finally, for p_2, we find

$$z_2 = t^2 - \left(t^2|p_0\right) p_0 - \left(t^2|p_1\right) p_1$$

$$= t^2 - \left(\int_{-1}^1 t^2 \frac{1}{\sqrt{2}} dt\right) \frac{1}{\sqrt{2}} - \left(\int_{-1}^1 t^2 \sqrt{\frac{3}{2}} t \, dt\right) \sqrt{\frac{3}{2}} t$$

$$= t^2 - \frac{1}{3}.$$

Thus,

$$p_2(t) = \frac{t^2 - \frac{1}{3}}{\left\| t^2 - \frac{1}{3} \right\|}$$

$$= \sqrt{\frac{45}{8}} \left(t^2 - \frac{1}{3} \right).$$

Example 3.3.11. A system of real equations $Ax = b$ can be interpreted geometrically as n equations

$$(a_1|x) = \beta_1,$$

$$\vdots \qquad \vdots$$

$$(a_n|x) = \beta_n,$$

where a_k is the kth row in A and β_k the kth coordinate for b. The solutions will be the intersection of the n hyperplanes $H_k = \{z : (a_k|z) = \beta_k\}$.

Example 3.3.12. We wish to show that the trigonometric functions

$$1 = \cos(0 \cdot t), \cos(t), \cos(2t), \ldots, \sin(t), \sin(2t), \ldots$$

are orthogonal in $C_{2\pi}^\infty(\mathbb{R}, \mathbb{R})$ with respect to the inner product

$$(f|g) = \frac{1}{2\pi} \int_{-\pi}^\pi f(t) g(t) \, dt.$$

First, observe that $\cos(mt)\sin(nt)$ is an odd function. This proves that

$$(\cos(mt) | \sin(nt)) = 0.$$

Thus, we are reduced to showing that each of the two sequences

$$1, \cos(t), \cos(2t), \ldots$$

$$\sin(t), \sin(2t), \ldots$$

are orthogonal. Using integration by parts, we see

$$(\cos(mt) \mid \cos(nt))$$

$$= \frac{1}{2\pi} \int_{-\pi}^{\pi} \cos(mt) \cos(nt) \, dt$$

$$= \frac{1}{2\pi} \frac{\sin(mt)}{m} \cos(nt) \Big|_{-\pi}^{\pi} - \frac{1}{2\pi} \int_{-\pi}^{\pi} \frac{\sin(mt)}{m}(-n) \sin(nt) \, dt$$

$$= \frac{n}{m} \frac{1}{2\pi} \int_{-\pi}^{\pi} \sin(mt) \sin(nt) \, dt$$

$$= \frac{n}{m} (\sin(mt) \mid \sin(nt))$$

$$= \frac{n}{m} \frac{1}{2\pi} \frac{-\cos(mt)}{m} \sin(nt) \Big|_{-\pi}^{\pi} - \frac{n}{m} \frac{1}{2\pi} \int_{-\pi}^{\pi} \frac{-\cos(mt)}{m} n \cos(nt) \, dt$$

$$= \left(\frac{n}{m}\right)^2 \frac{1}{2\pi} \int_{-\pi}^{\pi} \cos(mt) \cos(nt) \, dt$$

$$= \left(\frac{n}{m}\right)^2 (\cos(mt) \mid \cos(nt)).$$

When $n \neq m$ and $m > 0$, this clearly proves that $(\cos(mt) \mid \cos(nt)) = 0$ and in addition that $(\sin(mt) \mid \sin(nt)) = 0$. Finally, let us compute the norm of these functions. Clearly, $\|1\| = 1$. We just proved that $\|\cos(mt)\| = \|\sin(mt)\|$. This combined with the fact that

$$\sin^2(mt) + \cos^2(mt) = 1$$

shows that

$$\|\cos(mt)\| = \|\sin(mt)\| = \frac{1}{\sqrt{2}}$$

Example 3.3.13. Let us try to do Gram–Schmidt on $1, \cos t, \cos^2 t$ using the above inner product. We already know that the first two functions are orthogonal, so

$$e_1 = 1,$$

$$e_2 = \sqrt{2}\cos(t).$$

$$z_2 = \cos^2(t) - \left(\cos^2(t) \mid 1\right) 1 - \left(\cos^2(t) \mid \sqrt{2}\cos(t)\right) \sqrt{2}\cos(t)$$

$$= \cos^2(t) - \frac{1}{2\pi} \left(\int_{-\pi}^{\pi} \cos^2(t) \, dt\right) - \frac{2}{2\pi} \left(\int_{-\pi}^{\pi} \cos^2(t) \cos(t) \, dt\right) \cos t$$

$$= \cos^2(t) - \frac{1}{2} - \frac{1}{\pi} \left(\int_{-\pi}^{\pi} \cos^3(t)\, dt \right) \cos t$$

$$= \cos^2(t) - \frac{1}{2}$$

Thus, the third function is

$$e_3 = \frac{\cos^2(t) - \frac{1}{2}}{\left\| \cos^2(t) - \frac{1}{2} \right\|}$$

$$= 2\sqrt{2}\cos^2(t) - \sqrt{2}.$$

Exercises

1. Use Gram–Schmidt on the vectors

$$\begin{bmatrix} x_1 & x_2 & x_3 & x_4 & x_5 \end{bmatrix} = \begin{bmatrix} \sqrt{5} & -2 & 4 & e & 3 \\ 0 & 8 & \pi & 2 & -10 \\ 0 & 0 & 1+\sqrt{2} & 3 & -4 \\ 0 & 0 & 0 & -2 & 6 \\ 0 & 0 & 0 & 0 & 1 \end{bmatrix}$$

   to obtain an orthonormal basis for $\mathbb{F}^5$.
2. Find an orthonormal basis for $\mathbb{R}^3$ where the first vector is proportional to $(1,1,1)$.
3. Apply Gram–Schmidt to the collection $x_1 = (1,0,1,0)$, $x_2 = (1,1,1,0)$, $x_3 = (0,1,0,0)$.
4. Apply Gram–Schmidt to the collection $x_1 = (1,0,1,0)$, $x_2 = (0,1,1,0)$, $x_3 = (0,1,0,1)$ and complete to an orthonormal basis for $\mathbb{R}^4$.
5. Apply Gram–Schmidt to $\sin t$, $\sin^2 t$, $\sin^3 t$ using the inner product

$$(f|g) = \frac{1}{2\pi} \int_{-\pi}^{\pi} f(t)\, g(t)\, dt.$$

6. Given an arbitrary collection of vectors $x_1, \ldots, x_m$ in an inner product space V, show that it is possible to find orthogonal vectors $z_1, \ldots, z_n \in V$ such that

$$\begin{bmatrix} x_1 & \cdots & x_m \end{bmatrix} = \begin{bmatrix} z_1 & \cdots & z_n \end{bmatrix} A_{\text{ref}},$$

   where A_{ref} is an $n \times m$ matrix in row echelon form (see Sect. 1.13). Explain how this can be used to solve systems of the form

$$\begin{bmatrix} x_1 & \cdots & x_m \end{bmatrix} \begin{bmatrix} \xi_1 \\ \vdots \\ \xi_m \end{bmatrix} = b.$$

7. The goal of this exercise is to understand the dual basis to a basis $x_1, \ldots, x_n$ for an inner product space V. We say that $x_1^*, \ldots, x_n^*$ is dual to $x_1, \ldots, x_n$ if $(x_i | x_j^*) = \delta_{ij}$.

   (a) Show that each basis has a unique dual basis (you have to show it exists, that it is a basis, and that there is only one such basis).
   (b) Show that if $x_1, \ldots, x_n$ is a basis and $L : \mathbb{F}^n \to V$ is the usual coordinate isomorphism given by

   $$L\left(\begin{bmatrix} \alpha_1 \\ \vdots \\ \alpha_n \end{bmatrix}\right) = \begin{bmatrix} x_1 & \cdots & x_n \end{bmatrix} \begin{bmatrix} \alpha_1 \\ \vdots \\ \alpha_n \end{bmatrix}$$
   $$= \alpha_1 x_1 + \cdots + \alpha_n x_n,$$

   then its inverse is given by

   $$L^{-1}(x) = \begin{bmatrix} (x | x_1^*) \\ \vdots \\ (x | x_n^*) \end{bmatrix}.$$

   (c) Show that a basis is orthonormal if and only if it is *self-dual*, i.e., it is its own dual basis $x_i = x_i^*, i = 1, \ldots, n$.
   (d) Given $(1, 1, 0), (1, 0, 1), (0, 1, 1) \in \mathbb{R}^3$, find the dual basis.
   (e) Find the dual basis for $1, t, t^2 \in P_2$ with respect to the inner product

   $$(f | g) = \int_{-1}^1 f(t) g(t) \, dt$$

8. Using the inner product

   $$(f | g) = \int_0^1 f(t) g(t) \, dt$$

   on $\mathbb{R}[t]$, apply Gram–Schmidt to $1, t, t^2$ to find an orthonormal basis for P_2.

9. *(Legendre Polynomials)* Consider the inner product

   $$(f | g) = \int_a^b f(t) g(t) \, dt$$

   on $\mathbb{R}[t]$ and define

   $$q_{2n}(t) = (t - a)^n (t - b)^n,$$
   $$p_n(t) = \frac{d^n}{dt^n} (q_{2n}(t)).$$

(a) Show that

$$q_{2n}(a) = q_{2n}(b) = 0,$$

$$\vdots$$

$$\frac{d^{n-1}q_{2n}}{dt^{n-1}}(a) = \frac{d^{n-1}q_{2n}}{dt^{n-1}}(b) = 0.$$

(b) Show that p_n has degree n.
(c) Use induction on n to show that $p_n(t)$ is perpendicular to $1, t, \ldots, t^{n-1}$. Hint: Use integration by parts.
(d) Show that $p_0, p_1, \ldots, p_n, \ldots$ are orthogonal to each other.

10. *(Lagrange Interpolation)* Select $n+1$ distinct points $t_0, \ldots, t_n \in \mathbb{C}$ and consider

$$(p(t) | q(t)) = \sum_{i=0}^{n} p(t_i) \overline{q(t_i)}.$$

(a) Show that this defines an inner product on P_n but not on $\mathbb{C}[t]$.
(b) Consider

$$p_0(t) = \frac{(t - t_1)(t - t_2) \cdots (t - t_n)}{(t_0 - t_1)(t_0 - t_2) \cdots (t_0 - t_n)},$$

$$p_1(t) = \frac{(t - t_0)(t - t_2) \cdots (t - t_n)}{(t_1 - t_0)(t_1 - t_2) \cdots (t_1 - t_n)},$$

$$\vdots$$

$$p_n(t) = \frac{(t - t_0)(t - t_1) \cdots (t - t_{n-1})}{(t_n - t_0)(t_n - t_1) \cdots (t_n - t_{n-1})}.$$

Show that $p_i(t_j) = \delta_{ij}$ and that $p_0, \ldots, p_n$ form an orthonormal basis for P_n.
(c) Use $p_0, \ldots, p_n$ to solve the problem of finding a polynomial $p \in P_n$ such that $p(t_i) = b_i$.
(d) Let $\lambda_1, \ldots, \lambda_n \in \mathbb{C}$ (they may not be distinct) and $f : \mathbb{C} \to \mathbb{C}$ a function. Show that there is a polynomial $p(t) \in \mathbb{C}[t]$ such that $p(\lambda_1) = f(\lambda_1), \ldots, p(\lambda_n) = f(\lambda_n)$.

11. *(P. Enflo)* Let V be a finite-dimensional inner product space and $x_1, \ldots, x_n$, $y_1, \ldots, y_n \in V$. Show Enflo's inequality

$$\left(\sum_{i,j=1}^{n} |(x_i | y_j)|^2 \right)^2 \le \left(\sum_{i,j=1}^{n} |(x_i | x_j)|^2 \right) \left(\sum_{i,j=1}^{n} |(y_i | y_j)|^2 \right).$$

Hint: Use an orthonormal basis and start expanding on the left-hand side.

12. Let $L : V \to V$ be an operator on a finite-dimensional inner product space.

    (a) If λ is an eigenvalue for L, then

$$|\lambda| \le \|L\| .$$

    (b) Given examples of 2×2 matrices where strict inequality always holds.

13. Let $L : V_1 \to V_2$ and $K : V_2 \to V_3$ be linear maps between finite-dimensional inner product spaces. Show that

$$\|K \circ L\| \le \|K\| \, \|L\| .$$

14. Let $L, K : V \to V$ be operators on a finite-dimensional inner product space. If K is invertible, show that

$$\|L\| = \left\| K \circ L \circ K^{-1} \right\| .$$

15. Let $L, K : V \to W$ be linear maps between finite-dimensional inner product spaces. Show that

$$\|L + K\| \le \|L\| + \|K\| .$$

16. Let $A \in \mathrm{Mat}_{n \times m} (\mathbb{F})$. Show that

$$\left| \alpha_{ij} \right| \le \|A\| ,$$

    where $\|A\|$ is the operator norm of the linear map $A : \mathbb{F}^m \to \mathbb{F}^n$. Give examples where

$$\|A\| \ne \sqrt{\mathrm{tr} \, (AA^*)} = \sqrt{(A|A)}.$$

3.4   Orthogonal Complements and Projections

The goal of this section is to figure out if there is a best possible projection onto a subspace of a vector space. In general, there are quite a lot of projections, but if we have an inner product on the vector space, we can imagine that there should be a projection where the image of a vector is as close as possible to the original vector.

Let $M \subset V$ be a finite-dimensional subspace of an inner product space. From the previous section, we know that it is possible to find an orthonormal basis $e_1, \dots, e_m$ for M. Using that basis, we define $E : V \to V$ by

$$E (x) = (x|e_1) e_1 + \cdots + (x|e_m) e_m.$$

Note that $E (z) \in M$ for all $z \in V$. Moreover, if $x \in M$, then $E (x) = x$. Thus $E^2 (z) = E (z)$ for all $z \in V$. This shows that E is a projection whose image is M. Next, let us identify the kernel. If $x \in \ker (E)$, then

$$0 = E(x)$$
$$= (x|e_1)e_1 + \cdots + (x|e_m)e_m.$$

Since $e_1, \ldots, e_m$, is a basis this means that $(x|e_1) = \cdots = (x|e_m) = 0$. This in turn is equivalent to the condition

$$(x|z) = 0 \text{ for all } z \in M,$$

since any $z \in M$ is a linear combination of $e_1, \ldots, e_m$.

Definition 3.4.1. The set of all such vectors is denoted

$$M^{\perp} = \{x \in V : (x|z) = 0 \text{ for all } z \in M\}$$

and is called the *orthogonal complement* to M in V.

Given that $\ker(E) = M^{\perp}$, we have a formula for the kernel that does not depend on E. Thus, E is simply the projection of V onto M along $M^{\perp}$. The only problem with this characterization is that we do not know from the outset that $V = M \oplus M^{\perp}$. In case M is finite-dimensional, however, the existence of the projection E insures us that this must be the case as

$$x = E(x) + (1_V - E)(x)$$

and $(1_V - E)(x) \in \ker(E) = M^{\perp}$.

Definition 3.4.2. When there is an orthogonal direct sum decomposition: $V = M \oplus M^{\perp}$ we call the projection onto M along $M^{\perp}$ the *orthogonal projection* onto M and denote it by $\text{proj}_M : V \to V$.

The vector $\text{proj}_M(x)$ also solves our problem of finding the vector in M that is closest to x. To see why this is true, choose $z \in M$ and consider the triangle that has the three vectors x, $\text{proj}_M(x)$, and z as vertices. The sides are given by $x - \text{proj}_M(x)$, $\text{proj}_M(x) - z$, and $z - x$ (see Fig. 3.5). Since $\text{proj}_M(x) - z \in M$ and $x - \text{proj}_M(x) \in M^{\perp}$, these two vectors are perpendicular, and hence we have

$$\|x - \text{proj}_M(x)\|^2 \le$$
$$\|x - \text{proj}_M(x)\|^2 + \|\text{proj}_M(x) - z\|^2 = \|x - z\|^2,$$

where equality holds only when $\|\text{proj}_M(x) - z\|^2 = 0$, i.e., $\text{proj}_M(x)$ is the one and only point closest to x among all points in M.

Let us collect the above information in a theorem.

Theorem 3.4.3. (Orthogonal Sum Decomposition) *Let V be an inner product space and $M \subset V$ a finite-dimensional subspace. Then, $V = M \oplus M^{\perp}$ and for any orthonormal basis $e_1, \ldots, e_m$ for M, the projection onto M along $M^{\perp}$ is given by:*

$$\text{proj}_M(x) = (x|e_1)e_1 + \cdots + (x|e_m)e_m.$$

Fig. 3.5 Orthogonal
projection

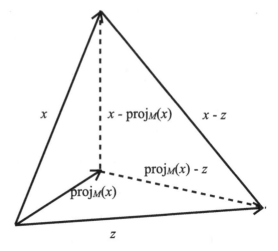

Corollary 3.4.4. *If V is a finite-dimensional inner product space and $M \subset V$ is a subspace, then*

$$V = M \oplus M^{\perp},$$

$$\left(M^{\perp}\right)^{\perp} = M^{\perp\perp} = M,$$

$$\dim V = \dim M + \dim M^{\perp}.$$

Proof. The first statement was proven in Theorem 3.4.3. The third statement now follows from Corollary 1.10.14. To prove the second statement, select an orthonormal basis $e_1, \ldots, e_k$ for M and $e_{k+1}, \ldots, e_n$ for $M^{\perp}$. Then, we see that $e_1, \ldots, e_k \in M^{\perp\perp}$ and consequently $M \subset M^{\perp\perp}$. On the other hand, note that, if we apply the first and third statements to $M^{\perp}$ instead if M, then we obtain

$$\dim V = \dim M^{\perp} + \dim M^{\perp\perp}.$$

In particular, we have $\dim M = \dim M^{\perp\perp}$ since $M \subset M^{\perp\perp}$, this proves the claim.
□

Orthogonal projections can also be characterized as follows.

Theorem 3.4.5. (Characterization of Orthogonal Projections) *Assume that V is a finite-dimensional inner product space and $E : V \to V$ a projection onto $M \subset V$. Then, the following conditions are equivalent:*

(1) $E = \text{proj}_M$
(2) $\text{im}\,(E)^{\perp} = \ker\,(E)$
(3) $\|E\,(x)\| \le \|x\|$ *for all* $x \in V$

Proof. We have already seen that the first two conditions are equivalent. These two conditions imply the third as $x = E\,(x) + (1_V - E)\,(x)$ is an orthogonal decomposition, and thus,

$$\|x\|^2 = \|E(x)\|^2 + \|(1_V - E)(x)\|^2$$
$$\geq \|E(x)\|^2.$$

It remains to be seen that the third condition implies that E is orthogonal. To prove this, choose $x \in \ker(E)^{\perp}$ and observe that $E(x) = x - (1_V - E)(x)$ is an orthogonal decomposition since $(1_V - E)(z) \in \ker(E)$ for all $z \in V$. Thus,

$$\|x\|^2 \geq \|E(x)\|^2$$
$$= \|x - (1_V - E)(x)\|^2$$
$$= \|x\|^2 + \|(1_V - E)(x)\|^2$$
$$\geq \|x\|^2$$

This means that $(1_V - E)(x) = 0$ and hence $x = E(x) \in \mathrm{im}(E)$. Thus, $\ker(E)^{\perp} \subset \mathrm{im}(E)$. We also know from the dimension formula (Theorem 1.11.7) and Corollary 3.4.4 that

$$\dim(\mathrm{im}(E)) = \dim(V) - \dim(\ker(E))$$
$$= \dim\left(\ker(E)^{\perp}\right).$$

This shows that $\ker(E)^{\perp} = \mathrm{im}(E)$. $\qquad\square$

Example 3.4.6. Let $V = \mathbb{R}^n$ and $M = \mathrm{span}\{(1,\ldots,1)\}$. Since $\|(1,\ldots,1)\|^2 = n$, we see that

$$\mathrm{proj}_M(x) = \mathrm{proj}_M\left(\begin{bmatrix} \alpha_1 \\ \vdots \\ \alpha_1 \end{bmatrix}\right)$$

$$= \frac{1}{n}\left(\begin{bmatrix} \alpha_1 \\ \vdots \\ \alpha_1 \end{bmatrix}\middle|\begin{bmatrix} 1 \\ \vdots \\ 1 \end{bmatrix}\right)\begin{bmatrix} 1 \\ \vdots \\ 1 \end{bmatrix}$$

$$= \frac{\alpha_1 + \cdots + \alpha_n}{n}\begin{bmatrix} 1 \\ \vdots \\ 1 \end{bmatrix}$$

$$= \bar{\alpha}\begin{bmatrix} 1 \\ \vdots \\ 1 \end{bmatrix},$$

where $\bar{\alpha}$ is the *average* or *mean* of the values $\alpha_1, \ldots, \alpha_n$. Since $\text{proj}_M(x)$ is the closest element in M to x, we get a geometric interpretation of the average of $\alpha_1, \ldots, \alpha_n$. If in addition we use that $\text{proj}_M(x)$ and $x - \text{proj}_M(x)$ are perpendicular, we arrive at a nice formula that helps compute the *variance*:

$$\text{Var}(\alpha_1, \ldots, \alpha_n) = \frac{1}{n-1} \sum_{i=1}^{n} |\alpha_i - \bar{\alpha}|^2,$$

where

$$\sum_{i=1}^{n} |\alpha_i - \bar{\alpha}|^2 = \|x - \text{proj}_M(x)\|^2$$

$$= \|x\|^2 - \|\text{proj}_M(x)\|^2$$

$$= \sum_{i=1}^{n} |\alpha_i|^2 - \sum_{i=1}^{n} |\bar{\alpha}|^2$$

$$= \left(\sum_{i=1}^{n} |\alpha_i|^2 \right) - n|\bar{\alpha}|^2$$

$$= \left(\sum_{i=1}^{n} |\alpha_i|^2 \right) - \frac{\left(\sum_{i=1}^{n} \alpha_i \right)^2}{n}.$$

Example 3.4.7. As above, let $M \subset V$ be a finite-dimensional subspace of an inner product space and $e_1, \ldots, e_m$ an orthonormal basis for M. Using the formula

$$\text{proj}_M(x) = (x|e_1)e_1 + \cdots + (x|e_m)e_m,$$

the inequality $\|\text{proj}_M(x)\| \le \|x\|$ translates into the *Bessel inequality*

$$|(x|e_1)|^2 + \cdots + |(x|e_m)|^2 \le \|x\|^2.$$

Exercises

1. Consider $\text{Mat}_{n \times n}(\mathbb{C})$ with the inner product $(A|B) = \text{tr}(AB^*)$. Describe the orthogonal complement to the space of all diagonal matrices.
2. Show that if $M = \text{span}\{z_1, \ldots, z_m\}$, then

$$M^{\perp} = \{x \in V : (x|z_1) = \cdots = (x|z_m) = 0\}.$$

3. Assume $V = M \oplus M^{\perp}$, show that

$$x = \text{proj}_M(x) + \text{proj}_{M^{\perp}}(x).$$

4. Find the element in span $\{1, \cos t, \sin t\}$ that is closest to $\sin^2 t$ using the inner product

$$(f|g) = \frac{1}{2\pi} \int_{-\pi}^{\pi} f(t) g(t) dt.$$

5. Assume $V = M \oplus M^\perp$ and that $L : V \to V$ is a linear operator. Show that both M and $M^\perp$ are L-invariant if and only if $\mathrm{proj}_M \circ L = L \circ \mathrm{proj}_M$.

6. Let $A \in \mathrm{Mat}_{m \times n}(\mathbb{R})$.

   a. Show that the row vectors of A are in the orthogonal complement of $\ker(A)$.

   b. Use this to show that the row rank and column rank of A are the same.

7. Let $M, N \subset V$ be subspaces of a finite-dimensional inner product space. Show that

$$(M + N)^\perp = M^\perp \cap N^\perp,$$
$$(M \cap N)^\perp = M^\perp + N^\perp.$$

8. Find the orthogonal projection onto span $\{(2, -1, 1), (1, -1, 0)\}$ by first computing the orthogonal projection onto the orthogonal complement.

9. Find the polynomial $p(t) \in P_2$ such that

$$\int_0^{2\pi} |p(t) - \cos t|^2 dt$$

is smallest possible.

10. Show that the decomposition into even and odd functions on $C^0([-a, a], \mathbb{C})$ is orthogonal if we use the inner product

$$(f|g) = \int_{-a}^{a} f(t) \overline{g(t)} dt.$$

11. Using the inner product

$$(f|g) = \int_0^1 f(t) g(t) dt,$$

find the orthogonal projection from $\mathbb{C}[t]$ onto span $\{1, t\} = P_1$. Given any $p \in \mathbb{C}[t]$, you should express the orthogonal projection in terms of the coefficients of p.

12. Using the inner product

$$(f|g) = \int_0^1 f(t) g(t) dt,$$

find the orthogonal projection from $\mathbb{C}[t]$ onto span $\{1, t, t^2\} = P_2$.

13. Compute the orthogonal projection onto the following subspaces:

(a)

$$\text{span} \left\{ \begin{bmatrix} 1 \\ 1 \\ 1 \\ 1 \end{bmatrix} \right\}$$

(b)

$$\text{span} \left\{ \begin{bmatrix} 1 \\ -1 \\ 0 \\ 1 \end{bmatrix}, \begin{bmatrix} 1 \\ 1 \\ 1 \\ 0 \end{bmatrix}, \begin{bmatrix} 2 \\ 0 \\ 1 \\ 1 \end{bmatrix} \right\}$$

(c)

$$\text{span} \left\{ \begin{bmatrix} 1 \\ i \\ 0 \\ 0 \end{bmatrix}, \begin{bmatrix} -i \\ 1 \\ 0 \\ 0 \end{bmatrix}, \begin{bmatrix} 0 \\ 1 \\ i \\ 0 \end{bmatrix} \right\}$$

14. *(Selberg)* Let $x, y_1, \ldots, y_n \in V$, where V is an inner product space. Show Selberg's generalization of Bessel's inequality

$$\sum_{i=1}^{n} \frac{|(x|y_i)|^2}{\sum_{j=1}^{n} |(y_i|y_j)|} \leq \|x\|^2.$$

Hint: It is a long calculation that comes from expanding the nonnegative quantity

$$\left\| x - \sum_{i=1}^{n} \frac{(x|y_i)}{\sum_{j=1}^{n} |(y_i|y_j)|} y_i \right\|^2.$$

3.5 Adjoint Maps

To introduce the concept of *adjoints* of linear maps, we start with the construction for matrices, i.e., linear maps $A : \mathbb{F}^m \to \mathbb{F}^n$, where $\mathbb{F} = \mathbb{R}$ or $\mathbb{C}$ and $\mathbb{F}^m$ and $\mathbb{F}^n$ are equipped with their standard inner products. We can write A as an $n \times m$ matrix and define the adjoint $A^* = \bar{A}^t$, i.e., A^* is the transposed and conjugate of A. In case $\mathbb{F} = \mathbb{R}$, conjugation is irrelevant, so $A^* = A^t$. Note that since A^* is an $m \times n$ matrix, it corresponds to a linear map $A^* : \mathbb{F}^n \to \mathbb{F}^m$. This matrix adjoint satisfies the crucial property

$$(Ax|y) = (x|A^*y)$$

for all $x \in V$ and $y \in W$. To see this, we simply think of x as an $m \times 1$ matrix, y as an $n \times 1$ matrix, and observe that

$$
\begin{aligned}
(Ax|y) &= (Ax)^t \, \bar{y} \\
&= x^t A^t \bar{y} \\
&= x^t \overline{(\bar{A}^t y)} \\
&= (x|A^* y).
\end{aligned}
$$

In the general case of a linear map $L : V \rightarrow W$ between finite-dimensional spaces, we can try to define the adjoint through matrix representations. To this end, select orthonormal bases for V and W so that we have a diagram

$$
\begin{array}{ccc}
V & \xrightarrow{L} & W \\
\updownarrow & & \updownarrow \\
\mathbb{F}^m & \xrightarrow{[L]} & \mathbb{F}^n
\end{array},
$$

where the vertical double arrows are isometries. Then, define $L^* : W \rightarrow V$ as the linear map whose matrix representation is $[L]^*$. In other words, $[L^*] = [L]^*$ and the following diagram commutes:

$$
\begin{array}{ccc}
V & \xleftarrow{L^*} & W \\
\updownarrow & & \updownarrow \\
\mathbb{F}^m & \xleftarrow{[L]^*} & \mathbb{F}^n
\end{array}.
$$

Because the vertical arrows are isometries we also have

$$
(Lx|y) = (x|L^* y).
$$

Proposition 3.5.1. *Let* $L : V \rightarrow W$ *be a linear map between finite-dimensional spaces. Then, there is a unique adjoint* $L^* : W \rightarrow V$ *with the property that*

$$
(Lx|y) = (x|L^* y)
$$

for all $x \in V$ *and* $y \in W$.

Proof. We saw already that such an adjoint exists, but we can give a similar construction of L^* that uses only an orthonormal basis $e_1, \ldots, e_m$ for V. To define $L^*(y)$, we need to know the inner products $(L^* y | e_j)$. The relationship $(Lx|y) = (x|L^* y)$ indicates that $(L^* y | e_j)$ can be calculated as

$$
\begin{aligned}
(L^* y | e_j) &= \overline{(e_j | L^* y)} \\
&= \overline{(Le_j | y)} \\
&= (y | Le_j).
\end{aligned}
$$

So let us define

$$L^* y = \sum_{j=1}^{m} (y | L e_j) \, e_j.$$

This clearly defines a linear map $L^* : W \to V$ satisfying

$$(L e_j | y) = (e_j | L^* y).$$

The more general condition

$$(Lx | y) = (x | L^* y)$$

follows immediately by writing x as a linear combination of $e_1, \ldots, e_m$ and using linearity in x on both sides of the equation.

Next, we address the issue of whether the adjoint is uniquely defined, i.e., could there be two linear maps $K_i : W \to V, i = 1, 2$ such that

$$(x | K_1 y) = (Lx | y) = (x | K_2 y)?$$

This would imply

$$0 = (x | K_1 y) - (x | K_2 y)$$
$$= (x | K_1 y - K_2 y).$$

If $x = K_1 y - K_2 y$, then

$$\| K_1 y - K_2 y \|^2 = 0,$$

and hence, $K_1 y = K_2 y$. This proves the claims. $\square$

The adjoint has the following useful elementary properties.

Proposition 3.5.2. *Let* $L, K : V \to W$, $L_1 : V_1 \to V_2$, *and* $L_2 : V_2 \to V_3$ *be linear maps between finite-dimensional inner product spaces. Then,*

(1) $(L + K)^* = L^* + K^*$.
(2) $L^{**} = L$
(3) $(\lambda 1_V)^* = \bar{\lambda} 1_V$.
(4) $(L_2 L_1)^* = L_1^* L_2^*$.
(5) *If L is invertible, then* $(L^{-1})^* = (L^*)^{-1}$.

Proof. The key to the proofs of these statements is the uniqueness statement in Proposition 3.5.1, i.e., any $L' : W \to V$ such that $(Lx | y) = (x | L' y)$ for all $x \in V$ and $y \in W$ must be the adjoint $L' = L^*$.

To check the first property, we calculate

$$(x | (L + K)^* y) = ((L + K) x | y)$$
$$= (Lx | y) + (Kx | y)$$

$$= \left(x|L^*y\right) + \left(x|K^*y\right)$$
$$= \left(x| \left(L^* + K^*\right) y\right).$$

The second is immediate from

$$(Lx|y) = \left(x|L^*y\right)$$
$$= \overline{(L^*y|x)}$$
$$= \overline{(y|L^{**}x)}$$
$$= \left(L^{**}x|y\right).$$

The third property follows from

$$(\lambda 1_V (x)\,|y) = (\lambda x|y)$$
$$= \left(x|\bar{\lambda}y\right)$$
$$= \left(x|\bar{\lambda}1_V (y)\right).$$

The fourth property

$$\left(x| (L_2L_1)^* y\right) = ((L_2L_1)(x)|y)$$
$$= (L_2 (L_1(x))|y)$$
$$= \left(L_1(x)|L_2^* (y)\right)$$
$$= \left(x|L_1^* (L_2^* (y))\right)$$
$$= \left(x| (L_1^*L_2^*)(y)\right).$$

And finally $1_V = L^{-1}L$ implies that

$$1_V = (1_V)^*$$
$$= \left(L^{-1}L\right)^*$$
$$= L^* \left(L^{-1}\right)^*$$

as desired.   □

Example 3.5.3. As an example, let us find the adjoint to

$$\begin{bmatrix} e_1 \cdots e_n \end{bmatrix} : \mathbb{F}^n \to V,$$

when $e_1, \ldots, e_n$ is an orthonormal basis. Recall that in Sect. 3.3, we already found a simple formula for the inverse

$$\begin{bmatrix} e_1 \cdots e_n \end{bmatrix}^{-1}(x) = \begin{bmatrix} (x|e_1) \\ \vdots \\ (x|e_n) \end{bmatrix}$$

and we proved that $[\, e_1 \, \cdots \, e_n \,]$ preserves inner products. If we let $x \in \mathbb{F}^n$ and $y \in V$, then we can write $y = [\, e_1 \, \cdots \, e_n \,](z)$ for some $z \in \mathbb{F}^n$. With that in mind, we can calculate

$$([\, e_1 \, \cdots \, e_n \,](x)\,|\,y) = ([\, e_1 \, \cdots \, e_n \,](x)\,|\,[\, e_1 \, \cdots \, e_n \,](z))$$

$$= (x|z)$$

$$= \left(x \,\middle|\, [\, e_1 \, \cdots \, e_n \,]^{-1}(y)\right).$$

Thus,

$$[\, e_1 \, \cdots \, e_n \,]^* = [\, e_1 \, \cdots \, e_n \,]^{-1}.$$

Below we shall generalize this relationship to all isomorphisms that preserve inner products, i.e., isometries.

The fact that

$$[\, e_1 \, \cdots \, e_n \,]^* = [\, e_1 \, \cdots \, e_n \,]^{-1}$$

simplifies the job of calculating matrix representations with respect to orthonormal bases. Assume that $L : V \to W$ is a linear map between finite-dimensional inner product spaces and that we have orthonormal bases $e_1, \ldots, e_m$ for V and $f_1, \ldots, f_n$ for W. Then,

$$L = [\, f_1 \, \cdots \, f_n \,][L][\, e_1 \, \cdots \, e_m \,]^*,$$

$$[L] = [\, f_1 \, \cdots \, f_n \,]^* L [\, e_1 \, \cdots \, e_m \,],$$

or in diagram form

$$\begin{array}{ccc} V & \xrightarrow{L} & W \\ {[\, e_1 \, \cdots \, e_m \,]^*} \downarrow & & \uparrow {[\, f_1 \, \cdots \, f_n \,]} \\ \mathbb{F}^m & \xrightarrow{[L]} & \mathbb{F}^n \end{array}$$

$$\begin{array}{ccc} V & \xrightarrow{L} & W \\ {[\, e_1 \, \cdots \, e_m \,]} \uparrow & & \downarrow {[\, f_1 \, \cdots \, f_n \,]^*} \\ \mathbb{F}^m & \xrightarrow{[L]} & \mathbb{F}^n \end{array}$$

From this, we see that the matrix definition of the adjoint is justified since the properties of the adjoint now tell us that:

$$L^* = \left([\, f_1 \, \cdots \, f_n \,][L][\, e_1 \, \cdots \, e_m \,]^*\right)^*$$

$$= [\, e_1 \, \cdots \, e_m \,][L]^* [\, f_1 \, \cdots \, f_n \,]^*.$$

A linear map and its adjoint have some remarkable relationships between their images and kernels. These properties are called the *Fredholm alternatives* and named after Fredholm who first used these properties to clarify when certain linear systems $L(x) = b$ can be solved (see also Sect. 4.9).

Theorem 3.5.4. (The Fredholm Alternative) *Let* $L : V \to W$ *be a linear map between finite-dimensional inner product spaces. Then,*

$$\ker(L) = \operatorname{im}\left(L^*\right)^{\perp},$$

$$\ker\left(L^*\right) = \operatorname{im}(L)^{\perp},$$

$$\ker(L)^{\perp} = \operatorname{im}\left(L^*\right),$$

$$\ker\left(L^*\right)^{\perp} = \operatorname{im}(L).$$

Proof. Since $L^{**} = L$ and $M^{\perp\perp} = M$, we see that all of the four statements are equivalent to each other. Thus. we need only prove the first. The two subspaces are characterized by

$$\ker(L) = \{x \in V : Lx = 0\},$$

$$\operatorname{im}\left(L^*\right)^{\perp} = \left\{x \in V : (x|L^*z) = 0 \text{ for all } z \in W\right\}.$$

Now, fix $x \in V$ and use that $(Lx|z) = (x|L^*z)$ for all $z \in V$. This implies first that if $x \in \ker(L)$, then also $x \in \operatorname{im}(L^*)^{\perp}$. Conversely, if $0 = (x|L^*z) = (Lx|z)$ for all $z \in W$, it must follow that $Lx = 0$ and hence $x \in \ker(L)$. □

Corollary 3.5.5. (The Rank Theorem) *Let* $L : V \to W$ *be a linear map between finite-dimensional inner product spaces. Then,*

$$\operatorname{rank}(L) = \operatorname{rank}\left(L^*\right).$$

Proof. Using the dimension formula (Theorem 1.11.7) for linear maps and that orthogonal complements have complementary dimension (Corollary 3.4.4) together with the Fredholm alternative, we see

$$\dim V = \dim(\ker(L)) + \dim(\operatorname{im}(L))$$
$$= \dim\left(\operatorname{im}\left(L^*\right)^{\perp}\right) + \dim(\operatorname{im}(L))$$
$$= \dim V - \dim\left(\operatorname{im}\left(L^*\right)\right) + \dim(\operatorname{im}(L)).$$

This implies the result. □

Next, we give another proof of the rank theorem for real and complex matrices (see Theorem 1.12.11).

Corollary 3.5.6. *For a real or complex $n \times m$ matrix A, the column rank equals the row rank.*

Proof. First, note that $\operatorname{rank}(B) = \operatorname{rank}(\bar{B})$ for all complex matrices B. Secondly, we know that the column rank of A is $\operatorname{rank}(A)$ is the same as the column rank, which by Corollary 3.5.5 equals $\operatorname{rank}(A^*)$ which in turn is the row rank of $\bar{A}$. This proves the result.   □

Corollary 3.5.7. *Let $L : V \to V$ be a linear operator on a finite-dimensional inner product space. Then, λ is an eigenvalue for L if and only if $\bar{\lambda}$ is an eigenvalue for L^*. Moreover, these eigenvalue pairs have the same geometric multiplicity:*

$$\dim\left(\ker\left(L - \lambda 1_V\right)\right) = \dim\left(\ker\left(L^* - \bar{\lambda}1_V\right)\right).$$

Proof. It suffices to prove the dimension statement. Note that $(L - \lambda 1_V)^* = L^* - \bar{\lambda}1_V$. Thus, the result follows if we can show

$$\dim\left(\ker\left(K\right)\right) = \dim\left(\ker\left(K^*\right)\right)$$

for $K : V \to V$. This comes from using the dimension formula (Theorem 1.11.7) and Corollary 3.5.5

$$\begin{aligned}
\dim\left(\ker\left(K\right)\right) &= \dim V - \dim\left(\operatorname{im}\left(K\right)\right) \\
&= \dim V - \dim\left(\operatorname{im}\left(K^*\right)\right) \\
&= \dim\left(\ker\left(K^*\right)\right).
\end{aligned}$$

                                                                                   □

Exercises

1. Let V and W be finite-dimensional inner product spaces.

   (a) Show that we can define an inner product on $\operatorname{Hom}_{\mathbb{F}}(V, W)$ by $(L|K) = \operatorname{tr}(LK^*) = \operatorname{tr}(K^*L)$.

   (b) Show that $(K|L) = (L^*|K^*)$.

   (c) If $e_1, \ldots, e_m$ is an orthonormal basis for V, show that

   $$(K|L) = (K(e_1)|L(e_1)) + \cdots + (K(e_m)|L(e_m)).$$

2. Assume that V is a complex inner product space. Recall from Exercise 6 in Sect. 1.4 that we have a vector space V^* with the same addition as in V but scalar multiplication is altered by conjugating the scalar. Show that the map $F : V^* \to \operatorname{Hom}(V, \mathbb{C})$ defined by $F(x) = (\cdot|x)$ is complex linear and an isomorphism when V is finite-dimensional. Use this to give another definition of the adjoint. Here

   $$F(x) = (\cdot|x) \in \operatorname{Hom}(V, \mathbb{C})$$

   is the linear map such that $(F(x))(z) = (z|x)$.

3. On $\text{Mat}_{n \times n}(\mathbb{C})$, use the inner product $(A|B) = \text{tr}(AB^*)$. For $A \in \text{Mat}_{n \times n}(\mathbb{C})$, consider the two linear operators on $\text{Mat}_{n \times n}(\mathbb{C})$ defined by $L_A(X) = AX$, $R_A(X) = XA$. Show that $(L_A)^* = L_{A^*}$ and $(R_A)^* = R_{A^*}$.

4. Let $x_1, \ldots, x_k \in V$, where V is a finite-dimensional inner product space.

   (a) Show that
   $$G(x_1, \ldots, x_k) = [x_1 \cdots x_k]^* [x_1 \cdots x_k],$$
   where $G(x_1, \ldots, x_k)$ is a $k \times k$ matrix whose ij entry is $(x_j|x_i)$. It is called the Gram matrix or Grammian.

   (b) Show that $G = G(x_1, \ldots, x_k)$ is nonnegative in the sense that $(Gx|x) \geq 0$ for all $x \in \mathbb{F}^k$.

   (c) Generalize part (a) to show that the composition
   $$[y_1 \cdots y_k]^* [x_1 \cdots x_k]$$
   is the matrix whose ij entry is $(x_j|y_i)$.

5. Find image and kernel for $A \in \text{Mat}_{3 \times 3}(\mathbb{R})$, where the ij entry is $\alpha_{ij} = (-1)^{i+j}$.

6. Find image and kernel for $A \in \text{Mat}_{3 \times 3}(\mathbb{C})$, where the kl entry is $\alpha_{kl} = (i)^{k+l}$.

7. Let $A \in \text{Mat}_{n \times n}(\mathbb{R})$ be symmetric, i.e., $A^* = A$, and assume A has rank $k \leq n$. Show that:

   (a) If the first k columns are linearly independent, then the principal $k \times k$ minor of A is invertible. The principal $k \times k$ minor of A is the $k \times k$ matrix one obtains by deleting the last $n - k$ columns and rows. Hint: Use a block decomposition
   $$A = \begin{bmatrix} B & C \\ C^t & D \end{bmatrix}$$
   and write
   $$\begin{bmatrix} C \\ D \end{bmatrix} = \begin{bmatrix} B \\ C^t \end{bmatrix} X, \ X \in \text{Mat}_{k \times (n-k)}(\mathbb{R})$$
   i.e., the last $n - k$ columns are linear combinations of the first k.

   (b) If rows $i_1, \ldots, i_k$ are linearly independent, then the $k \times k$ minor obtained by deleting all columns and rows not indexed by $i_1, \ldots, i_k$ is invertible. Hint: Note that $I_{kl} A I_{kl}$ is symmetric so one can use part a.

   (c) There are examples showing that (a) need not hold for $n \times n$ matrices in general.

8. Let $L : V \to V$ be a linear operator on a finite-dimensional inner product space. Show that:

   (a) If $M \subset V$ is an L-invariant subspace, then $M^\perp$ is L^*-invariant.

   (b) Show that there are examples where M is not L^*-invariant.

9. Consider two linear operators $K, L : V \to V$ and the commutator

$$[K, L] = K \circ L - L \circ K$$

   (a) Show that $[K, L]$ is skew-adjoint if K, L are both self-adjoint or both skew-adjoint.
   (b) Show that $[K, L]$ is self-adjoint if one of K, L is self-adjoint and the other skew-adjoint.

10. Let $L : V \to W$ be a linear operator between finite-dimensional vector spaces. Show that

   (a) L is one-to-one if and only if L^* is onto.
   (b) L^* is one-to-one if and only if L is onto.

11. Let $M, N \subset V$ be subspaces of a finite-dimensional inner product space and consider $L : M \times N \to V$ defined by $L(x, y) = x - y$.

   (a) Show that $L^*(z) = (\mathrm{proj}_M(z), -\mathrm{proj}_N(z))$.
   (b) Show that

$$\ker(L^*) = M^{\perp} \cap N^{\perp},$$
$$\mathrm{im}(L) = M + N.$$

   (c) Using the Fredholm alternative, show that

$$(M + N)^{\perp} = M^{\perp} \cap N^{\perp}.$$

   (d) Replace M and N by $M^{\perp}$ and $N^{\perp}$ and conclude

$$(M \cap N)^{\perp} = M^{\perp} + N^{\perp}.$$

12. Assume that $L : V \to W$ is a linear map between inner product spaces. Show that:

   (a) If both vector spaces are finite-dimensional, then

$$\dim(\ker(L)) - \dim(\mathrm{im}(L))^{\perp} = \dim V - \dim W.$$

   (b) If $V = W = \ell^2(\mathbb{Z})$, then for each integer $n \in \mathbb{Z}$, it is possible to find a linear operator L_n with finite-dimensional $\ker(L_n)$ and $(\mathrm{im}(L_n))^{\perp}$ so that

$$\mathrm{Ind}(L) = \dim(\ker(L)) - \dim(\mathrm{im}(L))^{\perp} = n.$$

   Hint: Consider linear maps that take (a_k) to (a_{k+l}) for some $l \in \mathbb{Z}$. An operator with finite-dimensional $\ker(L)$ and $(\mathrm{im}(L))^{\perp}$ is called a

Fredholm operator. The integer $\mathrm{Ind}\,(L) = \dim\,(\ker\,(L)) - \dim\,(\mathrm{im}\,(L))^{\perp}$ is the *index* of the operator and is an important invariant in functional analysis.

13. Let $L : V \to V$ be a linear operator on a finite-dimensional inner product space. Show that

$$\overline{\mathrm{tr}\,(L)} = \mathrm{tr}\left(L^{*}\right).$$

14. Let $L : V \to W$ be a linear map between inner product spaces. Show that

$$L : \ker\left(L^{*}L - \lambda 1_{V}\right) \to \ker\left(LL^{*} - \lambda 1_{V}\right)$$

and

$$L^{*} : \ker\left(LL^{*} - \lambda 1_{V}\right) \to \ker\left(L^{*}L - \lambda 1_{V}\right).$$

15. Let $L : V \to V$ be a linear operator on a finite-dimensional inner product space. Show that if $L\,(x) = \lambda x$, $L^{*}\,(y) = \mu y$, and $\lambda \neq \bar{\mu}$, then x and y are perpendicular.

16. Let V be a subspace of $C^{0}\,([0, 1], \mathbb{R})$ and consider the linear functionals $f_{t_0}\,(x) = x\,(t_0)$ and $f_{y}\,(x) = \int_{0}^{1} x\,(t)\,y\,(t)\,dt$. Show that:

   (a) If V is finite-dimensional, then $f_{t_0}|_V = f_y|_V$ for some $y \in V$.
   (b) If $V = P_2 = $ polynomials of degree ≤ 2, then there is an explicit $y \in V$ as in part (a).
   (c) If $V = C^{0}\,([0, 1], \mathbb{R})$, then there is no $y \in C^{0}\,([0, 1], \mathbb{R})$ such that $f_{t_0} = f_y$. The illusory function δ_{t_0} invented by Dirac to solve this problem is called Dirac's δ-function. It is defined as

$$\delta_{t_0}\,(t) = \begin{cases} 0 & \text{if } t \neq t_0 \\ \infty & \text{if } t = t_0 \end{cases}$$

$$\int_{0}^{1} \delta_{t_0}\,(t)\,dt = 1$$

so as to give the impression that

$$\int_{0}^{1} x\,(t)\,\delta_{t_0}\,(t)\,dt = x\,(t_0).$$

17. Find $q\,(t) \in P_2$ such that

$$p\,(5) = (p|q) = \int_{0}^{1} p\,(t)\,\overline{q\,(t)}\,dt$$

for all $p \in P_2$.

18. Find $f(t) \in \text{span}\{1, \sin(t), \cos(t)\}$ such that

$$(g|f) = \frac{1}{2\pi} \int_0^{2\pi} g(t) \overline{f(t)} dt$$

$$= \frac{1}{2\pi} \int_0^{2\pi} g(t) \left(1 + t^2\right) dt$$

for all $g \in \text{span}\{1, \sin(t), \cos(t)\}$.

3.6 Orthogonal Projections Revisited*

In this section, we shall give a new formula for an orthogonal projection. Instead of using Gram–Schmidt to create an orthonormal basis for the subspace, it gives a direct formula using an arbitrary basis for the subspace.

First, we need a new characterization of orthogonal projections using adjoints.

Lemma 3.6.1. (Characterization of Orthogonal Projections) *A projection* $E : V \to V$ *is orthogonal if and only if* $E = E^*$.

Proof. The Fredholm alternative (Theorem 3.5.4) tells us that $\text{im}(E) = \ker(E^*)^\perp$ so if $E = E^*$, we have shown that $\text{im}(E) = \ker(E)^\perp$, which implies that E is orthogonal (see Theorem 3.4.5).

Conversely, we can assume that $\text{im}(E) = \ker(E)^\perp$ since E is an orthogonal projection (see again Theorem 3.4.5). Using the Fredholm alternative again then tells us that

$$\text{im}(E) = \ker(E)^\perp = \text{im}\left(E^*\right),$$

$$\ker\left(E^*\right)^\perp = \text{im}(E) = \ker(E)^\perp.$$

As $(E^*)^2 = \left(E^2\right)^* = E^*$, it follows that E^* is a projection with the same image and kernel as E. Hence, $E = E^*$.   □

Using this characterization of orthogonal projections, it is possible to find a formula for proj_M using a general basis for $M \subset V$. Let $M \subset V$ be finite-dimensional with a basis $x_1, \ldots, x_m$. This yields an isomorphism

$$\left[x_1 \cdots x_m\right] : \mathbb{F}^m \to M$$

which can also be thought of as a one-to-one map $A : \mathbb{F}^m \to V$ whose image is M. This yields a linear map

$$A^*A : \mathbb{F}^m \to \mathbb{F}^m.$$

Since

$$\left(A^{*}Ay|y\right) = (Ay|Ay)$$
$$= \|Ay\|^{2},$$

the kernel satisfies

$$\ker\left(A^{*}A\right) = \ker(A) = \{0\}.$$

In particular, $A^{*}A$ is an isomorphism. This means that

$$E = A\left(A^{*}A\right)^{-1}A^{*}$$

defines linear operator $E : V \to V$. It is easy to check that $E = E^{*}$, and since

$$E^{2} = A\left(A^{*}A\right)^{-1}A^{*}A\left(A^{*}A\right)^{-1}A^{*}$$
$$= A\left(A^{*}A\right)^{-1}A^{*}$$
$$= E,$$

it is a projection. Finally, we must check that $\operatorname{im}(E) = M$. Since $(A^{*}A)^{-1}$ is an isomorphism and

$$\operatorname{im}\left(A^{*}\right) = (\ker(A))^{\perp} = (\{0\})^{\perp} = \mathbb{F}^{m},$$

we have

$$\operatorname{im}(E) = \operatorname{im}(A) = M$$

as desired.

To better understand this construction, we note that

$$A^{*}(x) = \begin{bmatrix} (x|x_{1}) \\ \vdots \\ (x|x_{m}) \end{bmatrix}.$$

This follows from

$$\left(\begin{bmatrix} \alpha_{1} \\ \vdots \\ \alpha_{m} \end{bmatrix}\middle|\begin{bmatrix} (x|x_{1}) \\ \vdots \\ (x|x_{m}) \end{bmatrix}\right) = \alpha_{1}\overline{(x|x_{1})} + \cdots + \alpha_{m}\overline{(x|x_{m})}$$

$$= \alpha_{1}(x_{1}|x) + \cdots + \alpha_{m}(x_{m}|x)$$
$$= (\alpha_{1}x_{1} + \cdots + \alpha_{m}x_{m}|x)$$

$$= \left(A\left(\begin{bmatrix} \alpha_{1} \\ \vdots \\ \alpha_{m} \end{bmatrix}\right)\middle|x\right)$$

The matrix form of A^*A can now be expressed as

$$A^*A = A^* \circ \left[x_1 \cdots x_m \right]$$
$$= \left[A^*(x_1) \cdots A^*(x_m) \right]$$
$$= \begin{bmatrix} (x_1|x_1) & \cdots & (x_m|x_1) \\ \vdots & \ddots & \vdots \\ (x_1|x_m) & \cdots & (x_m|x_m) \end{bmatrix}.$$

This is also called the *Gram matrix* of $x_1, \ldots, x_m$. This information specifies explicitly all of the components of the formula

$$E = A \left(A^*A \right)^{-1} A^*.$$

The only hard calculation is the inversion of A^*A. The calculation of $A\,(A^*A)^{-1}\,A^*$ should also be compared to using the Gram–Schmidt procedure for finding the orthogonal projection onto M.

Exercises

1. Using the inner product $\int_0^1 p(t)\bar{q}(t)\,dt$, find the orthogonal projection from $\mathbb{C}[t]$ onto span$\{1, t\} = P_1$. Given any $p \in \mathbb{C}[t]$, you should express the orthogonal projection in terms of the coefficients of p.
2. Using the inner product $\int_0^1 p(t)\bar{q}(t)\,dt$, find the orthogonal projection from $\mathbb{C}[t]$ onto span$\{1, t, t^2\} = P_2$.
3. Compute the orthogonal projection onto the following subspaces:

(a)
$$\text{span}\left\{ \begin{bmatrix} 1 \\ 1 \\ 1 \\ 1 \end{bmatrix} \right\} \subset \mathbb{R}^4$$

(b)
$$\text{span}\left\{ \begin{bmatrix} 1 \\ -1 \\ 0 \\ 1 \end{bmatrix}, \begin{bmatrix} 1 \\ 1 \\ 1 \\ 0 \end{bmatrix}, \begin{bmatrix} 2 \\ 0 \\ 1 \\ 1 \end{bmatrix} \right\} \subset \mathbb{R}^4$$

(c)
$$\text{span}\left\{ \begin{bmatrix} 1 \\ i \\ 0 \\ 0 \end{bmatrix}, \begin{bmatrix} -i \\ 1 \\ 0 \\ 0 \end{bmatrix}, \begin{bmatrix} 0 \\ 1 \\ i \\ 0 \end{bmatrix} \right\} \subset \mathbb{C}^4$$

4. Given an orthonormal basis $e_1, \ldots, e_k$ for the subspace $M \subset V$, show that the orthogonal projection onto M can be computed as

$$\mathrm{proj}_M = \begin{bmatrix} e_1 \cdots e_k \end{bmatrix} \begin{bmatrix} e_1 \cdots e_k \end{bmatrix}^*.$$

Hint: Show that

$$\begin{bmatrix} e_1 \cdots e_k \end{bmatrix}^* \begin{bmatrix} e_1 \cdots e_k \end{bmatrix} = 1_{\mathbb{F}^k}.$$

5. Show that if $M \subset V$ is an L-invariant subspace, then

$$(L|_M)^* = \mathrm{proj}_M \circ L^*|_M.$$

3.7 Matrix Exponentials*

In this section, we shall show that the initial value problem: $\dot{x} = Ax$, $x(t_0) = x_0$ where A is a square matrix with real or complex scalars as entries can be solved using matrix exponentials. More algebraic approaches are also available by using the Frobenius canonical form (Theorem 2.7.1) and the Jordan canonical form (Theorem 2.8.3). Later, we shall see how Schur's Theorem (4.8.1) also gives a very effective way of solving such systems.

Recall that in the one-dimensional situation, the solution is

$$x(t) = x_0 \exp\left(A(t - t_0)\right).$$

If we could make sense of this for square matrices A as well, we would have a possible way of writing down the solutions. The concept of operator norms introduced in Sect. 3.3 naturally leads to a norm of matrices as well. One key observation about this norm is that if $A = \begin{bmatrix} \alpha_{ij} \end{bmatrix}$, then $|\alpha_{ij}| \leq \|A\|$, i.e., the entries are bounded by the norm. Moreover, we also have that

$$\|AB\| \leq \|A\| \, \|B\|,$$

$$\|A + B\| \leq \|A\| + \|B\|$$

as

$$\|AB(x)\| \leq \|A\| \, \|B(x)\|$$

$$\leq \|A\| \, \|B\| \, \|x\|$$

and

$$\|(A + B)(x)\| \leq \|A(x)\| + \|B(x)\|.$$

Now, consider the series

$$\sum_{n=0}^{\infty} \frac{A^n}{n!}.$$

Since

$$\left\| \frac{A^n}{n!} \right\| \leq \frac{\|A\|^n}{n!}$$

and

$$\sum_{n=0}^{\infty} \frac{\|A\|^n}{n!}$$

is convergent, it follows that any given entry in

$$\sum_{n=0}^{\infty} \frac{A^n}{n!}$$

is bounded by a convergent series. Thus, the matrix series also converges. This means we can define

$$\exp(A) = \sum_{n=0}^{\infty} \frac{A^n}{n!}.$$

It is not hard to check that if $L \in \mathrm{Hom}\,(V, V)$, where V is a finite-dimensional inner product space, then we can similarly define

$$\exp(L) = \sum_{n=0}^{\infty} \frac{L^n}{n!}.$$

Now, consider the matrix-valued function

$$\exp(At) = \sum_{n=0}^{\infty} \frac{A^n t^n}{n!}$$

and with it the vector-valued function

$$x(t) = \exp(A(t - t_0)) x_0.$$

It still remains to be seen that this defines a differentiable function that solves $\dot{x} = Ax$. But it follows directly from the definition it has the correct initial value since $\exp(0) = 1_{\mathbb{F}^n}$. To check differentiability, we consider the matrix function $t \to \exp(At)$ and study $\exp(A(t + h))$. In fact, we claim that

$$\exp(A(t + h)) = \exp(At)\exp(Ah).$$

To establish this, we prove a more general version together with another useful fact.

Proposition 3.7.1. *Let* $L, K : V \rightarrow V$ *be linear operators on a finite-dimensional inner product space.*

(1) *If* $KL = LK$, *then* $\exp(K + L) = \exp(K) \circ \exp(L)$.
(2) *If* K *is invertible, then* $\exp(K \circ L \circ K^{-1}) = K \circ \exp(L) \circ K^{-1}$.

Proof. 1. This formula hinges on proving the binomial formula for commuting operators:

$$(L + K)^n = \sum_{k=0}^{n} \binom{n}{k} L^k K^{n-k},$$

$$\binom{n}{k} = \frac{n!}{(n-k)!k!}.$$

This formula is obvious for $n = 1$. Suppose that the formula holds for n. Using the conventions

$$\binom{n}{n+1} = 0,$$

$$\binom{n}{-1} = 0,$$

together with the formula from Pascal's triangle

$$\binom{n}{k-1} + \binom{n}{k} = \binom{n+1}{k},$$

it follows that

$$(L + K)^{n+1} = (L + K)^n (L + K)$$

$$= \left(\sum_{k=0}^{n} \binom{n}{k} L^k K^{n-k} \right) (L + K)$$

$$= \sum_{k=0}^{n} \binom{n}{k} L^k K^{n-k} L + \sum_{k=0}^{n} \binom{n}{k} L^k K^{n-k} K$$

$$= \sum_{k=0}^{n} \binom{n}{k} L^{k+1} K^{n-k} + \sum_{k=0}^{n} \binom{n}{k} L^k K^{n-k+1}$$

$$= \sum_{k=0}^{n+1} \binom{n}{k-1} L^k K^{n+1-k} + \sum_{k=0}^{n+1} \binom{n}{k} L^k K^{n+1-k}$$

$$= \sum_{k=0}^{n+1} \left(\binom{n}{k-1} + \binom{n}{k} \right) L^k K^{n+1-k}$$

$$= \sum_{k=0}^{n+1} \binom{n+1}{k} L^k K^{n+1-k}.$$

We can then compute

$$\sum_{n=0}^{N} \frac{(K+L)^n}{n!} = \sum_{n=0}^{N} \sum_{k=0}^{n} \frac{1}{n!} \binom{n}{k} L^k K^{n-k}$$

$$= \sum_{n=0}^{N} \sum_{k=0}^{n} \frac{1}{(n-k)!k!} L^k K^{n-k}$$

$$= \sum_{n=0}^{N} \sum_{k=0}^{n} \left(\frac{1}{k!} L^k \right) \left(\frac{1}{(n-k)!} K^{n-k} \right)$$

$$= \sum_{k,l=0, k+l \le N}^{N} \left(\frac{1}{k!} L^k \right) \left(\frac{1}{l!} K^l \right).$$

The last term is unfortunately not quite the same as the product

$$\sum_{k,l=0}^{N} \left(\frac{1}{k!} L^k \right) \left(\frac{1}{l!} K^l \right) = \left(\sum_{k=0}^{N} \frac{1}{k!} L^k \right) \left(\sum_{l=0}^{N} \frac{1}{l!} K^l \right).$$

However, the difference between these two sums can be estimated the following way:

$$\left\| \sum_{k,l=0}^{N} \left(\frac{1}{k!} L^k \right) \left(\frac{1}{l!} K^l \right) - \sum_{k,l=0, k+l \le N}^{N} \left(\frac{1}{k!} L^k \right) \left(\frac{1}{l!} K^l \right) \right\|$$

$$= \left\| \sum_{k,l=0, k+l > N}^{N} \left(\frac{1}{k!} L^k \right) \left(\frac{1}{l!} K^l \right) \right\|$$

$$\le \sum_{k,l=0, k+l > N}^{N} \left(\frac{1}{k!} \|L\|^k \right) \left(\frac{1}{l!} \|K\|^l \right)$$

$$\leq \sum_{k=0, l=N/2}^{N} \left(\frac{1}{k!} \|L\|^k \right) \left(\frac{1}{l!} \|K\|^l \right) + \sum_{l=0, k=N/2}^{N} \left(\frac{1}{k!} \|L\|^k \right) \left(\frac{1}{l!} \|K\|^l \right)$$

$$= \left(\sum_{k=0}^{N} \frac{1}{k!} \|L\|^k \right) \left(\sum_{l=N/2}^{N} \frac{1}{l!} \|K\|^l \right) + \left(\sum_{k=N/2}^{N} \frac{1}{k!} \|L\|^k \right) \left(\sum_{l=0}^{N} \frac{1}{l!} \|K\|^l \right)$$

$$\leq \exp(\|L\|) \left(\sum_{l=N/2}^{N} \frac{1}{l!} \|K\|^l \right) + \exp(\|K\|) \left(\sum_{k=N/2}^{N} \frac{1}{k!} \|L\|^k \right).$$

Since

$$\lim_{N \to \infty} \sum_{l=N/2}^{N} \frac{1}{l!} \|K\|^l = 0,$$

$$\lim_{N \to \infty} \sum_{k=N/2}^{N} \frac{1}{k!} \|L\|^k = 0,$$

it follows that

$$\lim_{N \to \infty} \left\| \sum_{n=0}^{N} \frac{(K+L)^n}{n!} - \left(\sum_{k=0}^{N} \frac{1}{k!} L^k \right) \left(\sum_{l=0}^{N} \frac{1}{l!} K^l \right) \right\| = 0.$$

Thus,

$$\sum_{n=0}^{\infty} \frac{(K+L)^n}{n!} = \sum_{k=0}^{\infty} \left(\frac{1}{k!} L^k \right) \sum_{l=0}^{\infty} \left(\frac{1}{l!} K^l \right)$$

as desired.

2. This is considerably simpler and uses that

$$\left(K \circ L \circ K^{-1} \right)^n = K \circ L^n \circ K^{-1}.$$

This is again proven by induction. First, observe it is trivial for $n = 1$ and then that

$$\left(K \circ L \circ K^{-1} \right)^{n+1} = \left(K \circ L \circ K^{-1} \right)^n \circ K \circ L \circ K^{-1}$$

$$= K \circ L^n \circ K^{-1} \circ K \circ L \circ K^{-1}$$

$$= K \circ L^n \circ L \circ K^{-1}$$

$$= K \circ L^{n+1} \circ K^{-1}.$$

Thus,

$$\sum_{n=0}^{N} \frac{\left(K \circ L \circ K^{-1}\right)^n}{n!} = \sum_{n=0}^{N} \frac{K \circ L^n \circ K^{-1}}{n!}$$

$$= K \circ \left(\sum_{n=0}^{N} \frac{L^n}{n!}\right) \circ K^{-1}.$$

By letting $N \to \infty$, we get the desired formula.   □

To calculate the derivative of $\exp(At)$, we observe that

$$\frac{\exp(A(t+h)) - \exp(At)}{h} = \frac{\exp(Ah)\exp(At) - \exp(At)}{h}$$

$$= \left(\frac{\exp(Ah) - 1_{\mathbb{F}^n}}{h}\right)\exp(At).$$

Using the definition of $\exp(Ah)$, it follows that

$$\frac{\exp(Ah) - 1_{\mathbb{F}^n}}{h} = \sum_{n=1}^{\infty} \frac{1}{h}\frac{A^n h^n}{n!}$$

$$= \sum_{n=1}^{\infty} \frac{A^n h^{n-1}}{n!}$$

$$= A + \sum_{n=2}^{\infty} \frac{A^n h^{n-1}}{n!}.$$

Since

$$\left\|\sum_{n=2}^{\infty} \frac{A^n h^{n-1}}{n!}\right\| \le \sum_{n=2}^{\infty} \frac{\|A\|^n |h|^{n-1}}{n!}$$

$$= \|A\| \sum_{n=2}^{\infty} \frac{\|A\|^{n-1} |h|^{n-1}}{n!}$$

$$= \|A\| \sum_{n=2}^{\infty} \frac{\|Ah\|^{n-1}}{n!}$$

$$\le \|A\| \sum_{n=1}^{\infty} \|Ah\|^n$$

$$= \|A\| \|Ah\| \frac{1}{1 - \|Ah\|}$$

$$\to 0 \text{ as } |h| \to 0,$$

we get that

$$\lim_{|h| \to 0} \frac{\exp(A(t+h)) - \exp(At)}{h} = \left(\lim_{|h| \to 0} \frac{\exp(Ah) - 1_{\mathbb{F}^n}}{h} \right) \exp(At)$$

$$= A \exp(At).$$

Therefore, if we define

$$x(t) = \exp(A(t - t_0)) x_0,$$

then

$$\dot{x} = A \exp(A(t - t_0)) x_0$$

$$= Ax.$$

The other problem we should solve at this point is uniqueness of solutions. To be more precise, if we have that both x and y solve the initial value problem $\dot{x} = Ax$, $x(t_0) = x_0$, then we wish to prove that $x = y$. Inner products can be used quite effectively to prove this as well. We consider the nonnegative function

$$\phi(t) = \|x(t) - y(t)\|^2$$

$$= (x_1 - y_1)^2 + \cdots + (x_n - y_n)^2.$$

In the complex situation, simply identify $\mathbb{C}^n = \mathbb{R}^{2n}$ and use the $2n$ real coordinates to define this norm. Recall that this norm comes from the usual inner product on Euclidean space. The derivative satisfies

$$\frac{d\phi}{dt}(t) = 2(\dot{x}_1 - \dot{y}_1)(x_1 - y_1) + \cdots + 2(\dot{x}_n - \dot{y}_n)(x_n - y_n)$$

$$= 2((\dot{x} - \dot{y}) \mid (x - y))$$

$$= 2(A(x - y) \mid (x - y))$$

$$\leq 2 \|A(x - y)\| \|x - y\|$$

$$\leq 2 \|A\| \|x - y\|^2$$

$$= 2 \|A\| \phi(t).$$

Thus, we have

$$\frac{d\phi}{dt}(t) - 2 \|A\| \phi(t) \leq 0.$$

If we multiply this by the positive integrating factor $\exp(-2 \|A\| (t - t_0))$ and use Leibniz' rule in reverse, we obtain

$$\frac{d}{dt}(\phi(t) \exp(-2 \|A\| (t - t_0))) \leq 0.$$

3 Inner Product Spaces

Together with the initial condition $\phi(t_0) = 0$, this yields

$$\phi(t) \exp(-2\,\|A\|\,(t - t_0)) \le 0, \text{ for } t \ge t_0.$$

Since the integrating factor is positive and ϕ is nonnegative, it must follow that $\phi(t) = 0$ for $t \ge t_0$. A similar argument using $-\exp(-2\,\|A\|\,(t - t_0))$ can be used to show that $\phi(t) = 0$ for $t \le t_0$. Altogether, we have established that the initial value problem $\dot{x} = Ax$, $x(t_0) = x_0$ always has a unique solution for matrices A with real (or complex) scalars as entries.

To explicitly solve these linear differential equations, it is often best to understand higher order equations first and then use the cyclic subspace decomposition from Sect. 2.6 to reduce systems to higher order equations. In Sect. 4.8.1, we shall give another method for solving systems of equations that does not use higher order equations.

Exercises

1. Let $f(z) = \sum_{n=0}^{\infty} a_n z^n$ define a power series and $A \in \mathrm{Mat}_{n\times n}(\mathbb{F})$. Show that one can define $f(A)$ as long as $\|A\| < $ radius of convergence.
2. Let $L : V \to V$ be an operator on a finite-dimensional inner product space. Show the following statements:

   (a) If $\|L\| < 1$, then $1_V + L$ has an inverse. Hint:

   $$(1_V + L)^{-1} = \sum_{n=1}^{\infty} (-1)^n L^n.$$

   (b) With L as above, show

   $$\|L^{-1}\| \le \frac{1}{1 - \|L\|},$$

   $$\left\| (1_V + L)^{-1} - 1_V \right\| \le \frac{\|L\|}{1 - \|L\|}.$$

   (c) If $\left\| L^{-1} \right\| \le \varepsilon^{-1}$ and $\|L - K\| < \varepsilon$, then K is invertible and

   $$\|K^{-1}\| \le \frac{\|L^{-1}\|}{1 - \|L^{-1}(K - L)\|},$$

   $$\|L^{-1} - K^{-1}\| \le \frac{\|L^{-1}\|^2}{(1 - \|L^{-1}\|\,\|L - K\|)^2}\,\|L - K\|.$$

3. Let $L : V \to V$ be an operator on a finite-dimensional inner product space.

   (a) Show that if λ is an eigenvalue for L, then

   $$|\lambda| \le \|L\| .$$

   (b) Give examples of 2×2 matrices where strict inequality always holds.

4. Show that

   $$x(t) = \left(\exp(A(t - t_0)) \int_{t_0}^{t} \exp(-A(s - t_0)) f(s) \, ds \right) x_0$$

   solves the initial value problem $\dot{x} = Ax + f(t), \; x(t_0) = x_0$.

5. Let $A = B + C \in \mathrm{Mat}_{n \times n}(\mathbb{R})$ where B is invertible and $\|C\|$ is very small compared to $\|B\|$.

   (a) Show that $B^{-1} - B^{-1} C B^{-1}$ is a good approximation to A^{-1}.

   (b) Use this to approximate the inverse to

   $$\begin{bmatrix} 1 & 0 & 1000 & 1 \\ 0 & -1 & 1 & 1000 \\ 2 & 1000 & -1 & 0 \\ 1000 & 3 & 2 & 0 \end{bmatrix}.$$

Chapter 4
Linear Operators on Inner Product Spaces

In this chapter, we are going to study linear operators on finite-dimensional inner product spaces. In the last chapter, we introduced adjoints of linear maps between possibly different inner product spaces. Here we shall see how the adjoint can be used to understand linear operators on a fixed inner product space. The important operators we study here are the self-adjoint, skew-adjoint, normal, orthogonal, and unitary operators. We shall spend several sections on the existence of eigenvalues, diagonalizability, and canonical forms for these special but important linear operators. Having done that, we go back to the study of general linear maps and operators and establish the singular value and polar decompositions. We also show Schur's theorem to the effect that complex linear operators have upper triangular matrix representations. This result does not depend on the spectral theorem. It is also possible to give a quick proof of the spectral theorem using only the material covered in Sect. 4.1. The chapter finishes with a section on quadratic forms and how they tie in with the theory of self-adjoint operators.

4.1 Self-Adjoint Maps

Definition 4.1.1. A linear operator $L : V \to V$ on a finite-dimensional vector space is called *self-adjoint* if $L^* = L$. Note that a real $m \times m$ matrix A is self-adjoint precisely when it is symmetric, i.e., $A = A^t$. The opposite of being self-adjoint is *skew-adjoint*: $L^* = -L$.

When V is a real inner product space, we also say that the operator is *symmetric* or *skew-symmetric*. In case the inner product is complex, these operators are also called *Hermitian* or *skew-Hermitian*.

Example 4.1.2. The following 2×2 matrices satisfy:

$$\begin{bmatrix} 0 & -\beta \\ \beta & 0 \end{bmatrix}$$

P. Petersen, *Linear Algebra*, Undergraduate Texts in Mathematics,
DOI 10.1007/978-1-4614-3612-6_4, © Springer Science+Business Media New York 2012

is skew-adjoint if β is real.

$$\begin{bmatrix} \alpha & -i\beta \\ i\beta & \alpha \end{bmatrix}$$

is self-adjoint if α and β are real.

$$\begin{bmatrix} i\alpha & -\beta \\ \beta & i\alpha \end{bmatrix}$$

is skew-adjoint if α and β are real.

In general, a complex 2×2 self-adjoint matrix looks like

$$\begin{bmatrix} \alpha & \beta + i\gamma \\ \beta - i\gamma & \delta \end{bmatrix}, \alpha, \beta, \gamma, \delta \in \mathbb{R}.$$

In general, a complex 2×2 skew-adjoint matrix looks like

$$\begin{bmatrix} i\alpha & i\beta - \gamma \\ i\beta + \gamma & i\delta \end{bmatrix}, \alpha, \beta, \gamma, \delta \in \mathbb{R}.$$

Example 4.1.3. We saw in Sect. 3.6 that self-adjoint projections are precisely the orthogonal projections.

Example 4.1.4. If $L : V \to W$ is a linear map we can create two self-adjoint maps $L^*L : V \to V$ and $LL^* : W \to W$.

Example 4.1.5. Consider the space of periodic functions $C_{2\pi}^{\infty}$ $(\mathbb{R}, \mathbb{C})$ with the inner product

$$(x|y) = \frac{1}{2\pi} \int_0^{2\pi} x(t) \overline{y(t)} dt.$$

The linear operator

$$D(x) = \frac{dx}{dt}$$

can be seen to be skew-adjoint even though we have not defined the adjoint of maps on infinite-dimensional spaces. In general, we say that a map is self-adjoint or skew-adjoint if

$$(L(x)|y) = (x|L(y)) \text{ or}$$
$$(L(x)|y) = -(x|L(y))$$

for all x, y. Using that definition, we note that integration by parts and the fact that the functions are periodic imply our claim:

$$(D(x)|y) = \frac{1}{2\pi} \int_0^{2\pi} \left(\frac{dx}{dt}(t) \right) \overline{y(t)} dt$$

$$= \frac{1}{2\pi} x(t) \overline{y(t)}\big|_0^{2\pi} - \frac{1}{2\pi} \int_0^{2\pi} x(t) \overline{\frac{dy}{dt}}(t) dt$$

$$= -\frac{1}{2\pi} \int_0^{2\pi} x(t) \overline{\frac{dy}{dt}}(t) dt$$

$$= -(x|D(y)).$$

In quantum mechanics, one often makes D self-adjoint by instead considering iD. In analogy with the formulae

$$\exp(x) = \frac{\exp(x) + \exp(-x)}{2} + \frac{\exp(x) - \exp(-x)}{2}$$

$$= \cosh(x) + \sinh(x),$$

we have

$$L = \frac{1}{2}(L + L^*) + \frac{1}{2}(L - L^*),$$

$$L^* = \frac{1}{2}(L + L^*) - \frac{1}{2}(L - L^*),$$

where $\frac{1}{2}(L + L^*)$ is self-adjoint and $\frac{1}{2}(L - L^*)$ is skew-adjoint. In the complex case, we also have

$$\exp(ix) = \frac{\exp(ix) + \exp(-ix)}{2} + \frac{\exp(ix) - \exp(-ix)}{2}$$

$$= \frac{\exp(ix) + \exp(-ix)}{2} + i\frac{\exp(ix) - \exp(-ix)}{2i}$$

$$= \cos(x) + i\sin(x),$$

which is a nice analogy for

$$L = \frac{1}{2}(L + L^*) + i\frac{1}{2i}(L - L^*),$$

$$L^* = \frac{1}{2}(L + L^*) - i\frac{1}{2i}(L - L^*),$$

where now also $\frac{1}{2i}(L - L^*)$ is self-adjoint. The idea behind this formula is that multiplication by i takes skew-adjoint maps to self-adjoints maps and vice versa.

Self- and skew-adjoint maps are clearly quite special by virtue of their definitions. The above decomposition which has quite a lot in common with dividing functions into odd and even parts or dividing complex numbers into real and imaginary parts seems to give some sort of indication that these maps could be central to the understanding of general linear maps. This is not quite true, but we shall be able to get a grasp on quite a lot of different maps.

Aside from these suggestive properties, self- and skew-adjoint maps are *completely reducible.*

Definition 4.1.6. A linear map $L : V \to V$ is said to be completely reducible if every invariant subspace has a complementary invariant subspace.

Recall that maps like

$$L = \begin{bmatrix} 0 & 1 \\ 0 & 0 \end{bmatrix} : \mathbb{R}^2 \to \mathbb{R}^2$$

can have invariant subspaces without having complementary subspaces that are invariant.

Proposition 4.1.7. (Reducibility of Self- or Skew-Adjoint Operators) *Let $L : V \to V$ be a linear operator on a finite-dimensional inner product space. If L is self- or skew-adjoint, then for each invariant subspace $M \subset V$ the orthogonal complement is also invariant, i.e., if $L(M) \subset M$, then $L(M^\perp) \subset M^\perp$.*

Proof. Assume that $L(M) \subset M$. Let $x \in M$ and $z \in M^\perp$. Since $L(x) \in M$, we have

$$0 = (z|L(x))$$
$$= (L^*(z)|x)$$
$$= \pm(L(z)|x).$$

As this holds for all $x \in M$, it follows that $L(z) \in M^\perp$.   □

Remark 4.1.8. This property almost tells us that these operators are diagonalizable. Certainly in the case where we have complex scalars, we can use induction on dimension to show that such maps are diagonalizable. In the case of real scalars, the problem is that it is not clear that self- and/or skew-adjoint maps have any invariant subspaces whatsoever. The map which is rotation by 90° in the plane is clearly skew-symmetric, but it has no non-trivial invariant subspaces. Thus, we cannot make the map any simpler. We shall see below that this is basically the worst case scenario for such maps.

Exercises

1. Let $L : P_n \to P_n$ be a linear map on the space of real polynomials of degree $\leq n$ such that $[L]$ with respect to the standard basis $1, t, \ldots, t^n$ is self-adjoint. Is L self-adjoint if we use the inner product

$$(p|q) = \int_a^b p(t) q(t) \, dt \ ?$$

2. If V is finite-dimensional, show that the three subsets of $\mathrm{Hom}(V, V)$ defined by

$$M_1 = \mathrm{span}\{1_V\},$$

$$M_2 = \{L : L \text{ is skew-adjoint}\},$$

$$M_3 = \{L : \mathrm{tr} L = 0 \text{ and } L \text{ is self-adjoint}\}$$

are subspaces over $\mathbb{R}$, are mutually orthogonal with respect to the real inner product $\mathrm{Re}(L, K) = \mathrm{Re}(\mathrm{tr}(L^* K))$, and yield a direct sum decomposition of $\mathrm{Hom}(V, V)$.

3. Let E be an orthogonal projection and L a linear operator. Recall from Exercise 11 in Sect. 1.11 and Exercise 5 in Sect. 3.4 that L leaves $M = \mathrm{im}(E)$ invariant if and only if $ELE = LE$ and that $M \oplus M^{\perp}$ reduces L if and only if $EL = LE$. Show that if L is skew- or self-adjoint and $ELE = LE$, then $EL = LE$.

4. Let V be a finite-dimensional complex inner product space. Show that both the space of self-adjoint and skew-adjoint maps form a real vector space. Show that multiplication by i yields an $\mathbb{R}$-linear isomorphism between these spaces.

5. Show that $D^{2k} : C_{2\pi}^{\infty}(\mathbb{R}, \mathbb{C}) \to C_{2\pi}^{\infty}(\mathbb{R}, \mathbb{C})$ is self-adjoint and that $D^{2k+1} : C_{2\pi}^{\infty}(\mathbb{R}, \mathbb{C}) \to C_{2\pi}^{\infty}(\mathbb{R}, \mathbb{C})$ is skew-adjoint.

6. Let $x_1, \ldots, x_k$ be vectors in an inner product space V. Show that the $k \times k$ matrix $G(x_1, \ldots, x_k)$ whose ij entry is $(x_j | x_i)$ is self-adjoint and that all its eigenvalues are nonnegative.

7. Let $L : V \to V$ be a self-adjoint operator on a finite-dimensional inner product space and $p \in \mathbb{R}[t]$ a real polynomial. Show that $p(L)$ is also self-adjoint.

8. Assume that $L : V \to V$ is self-adjoint and $\lambda \in \mathbb{R}$. Show:

   (a) $\ker(L) = \ker(L^k)$ for any $k \geq 1$. Hint: Start with $k = 2$.
   (b) $\mathrm{im}(L) = \mathrm{im}(L^k)$ for any $k \geq 1$.
   (c) $\ker(L - \lambda 1_V) = \ker\left((L - \lambda 1_V)^k\right)$ for any $k \geq 1$.
   (d) Show that the eigenvalues of L are real.
   (d) Show that $\mu_L(t)$ has no multiple roots.

9. Let $L : V \to V$ be a self-adjoint operator on a finite-dimensional vector space.

   (a) Show that the eigenvalues of L are real.
   (b) In case V is complex, show that L has an eigenvalue.

(c) In case V is real, show that L has an eigenvalue. Hint: Choose an orthonormal basis and observe that $[L] \in \mathrm{Mat}_{n \times n}(\mathbb{R}) \subset \mathrm{Mat}_{n \times n}(\mathbb{C})$ is also self-adjoint as a complex matrix. Thus, all roots of $\chi_{[L]}(t)$ must be real by (a).

10. Assume that $L_1, L_2 : V \to V$ are both self-adjoint or skew-adjoint.

   (a) Show that $L_1 L_2$ is skew-adjoint if and only if $L_1 L_2 + L_2 L_1 = 0$.
   (b) Show that $L_1 L_2$ is self-adjoint if and only if $L_1 L_2 = L_2 L_1$.
   (c) Give an example where $L_1 L_2$ is neither self-adjoint nor skew-adjoint.

4.2   Polarization and Isometries

The idea of *polarization* is that many bilinear expressions such as $(x|y)$ can be expressed as a sum of quadratic terms $\|z\|^2 = (z|z)$ for suitable z.

Let us start with a real inner product on V. Then,

$$(x + y|x + y) = (x|x) + 2(x|y) + (y|y),$$

so

$$
(x|y) = \frac{1}{2}((x+y|x+y) - (x|x) - (y|y))
$$
$$
= \frac{1}{2}\left(\|x+y\|^2 - \|x\|^2 - \|y\|^2\right).
$$

Since complex inner products are only conjugate symmetric, we only get

$$(x + y|x + y) = (x|x) + 2\mathrm{Re}\,(x|y) + (y|y),$$

which implies

$$\mathrm{Re}\,(x|y) = \frac{1}{2}\left(\|x+y\|^2 - \|x\|^2 - \|y\|^2\right).$$

Nevertheless, the real part of the complex inner product determines the entire inner product as

$$
\mathrm{Re}\,(x|iy) = \mathrm{Re}\,(-i\,(x|y))
$$
$$
= \mathrm{Im}\,(x|y).
$$

In particular, we have

$$\mathrm{Im}\,(x|y) = \frac{1}{2}\left(\|x+iy\|^2 - \|x\|^2 - \|iy\|^2\right).$$

We can use these ideas to check when linear operators $L : V \to V$ are zero. First, we note that $L = 0$ if and only if $(L(x)|y) = 0$ for all $x, y \in V$. To check the "if" part, just let $y = L(x)$ to see that $\|L(x)\|^2 = 0$ for all $x \in V$. When L is self-adjoint, this can be improved.

Proposition 4.2.1. *Let $L : V \to V$ be self-adjoint on an inner product space. Then, $L = 0$ if and only if $(L(x)|x) = 0$ for all $x \in V$.*

Proof. There is nothing to prove when $L = 0$.

Conversely, assume that $(L(x)|x) = 0$ for all $x \in V$. The polarization trick from above implies

$$
\begin{aligned}
0 &= (L(x + y)|x + y)\\
&= (L(x)|x) + (L(x)|y) + (L(y)|x) + (L(y)|y)\\
&= (L(x)|y) + (y|L^*(x))\\
&= (L(x)|y) + (y|L(x))\\
&= 2\mathrm{Re}\,(L(x)|y)\,.
\end{aligned}
$$

Next, insert $y = L(x)$ to see that

$$
\begin{aligned}
0 &= \mathrm{Re}\,(L(x)|L(x))\\
&= \|L(x)\|^2
\end{aligned}
$$

as desired.   □

If L is not self-adjoint, there is no reason to think that such a result should hold. For instance, when V is a real inner product space and L is skew-symmetric, then we have

$$
\begin{aligned}
(L(x)|x) &= -(x|L(x))\\
&= -(L(x)|x)
\end{aligned}
$$

so $(L(x)|x) = 0$ for all x. Therefore, it is somewhat surprising that we can use the complex polarization trick to prove the next result.

Proposition 4.2.2. *Let $L : V \to V$ be a linear operator on a complex inner product space. Then, $L = 0$ if and only if $(L(x)|x) = 0$ for all $x \in V$.*

Proof. There is nothing to prove when $L = 0$.

Conversely, assume that $(L(x)|x) = 0$ for all $x \in V$. We use the complex polarization trick from above for fixed $x, y \in V$:

$$
\begin{aligned}
0 &= (L(x + y)|x + y)\\
&= (L(x)|x) + (L(x)|y) + (L(y)|x) + (L(y)|y)\\
&= (L(x)|y) + (L(y)|x)
\end{aligned}
$$

$$0 = (L\,(x + iy)\,|x + iy)$$
$$= (L\,(x)\,|x) + (L\,(x)\,|iy) + (L\,(iy)\,|x) + (L\,(iy)\,|iy)$$
$$= -i\,(L\,(x)\,|y) + i\,(L\,(y)\,|x)\,.$$

This yields a system

$$\begin{bmatrix} 1 & 1 \\ -i & i \end{bmatrix} \begin{bmatrix} (L\,(x)\,|y) \\ (L\,(y)\,|x) \end{bmatrix} = \begin{bmatrix} 0 \\ 0 \end{bmatrix}.$$

Since the columns of $\left[\begin{smallmatrix} 1 & 1 \\ -i & i \end{smallmatrix}\right]$ are linearly independent, the only solution is the trivial one. In particular, $(L\,(x)\,|y) = 0$. $\qquad\square$

Polarization can also be used to give a nice characterization of isometries (see also Sect. 3.3). These properties tie in nicely with our observation that

$$\left[\,e_1 \cdots e_n\,\right]^* = \left[\,e_1 \cdots e_n\,\right]^{-1}$$

when $e_1, \dots, e_n$ is an orthonormal basis.

Proposition 4.2.3. (Characterization of Isometries) *Let* $L\ :\ V\ \to\ W$ *be a linear map between finite-dimensional inner product spaces, then the following are equivalent:*

(1) $\|L\,(x)\| = \|x\|$ *for all* $x \in V$.
(2) $(L\,(x)\,|L\,(y)) = (x|y)$ *for all* $x, y \in V$.
(3) $L^*L = 1_V$
(4) L *takes orthonormal sets of vectors to orthonormal sets of vectors.*

Proof. $(1) \Rightarrow (2)$: Depending on whether we are in the complex or real case, simply write $(L\,(x)\,|L\,(y))$ and $(x|y)$ in terms of norms and use (1) to see that both terms are the same.

$(2) \Rightarrow (3)$: Just use that $(L^*L\,(x)\,|y) = (L\,(x)\,|L\,(y)) = (x|y)$ for all $x, y \in V$.

$(3) \Rightarrow (4)$: We are assuming $(x|y) = (L^*L\,(x)\,|y) = (L\,(x)\,|L\,(y))$, which immediately implies (4).

$(4) \Rightarrow (1)$: Evidently, L takes unit vectors to unit vectors. So (1) holds if $\|x\| = 1$. Now, use the scaling property of norms to finish the argument. $\qquad\square$

Recall the definition of the operator norm for linear maps $L : V \to W$

$$\|L\| = \max_{\|x\|=1} \|L\,(x)\|\,.$$

It was shown in Theorem 3.3.8 that this norm is finite when V is finite-dimensional. It is important to realize that this operator norm is not the same as the norm we get from the inner product $(L|K) = \mathrm{tr}\,(LK^*)$ defined on $\mathrm{Hom}\,(V, W)$. To see this, it suffices to consider 1_V. Clearly, $\|1_V\| = 1$, but $(1_V|1_V) = \mathrm{tr}\,(1_V 1_V) = \dim\,(V)$.

Remark 4.2.4. Note that if $L : V \to W$ satisfies the conditions in Proposition 4.2.3, then $\|L\| = 1$.

We also obtain

Corollary 4.2.5. (Characterization of Isometries) *Let $L : V \to W$ be an isomorphism between finite-dimensional inner product spaces, then L is an isometry if and only if $L^* = L^{-1}$.*

Proof. If L is an isometry, then it satisfies all of the above four conditions. In particular, $L^* L = 1_V$. Since L is invertible, it must follow that $L^{-1} = L^*$.

Conversely, if $L^{-1} = L^*$, then $L^* L = 1_V$, and it follows from Proposition 4.2.3 that $(L(x)|L(y)) = (x|y)$ so L is an isometry. □

Just as for self-adjoint and skew-adjoint operators, we have that isometries are completely reducible.

Corollary 4.2.6. (Reducibility of Isometries) *Let $L : V \to V$ be a linear operator on a finite-dimensional inner product space that is also an isometry. If $M \subset V$ is L-invariant, then so is $M^\perp$.*

Proof. If $x \in M$ and $y \in M^\perp$, then we note that

$$0 = (L(x)|y) = \left(x|L^*(y)\right).$$

Therefore, $L^*(y) = L^{-1}(y) \in M^\perp$ for all $y \in M^\perp$. Now observe that $L^{-1}|_{M^\perp} : M^\perp \to M^\perp$ must be an isomorphism as its kernel is trivial. This implies that each $z \in M^\perp$ is of the form $z = L^{-1}(y)$ for $y \in M^\perp$. Thus, $L(z) = y \in M^\perp$, and hence, $M^\perp$ is L-invariant. □

Definition 4.2.7. In the special case where $V = W = \mathbb{R}^n$, we call the linear isometries *orthogonal* matrices. The collection of orthogonal matrices is denoted O_n.

Note that these matrices are a *subgroup* of $Gl_n(\mathbb{R})$, i.e., if $A, B \in O_n$, then $AB \in O_n$. In particular, we see that O_n is itself a group. Similarly, when $V = W = \mathbb{C}^n$, we have the subgroup of *unitary* matrices $U_n \subset Gl_n(\mathbb{C})$ consisting of complex matrices that are also isometries.

Exercises

1. On $\text{Mat}_{n \times n}(\mathbb{R})$, use the inner product $(A|B) = \text{tr}(AB^t)$. Consider the linear operator $L(X) = X^t$. Show that L is orthogonal. Is it skew- or self-adjoint?
2. On $\text{Mat}_{n \times n}(\mathbb{C})$, use the inner product $(A|B) = \text{tr}(AB^*)$. For $A \in \text{Mat}_{n \times n}(\mathbb{C})$, consider the two linear operators on $\text{Mat}_{n \times n}(\mathbb{C})$ defined by $L_A(X) = AX$, $R_A(X) = XA$ (see also Exercise 3 in Sect. 3.5) Show that

   (a) L_A and R_A are unitary if A is unitary.
   (b) L_A and R_A are self- or skew-adjoint if A is self- or skew-adjoint.

3. Show that the operator D defines an isometry on both

$$\text{span}_{\mathbb{C}} \{\exp{(it)}, \exp{(-it)}\}$$

and

$$\text{span}_{\mathbb{R}} \{\cos{(t)}, \sin{(t)}\}$$

provided we use the inner product

$$(f|g) = \frac{1}{2\pi} \int_{-\pi}^{\pi} f(t) g(t) \, dt$$

inherited from $C_{2\pi}^{\infty} (\mathbb{R}, \mathbb{C})$.

4. Let $L : V \to V$ be a linear operator on a complex inner product space. Show that L is self-adjoint if and only if $(L(x)|x)$ is real for all $x \in V$.

5. Let $L : V \to V$ be a linear operator on a real inner product space. Show that L is skew-adjoint if and only if $(L(x)|x) = 0$ for all $x \in V$.

6. Let $e_1, \ldots, e_n$ be an orthonormal basis for V and assume that $L : V \to W$ has the property that $L(e_1), \ldots, L(e_n)$ is an orthonormal basis for W. Show that L is an isometry.

7. Let $L : V \to V$ be a linear operator on a finite-dimensional inner product space. Show that if $L \circ K = K \circ L$ for all isometries $K : V \to V$, then $L = \lambda 1_V$.

8. Let $L : V \to V$ be a linear operator on an inner product space such that $(L(x)|L(y)) = 0$ if $(x|y) = 0$.

   (a) Show that if $\|x\| = \|y\|$ and $(x|y) = 0$, then $\|L(x)\| = \|L(y)\|$. Hint: Use and show that $x + y$ and $x - y$ are perpendicular.
   (b) Show that $L = \lambda U$, where U is an isometry.

9. Let V be a finite-dimensional real inner product space and $F : V \to V$ be a bijective map that preserves distances, i.e., for all $x, y \in V$

$$\|F(x) - F(y)\| = \|x - y\|.$$

   (a) Show that $G(x) = F(x) - F(0)$ also preserves distances and that $G(0) = 0$.
   (b) Show that $\|G(x)\| = \|x\|$ for all $x \in V$.
   (c) Using polarization to show that $(G(x)|G(y)) = (x|y)$ for all $x, y \in V$. (See also next the exercise for what can happen in the complex case.)
   (d) If $e_1, \ldots, e_n$ is an orthonormal basis, then show that $G(e_1), \ldots, G(e_n)$ is also an orthonormal basis.
   (e) Show that

$$G(x) = (x|e_1) G(e_1) + \cdots + (x|e_n) G(e_n),$$

   and conclude that G is linear.
   (f) Conclude that $F(x) = L(x) + F(0)$ for a linear isometry L.

10. On $\text{Mat}_{n \times n}(\mathbb{C})$, use the inner product $(A|B) = \text{tr}(AB^*)$. Consider the map $L(X) = X^*$.

   (a) Show that L is real linear but not complex linear.
   (b) Show that

$$\|L(X) - L(Y)\| = \|X - Y\|$$

   for all X, Y but that

$$(L(X)|L(Y)) \neq (X|Y)$$

   for some choices of X, Y.

4.3 The Spectral Theorem

We are now ready to present and prove the most important theorem about when it is possible to find a basis that diagonalizes a special class of operators. This is the spectral theorem and it states that a self-adjoint linear operator on a finite-dimensional inner product space is diagonalizable with respect to an orthonormal basis. There are several reasons for why this particular result is important. Firstly, it forms the foundation for all of our other results for linear maps between inner product spaces, including isometries, skew-adjoint maps, and general linear maps between inner product spaces. Secondly, it is the one result of its type that has a truly satisfying generalization to infinite-dimensional spaces. In the infinite-dimensional setting, it becomes a cornerstone for several developments in analysis, functional analysis, partial differential equations, representation theory, and much more.

First, we revisit some material from Sect. 2.5. Our general goal for linear operators $L : V \to V$ is to find a basis such that the matrix representation for L is as simple as possible. Since the simplest matrices are the diagonal matrices, one might well ask if it is always possible to find a basis $x_1, \ldots, x_m$ that diagonalizes L, i.e., $L(x_1) = \lambda_1 x_1, \ldots, L(x_m) = \lambda_m x_m$? The central idea behind finding such a basis is quite simple and reappears in several proofs in this chapter. Given some special information about the linear operator L on V, we show that L^* has an eigenvector $x \neq 0$ and that the orthogonal complement to x in V is L-invariant. The existence of this invariant subspace of V then indicates that the procedure for establishing a particular result about exhibiting a nice matrix representation for L is a simple induction on the dimension of the vector space.

Example 4.3.1. A rotation by $90°$ in $\mathbb{R}^2$ does not have a basis of eigenvectors. However, if we interpret it as a complex map on $\mathbb{C}$, it is just multiplication by i and therefore already diagonalized. We could also view the 2×2 matrix as a map on $\mathbb{C}^2$. As such, we can also diagonalize it by using $x_1 = (i, 1)$ and $x_2 = (-i, 1)$ so that x_1 is mapped to $i x_1$ and x_2 to $-i x_2$.

Example 4.3.2. A much worse example is the linear map represented by

$$A = \begin{bmatrix} 0 & 1 \\ 0 & 0 \end{bmatrix}.$$

Here $x_1 = (1,0)$ does have the property that $Ax_1 = 0$, but it is not possible to find x_2 linearly independent from x_1 so that $Ax_2 = \lambda x_2$. In case $\lambda = 0$, we would just have $A = 0$ which is not true. So $\lambda \neq 0$, but then $x_2 \in \text{im}(A) = \text{span}\{x_1\}$. Note that using complex scalars cannot alleviate this situation due to the very general nature of the argument.

At this point, it should be more or less clear that the first goal is to show that self-adjoint operators have eigenvalues. Recall that in Sects. 2.3 and also 2.7, we constructed a characteristic polynomial for L with the property that any eigenvalue must be a root of this polynomial. This is fine if we work with complex scalars as we can then appeal to the Fundamental theorem of algebra 2.1.8 in order to find roots. But this is less satisfactory if we use real scalars. Although it is in fact not hard to deal with by passing to suitable matrix representations (see Exercise 9 in Sect. 4.1), Lagrange gave a very elegant proof (and most likely the first proof) that self-adjoint operators have real eigenvalues using Lagrange multipliers. We shall give a similar proof here that does not require quite as much knowledge of multivariable calculus.

Theorem 4.3.3. (Existence of Eigenvalues for Self-Adjoint Operators) *Let $L : V \to V$ be self-adjoint on a finite-dimensional inner product space. Then, L has a real eigenvalue.*

Proof. We use the compact set $S = \{x \in V : (x|x) = 1\}$ and the real-valued function $x \to (Lx|x)$ on S. Select $x_1 \in S$ so that

$$(Lx|x) \leq (Lx_1|x_1)$$

for all $x \in S$. If we define $\lambda_1 = (Lx_1|x_1)$, then this implies that

$$(Lx|x) \leq \lambda_1, \text{ for all } x \in S.$$

Consequently,

$$(Lx|x) \leq \lambda_1 (x|x) = \lambda_1 \|x\|^2, \text{ for all } x \in V.$$

This shows that the real-valued function

$$f(x) = \frac{(Lx|x)}{\|x\|^2}$$

has a maximum at $x = x_1$ and that the value there is λ_1.

This implies that for any $y \in V$, the function $\phi(t) = f(x_1 + ty)$ has a maximum at $t = 0$ and hence the derivative at $t = 0$ is zero. To be able to use this, we need to compute the derivative of

$$\phi(t) = \frac{(L(x_1 + ty) | x_1 + ty)}{\|x_1 + ty\|^2}$$

at $t = 0$. We start by computing the derivative of the numerator at $t = 0$ using the definition of a derivative

$$\lim_{h \to 0} \frac{(L(x_1 + hy) | x_1 + hy) - (L(x_1) | x_1)}{h}$$

$$= \lim_{h \to 0} \frac{(L(hy) | x_1) + (L(x_1) | hy) + (L(hy) | hy)}{h}$$

$$= \lim_{h \to 0} \frac{(hy | L(x_1)) + (L(x_1) | hy) + (L(hy) | hy)}{h}$$

$$= (y | L(x_1)) + (L(x_1) | y) + \lim_{h \to 0} (L(y) | hy)$$

$$= 2\text{Re}(L(x_1) | y).$$

The derivative of the denominator is computed the same way simply observing that we can let $L = 1_V$. The derivative of the quotient at $t = 0$ can now be calculated using that $\|x_1\| = 1$, $\lambda_1 = (Lx_1 | x_1)$, and the fact that $\lambda_1 \in \mathbb{R}$

$$\phi'(0) = \frac{2\text{Re}(L(x_1) | y) \|x_1\|^2 - 2\text{Re}(x_1 | y)(L(x_1) | x_1)}{\|x_1\|^4}$$

$$= 2\text{Re}(L(x_1) | y) - 2\text{Re}(x_1 | y)\lambda_1$$

$$= 2\text{Re}(L(x_1) - \lambda_1 x_1 | y).$$

Since $\phi'(0) = 0$ for any choice of y, we note that by using $y = L(x_1) - \lambda_1 x_1$, we obtain

$$0 = \phi'(0) = 2\|L(x_1) - \lambda_1 x_1\|^2.$$

This shows that λ_1 and x_1 form an eigenvalue/vector pair. $\qquad\square$

We can now prove.

Theorem 4.3.4. (The Spectral Theorem) *Let $L : V \to V$ be a self-adjoint operator on a finite-dimensional inner product space. Then, there exists an orthonormal basis $e_1, \ldots, e_n$ of eigenvectors, i.e., $L(e_1) = \lambda_1 e_1, \ldots, L(e_n) = \lambda_n e_n$. Moreover, all eigenvalues $\lambda_1, \ldots, \lambda_n$ are real.*

Proof. We just proved that we can find an eigenvalue/vector pair $L(e_1) = \lambda_1 e_1$. Recall that λ_1 was real and we can, if necessary, multiply e_1 by a suitable scalar to make it a unit vector.

Next, we use again that L is self-adjoint to see that L leaves the orthogonal complement to e_1 invariant (see also Proposition 4.1.7), i.e., $L(M) \subset M$, where $M = \{x \in V : (x|e_1) = 0\}$. To show this, let $x \in M$ and calculate

$$(L(x)|e_1) = (x|L^*(e_1))$$
$$= (x|L(e_1))$$
$$= (x|\lambda_1 e_1)$$
$$= \bar{\lambda}_1 (x|e_1)$$
$$= 0.$$

Now, we have a new operator $L : M \to M$ on a space of dimension $\dim M = \dim V - 1$. We note that this operator is also self-adjoint. Thus, we can use induction on $\dim V$ to prove the theorem. Alternatively, we can extract an eigenvalue/vector pair $L(e_2) = \lambda_2 e_2$, where $e_2 \in M$ is a unit vector and then pass down to the orthogonal complement of e_2 inside M. This procedure will end in $\dim V$ steps and will also generate an orthonormal basis of eigenvectors as the vectors are chosen successively to be orthogonal to each other.   □

In terms of matrix representations (see Sects. 1.7 and 3.5), we have proven the following:

Corollary 4.3.5. *Let $L : V \to V$ be a self-adjoint operator on a finite-dimensional inner product space. Then, there exists an orthonormal basis $e_1, \ldots, e_n$ of eigenvectors and a real $n \times n$ diagonal matrix D such that*

$$L = \begin{bmatrix} e_1 \cdots e_n \end{bmatrix} D \begin{bmatrix} e_1 \cdots e_n \end{bmatrix}^*$$

$$= \begin{bmatrix} e_1 \cdots e_n \end{bmatrix} \begin{bmatrix} \lambda_1 & \cdots & 0 \\ \vdots & \ddots & \vdots \\ 0 & \cdots & \lambda_n \end{bmatrix} \begin{bmatrix} e_1 \cdots e_n \end{bmatrix}^*.$$

The same eigenvalue can apparently occur several times, just think of the operator 1_V. Recall that the geometric multiplicity of an eigenvalue is $\dim(\ker(L - \lambda 1_V))$. This is clearly the same as the number of times it occurs in the above diagonal form of the operator. Thus, the basis vectors that correspond to λ in the diagonalization yield a basis for $\ker(L - \lambda 1_V)$. With this in mind, we can rephrase the spectral theorem. In the form stated below, it is also known as the *spectral resolution* of L with respect to 1_V as both of these operators are resolved according to the eigenspaces for L.

Theorem 4.3.6. (The Spectral Resolution of Self-Adjoint Operators) *Let $L : V \to V$ be a self-adjoint operator on a finite-dimensional inner product space and $\lambda_1, \ldots, \lambda_k$ the distinct eigenvalues for L. Then,*

$$1_V = \text{proj}_{\ker(L - \lambda_1 1_V)} + \cdots + \text{proj}_{\ker(L - \lambda_k 1_V)}$$

and

$$L = \lambda_1 \text{proj}_{\ker(L - \lambda_1 1_V)} + \cdots + \lambda_k \text{proj}_{\ker(L - \lambda_k 1_V)}.$$

Proof. The proof relies on showing that eigenspaces are mutually orthogonal to each other. This actually follows from our constructions in the proof of the spectral theorem. Nevertheless, it is desirable to have a direct proof of this. Let $L(x) = \lambda x$ and $L(y) = \mu y$, then

$$\begin{aligned}
\lambda(x|y) &= (L(x)|y) \\
&= (x|L(y)) \\
&= (x|\mu y) \\
&= \mu(x|y) \text{ since } \mu \text{ is real.}
\end{aligned}$$

If $\lambda \neq \mu$, then we get

$$(\lambda - \mu)(x|y) = 0,$$

which implies $(x|y) = 0$.

With this in mind, we can now see that if $x_i \in \ker(L - \lambda_i 1_V)$, then

$$\text{proj}_{\ker(L - \lambda_j 1_V)}(x_i) = \begin{cases} x_j & \text{if } i = j \\ 0 & \text{if } i \neq j \end{cases}$$

as x_i is perpendicular to $\ker(L - \lambda_j 1_V)$ in case $i \neq j$. Since we can write $x = x_1 + \cdots + x_k$, where $x_i \in \ker(L - \lambda_i 1_V)$, we have

$$\text{proj}_{\ker(L - \lambda_i 1_V)}(x) = x_i.$$

This shows that

$$x = \text{proj}_{\ker(L - \lambda_1 1_V)}(x) + \cdots + \text{proj}_{\ker(L - \lambda_k 1_V)}(x)$$

as well as

$$L(x) = \left(\lambda_1 \text{proj}_{\ker(L - \lambda_1 1_V)} + \cdots + \lambda_k \text{proj}_{\ker(L - \lambda_k 1_V)} \right)(x). \qquad \square$$

The fact that we can diagonalize self-adjoint operators has an immediate consequence for complex skew-adjoint operators as they become self-adjoint after multiplying them by $i = \sqrt{-1}$. Thus, we have

Corollary 4.3.7 (The Spectral Theorem for Complex Skew-Adjoint Operators).
Let $L : V \to V$ be a skew-adjoint linear operator on a complex finite-dimensional inner product space. Then, we can find an orthonormal basis such that $L(e_1) = i\mu_1 e_1, \ldots, L(e_n) = i\mu_n e_n$, where $\mu_1, \ldots, \mu_n \in \mathbb{R}$.

It is worth pondering this statement. Apparently, we have not said anything about skew-adjoint real linear operators. The statement, however, does cover both real and complex matrices as long as we view them as maps on $\mathbb{C}^n$. It just so happens that the corresponding diagonal matrix has purely imaginary entries, unless they are 0, and hence is forced to be complex.

Before doing some examples, it is worthwhile trying to find a way of remembering the formula

$$L = \begin{bmatrix} e_1 & \cdots & e_n \end{bmatrix} D \begin{bmatrix} e_1 & \cdots & e_n \end{bmatrix}^* .$$

If we solve it for D instead, it reads

$$D = \begin{bmatrix} e_1 & \cdots & e_n \end{bmatrix}^* L \begin{bmatrix} e_1 & \cdots & e_n \end{bmatrix} .$$

This is quite natural as

$$L \begin{bmatrix} e_1 & \cdots & e_n \end{bmatrix} = \begin{bmatrix} \lambda_1 e_1 & \cdots & \lambda_n e_n \end{bmatrix}$$

and then observing that

$$\begin{bmatrix} e_1 & \cdots & e_n \end{bmatrix}^* \begin{bmatrix} \lambda_1 e_1 & \cdots & \lambda_n e_n \end{bmatrix}$$

is the matrix whose ij entry is $(\lambda_j e_j | e_i)$ as the rows $\begin{bmatrix} e_1 & \cdots & e_n \end{bmatrix}^*$ correspond to the columns in $\begin{bmatrix} e_1 & \cdots & e_n \end{bmatrix}$. This gives a quick check for whether we have the change of basis matrices in the right places.

Example 4.3.8. Let

$$A = \begin{bmatrix} 0 & -i \\ i & 0 \end{bmatrix} .$$

Then, A is both self-adjoint and unitary. This shows that ± 1 are the only possible eigenvalues. We can easily find nontrivial solutions to both equations $(A \mp 1_{\mathbb{C}^2})(x) = 0$ by observing that

$$(A - 1_{\mathbb{C}^2}) \begin{bmatrix} -i \\ 1 \end{bmatrix} = \begin{bmatrix} -1 & -i \\ i & -1 \end{bmatrix} \begin{bmatrix} -i \\ 1 \end{bmatrix} = 0$$

$$(A + 1_{\mathbb{C}^2}) \begin{bmatrix} 1 \\ i \end{bmatrix} = \begin{bmatrix} 1 & -i \\ i & 1 \end{bmatrix} \begin{bmatrix} i \\ 1 \end{bmatrix} = 0$$

The vectors

$$z_1 = \begin{bmatrix} -i \\ 1 \end{bmatrix}, z_2 = \begin{bmatrix} i \\ 1 \end{bmatrix}$$

form an orthogonal set that we can normalize to an orthonormal basis of eigenvectors

$$x_1 = \begin{bmatrix} \frac{-i}{\sqrt{2}} \\ \frac{1}{\sqrt{2}} \end{bmatrix}, x_2 = \begin{bmatrix} \frac{i}{\sqrt{2}} \\ \frac{1}{\sqrt{2}} \end{bmatrix} .$$

This means that

$$A = \begin{bmatrix} x_1 & x_2 \end{bmatrix} \begin{bmatrix} 1 & 0 \\ 0 & -1 \end{bmatrix} \begin{bmatrix} x_1 & x_2 \end{bmatrix}^{-1}$$

or more concretely that

$$\begin{bmatrix} 0 & -i \\ i & 0 \end{bmatrix} = \begin{bmatrix} \frac{-i}{\sqrt{2}} & \frac{i}{\sqrt{2}} \\ \frac{1}{\sqrt{2}} & \frac{1}{\sqrt{2}} \end{bmatrix} \begin{bmatrix} 1 & 0 \\ 0 & -1 \end{bmatrix} \begin{bmatrix} \frac{i}{\sqrt{2}} & \frac{1}{\sqrt{2}} \\ \frac{-i}{\sqrt{2}} & \frac{1}{\sqrt{2}} \end{bmatrix}.$$

Example 4.3.9. Let

$$B = \begin{bmatrix} 0 & -1 \\ 1 & 0 \end{bmatrix}.$$

The corresponding self-adjoint matrix is

$$\begin{bmatrix} 0 & -i \\ i & 0 \end{bmatrix}.$$

Using the identity

$$\begin{bmatrix} 0 & -i \\ i & 0 \end{bmatrix} = \begin{bmatrix} \frac{-i}{\sqrt{2}} & \frac{i}{\sqrt{2}} \\ \frac{1}{\sqrt{2}} & \frac{1}{\sqrt{2}} \end{bmatrix} \begin{bmatrix} 1 & 0 \\ 0 & -1 \end{bmatrix} \begin{bmatrix} \frac{i}{\sqrt{2}} & \frac{1}{\sqrt{2}} \\ \frac{-i}{\sqrt{2}} & \frac{1}{\sqrt{2}} \end{bmatrix}$$

and then multiplying by $-i$ to get back to

$$\begin{bmatrix} 0 & -1 \\ 1 & 0 \end{bmatrix},$$

we obtain

$$\begin{bmatrix} 0 & -1 \\ 1 & 0 \end{bmatrix} = \begin{bmatrix} \frac{-i}{\sqrt{2}} & \frac{i}{\sqrt{2}} \\ \frac{1}{\sqrt{2}} & \frac{1}{\sqrt{2}} \end{bmatrix} \begin{bmatrix} -i & 0 \\ 0 & i \end{bmatrix} \begin{bmatrix} \frac{i}{\sqrt{2}} & \frac{1}{\sqrt{2}} \\ \frac{-i}{\sqrt{2}} & \frac{1}{\sqrt{2}} \end{bmatrix}.$$

It is often more convenient to find the eigenvalues using the characteristic polynomial; to see why this is, let us consider some more complicated examples.

Example 4.3.10. We consider the real symmetric operator

$$A = \begin{bmatrix} \alpha & \beta \\ \beta & \alpha \end{bmatrix}, \alpha, \beta \in \mathbb{R}.$$

This time, one can more or less readily see that

$$x_1 = \begin{bmatrix} 1 \\ 1 \end{bmatrix}, x_2 = \begin{bmatrix} 1 \\ -1 \end{bmatrix}$$

are eigenvectors and that the corresponding eigenvalues are $(\alpha \pm \beta)$. But if one did not guess that, then computing the characteristic polynomial is clearly the way to go.

Even with relatively simple examples such as

$$A = \begin{bmatrix} 1 & 1 \\ 1 & 2 \end{bmatrix}$$

things quickly get out of hand. Clearly, the method of using Gauss elimination on the system $A - \lambda 1_{\mathbb{C}^n}$ and then finding conditions on λ that ensure that we have nontrivial solutions is more useful in finding all eigenvalues/vectors.

Example 4.3.11. Let us try this with

$$A = \begin{bmatrix} 1 & 1 \\ 1 & 2 \end{bmatrix}.$$

Thus, we consider

$$\begin{bmatrix} 1-\lambda & 1 & 0 \\ 1 & 2-\lambda & 0 \end{bmatrix}$$

$$\begin{bmatrix} 1 & 2-\lambda & 0 \\ 1-\lambda & 1 & 0 \end{bmatrix}$$

$$\begin{bmatrix} 1 & (2-\lambda) & 0 \\ 0 & -(1-\lambda)(2-\lambda)+1 & 0 \end{bmatrix}$$

Thus, there is a nontrivial solution precisely when

$$-(1-\lambda)(2-\lambda)+1 = -1 + 3\lambda - \lambda^2 = 0.$$

The roots of this polynomial are $\lambda_{1,2} = \frac{3}{2} \pm \frac{1}{2}\sqrt{5}$. The corresponding eigenvectors are found by inserting the root and then finding a nontrivial solution. Thus, we are trying to solve

$$\begin{bmatrix} 1 & (2-\lambda_{1,2}) & 0 \\ 0 & 0 & 0 \end{bmatrix}$$

which means that

$$x_{1,2} = \begin{bmatrix} \lambda_{1,2} - 2 \\ 1 \end{bmatrix}.$$

We should normalize this to get a unit vector

$$e_{1,2} = \frac{1}{\sqrt{5 - 4\lambda_{1,2} + (\lambda_{1,2})^2}} \begin{bmatrix} \lambda_{1,2} - 2 \\ 1 \end{bmatrix}$$

$$= \frac{1}{\sqrt{\left(\frac{5}{2} \mp \frac{1}{2}\sqrt{5}\right)}} \begin{bmatrix} -1 \pm \sqrt{5} \\ 1 \end{bmatrix}$$

Exercises

1. Let L be self- or skew-adjoint on a complex finite-dimensional inner product space.

   (a) Show that $L = K^2$ for some $K : V \to V$.
   (b) Show by example that K need not be self-adjoint if L is self-adjoint.
   (c) Show by example that K need not be skew-adjoint if L is skew-adjoint.

2. Diagonalize the matrix that is zero everywhere except for 1s on the anti-diagonal.

$$\begin{bmatrix} 0 & \cdots & 0 & 1 \\ \vdots & & 1 & 0 \\ 0 & & & \vdots \\ 1 & 0 & \cdots & 0 \end{bmatrix}$$

3. Diagonalize the real matrix that has αs on the diagonal and βs everywhere else.

$$\begin{bmatrix} \alpha & \beta & \cdots & \beta \\ \beta & \alpha & & \beta \\ \vdots & & \ddots & \vdots \\ \beta & \beta & \cdots & \alpha \end{bmatrix}$$

4. Let $K, L : V \to V$ be self-adjoint operators on a finite-dimensional vector space. If $KL = LK$, then show that there is an orthonormal basis diagonalizing both K and L.

5. Let $L : V \to V$ be self-adjoint. If there is a unit vector $x \in V$ such that

$$\|L(x) - \mu x\| \le \varepsilon,$$

   then L has an eigenvalue λ so that $|\lambda - \mu| \le \varepsilon$.

6. Let $L : V \to V$ be self-adjoint on a finite-dimensional inner product space. Show that either $\|L\|$ or $-\|L\|$ are eigenvalues for L.

7. If an operator $L : V \to V$ on a finite-dimensional inner product space satisfies one of the following four conditions, then it is said to be *positive*. Show that these conditions are equivalent:

   (a) L is self-adjoint with positive eigenvalues.
   (b) L is self-adjoint and $(L(x)|x) > 0$ for all $x \in V - \{0\}$.
   (c) $L = K^* \circ K$ for an injective operator $K : V \to W$, where W is also an inner product space.
   (d) $L = K \circ K$ for an invertible self-adjoint operator $K : V \to V$.

8. Let P, Q be two positive operators on a finite-dimensional inner product space. If $P^2 = Q^2$, then show that $P = Q$.

9. Let P be a nonnegative operator on a finite-dimensional inner product space, i.e., self-adjoint with nonnegative eigenvalues.

   (a) Show that $\operatorname{tr} P \geq 0$.
   (b) Show that $P = 0$ if and only if $\operatorname{tr} P = 0$.

10. Let $L : V \to V$ be a linear operator on a finite-dimensional inner product space.

    (a) If L is self-adjoint, show that L^2 is self-adjoint and has nonnegative eigenvalues.
    (b) If L is skew-adjoint, show that L^2 is self-adjoint and has nonpositive eigenvalues.

11. Consider the *Killing form* on $\operatorname{Hom}(V, V)$, where V is a finite-dimensional vector space of dimension > 1, defined by

$$K(L, K) = \operatorname{tr} L \operatorname{tr} K - \operatorname{tr}(LK).$$

    (a) Show that $K(L, K) = K(K, L)$.
    (b) Show that $K \to K(L, K)$ is linear.
    (c) Assume in addition that V is an inner product space. Show that $K(L, L) > 0$ if L is skew-adjoint and $L \neq 0$.
    (d) Show that $K(L, L) < 0$ if L is self-adjoint and $L \neq 0$.
    (e) Show that K is nondegenerate, i.e., if $L \neq 0$, then we can find $K \neq 0$, so that $K(L, K) \neq 0$. Hint: Let $K = \frac{1}{2}(L + L^*)$ or $K = \frac{1}{2}(L - L^*)$ depending on the value of $\operatorname{tr}\left(\frac{1}{2}(L + L^*)\frac{1}{2}(L - L^*)\right)$.

4.4　Normal Operators

The concept of a normal operator is somewhat more general than the previous special types of operators we have encountered. The definition is quite simple and will be motivated below.

Definition 4.4.1. We say that a linear operator $L : V \to V$ on a finite-dimensional inner product space is *normal* if $LL^* = L^*L$.

With this definition, it is clear that all self-adjoint, skew-adjoint, and isometric operators are normal.

First, let us show that any operator that is diagonalizable with respect to an orthonormal basis must be normal. Suppose that L is diagonalized in the orthonormal basis $e_1, \ldots, e_n$ and that D is the diagonal matrix representation in this basis, then

$$L = \begin{bmatrix} e_1 \cdots e_n \end{bmatrix} D \begin{bmatrix} e_1 \cdots e_n \end{bmatrix}^*$$

$$= \begin{bmatrix} e_1 \cdots e_n \end{bmatrix} \begin{bmatrix} \lambda_1 & \cdots & 0 \\ \vdots & \ddots & \vdots \\ 0 & \cdots & \lambda_n \end{bmatrix} \begin{bmatrix} e_1 \cdots e_n \end{bmatrix}^*,$$

and

$$L^* = \begin{bmatrix} e_1 \cdots e_n \end{bmatrix} D^* \begin{bmatrix} e_1 \cdots e_n \end{bmatrix}^*$$

$$= \begin{bmatrix} e_1 \cdots e_n \end{bmatrix} \begin{bmatrix} \bar{\lambda}_1 & \cdots & 0 \\ \vdots & \ddots & \vdots \\ 0 & \cdots & \bar{\lambda}_n \end{bmatrix} \begin{bmatrix} e_1 \cdots e_n \end{bmatrix}^*.$$

Thus,

$$LL^* = \begin{bmatrix} e_1 \cdots e_n \end{bmatrix} \begin{bmatrix} \lambda_1 & \cdots & 0 \\ \vdots & \ddots & \vdots \\ 0 & \cdots & \lambda_n \end{bmatrix} \begin{bmatrix} \bar{\lambda}_1 & \cdots & 0 \\ \vdots & \ddots & \vdots \\ 0 & \cdots & \bar{\lambda}_n \end{bmatrix} \begin{bmatrix} e_1 \cdots e_n \end{bmatrix}^*$$

$$= \begin{bmatrix} e_1 \cdots e_n \end{bmatrix} \begin{bmatrix} |\lambda_1|^2 & \cdots & 0 \\ \vdots & \ddots & \vdots \\ 0 & \cdots & |\lambda_n|^2 \end{bmatrix} \begin{bmatrix} e_1 \cdots e_n \end{bmatrix}^*$$

$$= L^*L$$

since $DD^* = D^*D$.

For real operators, the spectral theorem tells us that they must be self-adjoint in order to be diagonalizable with respect to an orthonormal basis. For complex operators, things are a little different as also skew-adjoint operators are diagonalizable with respect to an orthonormal basis. Below we shall generalize the spectral theorem to normal operators and show that in the complex case, these are precisely the operators that can be diagonalized with respect to an orthonormal basis. Another very simple normal operator that is not necessarily of those three types is the homothety $\lambda 1_V$ for all $\lambda \in \mathbb{C}$. The canonical form for real normal operators is somewhat more complicated and will be studied in Sect. 4.6.

Example 4.4.2. We note that

$$\begin{bmatrix} 1 & 1 \\ 0 & 2 \end{bmatrix}$$

is not normal as

$$\begin{bmatrix} 1 & 1 \\ 0 & 2 \end{bmatrix}\begin{bmatrix} 1 & 0 \\ 1 & 2 \end{bmatrix} = \begin{bmatrix} 2 & 2 \\ 2 & 4 \end{bmatrix},$$

$$\begin{bmatrix} 1 & 0 \\ 1 & 2 \end{bmatrix}\begin{bmatrix} 1 & 1 \\ 0 & 2 \end{bmatrix} = \begin{bmatrix} 1 & 1 \\ 1 & 5 \end{bmatrix}.$$

Nevertheless, it is diagonalizable with respect to the basis

$$x_1 = \begin{bmatrix} 1 \\ 0 \end{bmatrix}, x_2 = \begin{bmatrix} 1 \\ 1 \end{bmatrix}$$

as

$$\begin{bmatrix} 1 & 1 \\ 0 & 2 \end{bmatrix}\begin{bmatrix} 1 \\ 0 \end{bmatrix} = \begin{bmatrix} 1 \\ 0 \end{bmatrix},$$

$$\begin{bmatrix} 1 & 1 \\ 0 & 2 \end{bmatrix}\begin{bmatrix} 1 \\ 1 \end{bmatrix} = \begin{bmatrix} 2 \\ 2 \end{bmatrix} = 2\begin{bmatrix} 1 \\ 1 \end{bmatrix}.$$

While we can normalize x_2 to be a unit vector, there is nothing we can do about x_1 and x_2 not being perpendicular.

Example 4.4.3. Let

$$A = \begin{bmatrix} \alpha & \gamma \\ \beta & \delta \end{bmatrix} : \mathbb{C}^2 \to \mathbb{C}^2.$$

Then,

$$AA^* = \begin{bmatrix} \alpha & \gamma \\ \beta & \delta \end{bmatrix}\begin{bmatrix} \bar\alpha & \bar\beta \\ \bar\gamma & \bar\delta \end{bmatrix} = \begin{bmatrix} |\alpha|^2 + |\gamma|^2 & \alpha\bar\beta + \gamma\bar\delta \\ \bar\alpha\beta + \bar\gamma\delta & |\beta|^2 + |\delta|^2 \end{bmatrix}$$

$$A^*A = \begin{bmatrix} \bar\alpha & \bar\beta \\ \bar\gamma & \bar\delta \end{bmatrix}\begin{bmatrix} \alpha & \gamma \\ \beta & \delta \end{bmatrix} = \begin{bmatrix} |\alpha|^2 + |\beta|^2 & \bar\alpha\gamma + \bar\beta\delta \\ \alpha\bar\gamma + \beta\bar\delta & |\gamma|^2 + |\delta|^2 \end{bmatrix}.$$

So the conditions for A to be normal are

$$|\beta|^2 = |\gamma|^2,$$
$$\alpha\bar\gamma + \beta\bar\delta = \bar\alpha\beta + \bar\gamma\delta.$$

The last equation is easier to remember if we note that it means that the columns of A must have the same inner product as the columns of A^*.

Proposition 4.4.4. (Characterization of Normal Operators) *Let $L : V \to V$ be an operator on a finite-dimensional inner product space. Then, the following conditions are equivalent:*

(1) $LL^* = L^*L$.
(2) $\|L(x)\| = \|L^*(x)\|$ *for all* $x \in V$.
(3) $AB = BA$, *where* $A = \frac{1}{2}(L + L^*)$ *and* $B = \frac{1}{2}(L - L^*)$.

Proof. $(1) \Leftrightarrow (2)$: Note that for all $x \in V$, we have

$$
\begin{aligned}
\|L(x)\| = \|L^*(x)\| &\Leftrightarrow \|L(x)\|^2 = \|L^*(x)\|^2 \\
&\Leftrightarrow (L(x)|L(x)) = (L^*(x)|L^*(x)) \\
&\Leftrightarrow (x|L^*L(x)) = (x|LL^*(x)) \\
&\Leftrightarrow (x|(L^*L - LL^*)(x)) = 0 \\
&\Leftrightarrow L^*L - LL^* = 0
\end{aligned}
$$

The last implication is a consequence of the fact that $L^*L - LL^*$ is self-adjoint (see Proposition 4.2.1).

$(3) \Leftrightarrow (1)$: We note that

$$
\begin{aligned}
AB &= \frac{1}{2}(L + L^*)\frac{1}{2}(L - L^*) \\
&= \frac{1}{4}(L + L^*)(L - L^*) \\
&= \frac{1}{4}\left(L^2 - (L^*)^2 + L^*L - LL^*\right), \\
BA &= \frac{1}{4}(L - L^*)(L + L^*) \\
&= \frac{1}{4}\left(L^2 - (L^*)^2 - L^*L + LL^*\right).
\end{aligned}
$$

So $AB = BA$ if and only if $L^*L - LL^* = -L^*L + LL^*$ which is the same as saying that $LL^* = L^*L$. $\square$

We also need a general result about invariant subspaces.

Lemma 4.4.5. *Let* $L : V \to V$ *be a linear operator on a finite-dimensional inner product space. If* $M \subset V$ *is an* L- *and* L^*-*invariant subspace, then* $M^{\perp}$ *is also* L- *and* L^*-*invariant. In particular,*

$$
(L|_{M^{\perp}})^* = L^*|_{M^{\perp}}.
$$

Proof. Let $x \in M$ and $y \in M^{\perp}$. We have to show that

$$
(x|L(y)) = 0,
$$
$$
(x|L^*(y)) = 0.
$$

For the first identity, use that

$$
(x|L(y)) = (L^*(x)|y) = 0
$$

as $L^*(x) \in M$. Similarly, for the second use that

$$\left(x|L^*(y)\right) = (L(x)|y) = 0$$

as $L(x) \in M$. $\square$

We are now ready to prove the spectral theorem for normal operators.

Theorem 4.4.6 (The Spectral Theorem for Normal Operators). *Let $L : V \to V$ be a normal operator on a finite-dimensional complex inner product space; then, there is an orthonormal basis $e_1, \ldots, e_n$ of eigenvectors, i.e., $L(e_1) = \lambda_1 e_1, \ldots, L(e_n) = \lambda_n e_n$.*

Proof. As with the spectral theorem (see Theorem 4.3.4), the proof depends on showing that we can find an eigenvalue and that the orthogonal complement to an eigenvector is invariant.

Rather than appealing to the Fundamental Theorem of Algebra 2.1.8 in order to find an eigenvalue for L, we shall use what we know about self-adjoint operators. This has the advantage of also yielding a proof that works in the real case (see Sect. 4.6). First, decompose $L = A + iC$ where $A = \frac{1}{2}(L + L^*)$ and $C = \frac{1}{i}B = \frac{1}{2i}(L - L^*)$ are both self-adjoint (compare with Proposition 4.4.4). Then, use Theorem 4.3.3 to find $\alpha \in \mathbb{R}$ such that $\ker(A - \alpha 1_V) \neq \{0\}$. If $x \in \ker(A - \alpha 1_V)$, then

$$
\begin{aligned}
(A - \alpha 1_V)(C(x)) &= AC(x) - \alpha C(x) \\
&= CA(x) - C(\alpha x) \\
&= C((A - \alpha 1_V)(x)) \\
&= 0.
\end{aligned}
$$

Thus, $\ker(A - \alpha 1_V)$ is a C-invariant subspace. Using that C and hence also its restriction to $\ker(A - \alpha 1_V)$ is self-adjoint, we can find $x \in \ker(A - \alpha 1_V)$ so that $C(x) = \beta x$, with $\beta \in \mathbb{R}$ (see Theorem 4.3.3). This means that

$$
\begin{aligned}
L(x) &= A(x) + iC(x) \\
&= \alpha x + i\beta x \\
&= (\alpha + i\beta)x.
\end{aligned}
$$

Hence, we have found an eigenvalue $\alpha + i\beta$ for L with a corresponding eigenvector x. It follows from Proposition 3.5.2 that

$$
\begin{aligned}
L^*(x) &= A(x) - iC(x) \\
&= (\alpha - i\beta)x.
\end{aligned}
$$

Thus, span $\{x\}$ is both L- and L^*-invariant. Lemma 4.4.5 then shows that $M = (\text{span}\{x\})^{\perp}$ is also L- and L^*-invariant. Hence, $(L|_M)^* = L^*|_M$ showing that $L|_M : M \to M$ is also normal. We can then use induction as in the spectral theorem to finish the proof. $\qquad \square$

As an immediate consequence, we get a result for unitary operators.

Theorem 4.4.7 (The Spectral Theorem for Unitary Operators). *Let $L : V \to V$ be unitary; then, there is an orthonormal basis $e_1, \ldots, e_n$ such that $L(e_1) = e^{i\theta_1}e_1, \ldots, L(e_n) = e^{i\theta_n}e_n$, where $\theta_1, \ldots, \theta_n \in \mathbb{R}$.*

We also have the resolution version of the spectral theorem.

Theorem 4.4.8 (The Spectral Resolution of Normal Operators). *Let $L : V \to V$ be a normal operator on a complex finite-dimensional inner product space and $\lambda_1, \ldots, \lambda_k$ the distinct eigenvalues for L. Then,*

$$1_V = \text{proj}_{\ker(L-\lambda_1 1_V)} + \cdots + \text{proj}_{\ker(L-\lambda_k 1_V)}$$

and

$$L = \lambda_1 \text{proj}_{\ker(L-\lambda_1 1_V)} + \cdots + \lambda_k \text{proj}_{\ker(L-\lambda_k 1_V)}.$$

Let us see what happens in some examples.

Example 4.4.9. Let

$$L = \begin{bmatrix} \alpha & \beta \\ -\beta & \alpha \end{bmatrix}, \alpha, \beta \in \mathbb{R};$$

then L is normal. When $\alpha = 0$, it is skew-adjoint; when $\beta = 0$, it is self-adjoint; and when $\alpha^2 + \beta^2 = 1$, it is an orthogonal transformation. The decomposition $L = A + iC$ looks like

$$\begin{bmatrix} \alpha & \beta \\ -\beta & \alpha \end{bmatrix} = \begin{bmatrix} \alpha & 0 \\ 0 & \alpha \end{bmatrix} + i \begin{bmatrix} 0 & -i\beta \\ i\beta & 0 \end{bmatrix}.$$

Here

$$\begin{bmatrix} \alpha & 0 \\ 0 & \alpha \end{bmatrix}$$

has α as an eigenvalue and

$$\begin{bmatrix} 0 & -i\beta \\ i\beta & 0 \end{bmatrix}$$

has $\pm\beta$ as eigenvalues. Thus, L has eigenvalues $(\alpha \pm i\beta)$.

Example 4.4.10. The matrix

$$\begin{bmatrix} 0 & 1 & 0 \\ -1 & 0 & 0 \\ 0 & 0 & 1 \end{bmatrix}$$

is normal and has 1 as an eigenvalue. We are then reduced to looking at

$$\begin{bmatrix} 0 & 1 \\ -1 & 0 \end{bmatrix}$$

which has $\pm i$ as eigenvalues.

Exercises

1. Consider $L_A(X) = AX$ and $R_A(X) = XA$ as linear operators on $\mathrm{Mat}_{n \times n}(\mathbb{C})$. What conditions do you need on A in order for these maps to be normal (see also Exercise 3 in Sect. 3.5)?

2. Assume that $L : V \to V$ is normal and that $p \in \mathbb{F}[t]$. Show that $p(L)$ is also normal.

3. Assume that $L : V \to V$ is normal. Without using the spectral theorem, show:

   (a) $\ker(L) = \ker(L^*)$.
   (b) $\ker(L - \lambda 1_V) = \ker(L^* - \bar{\lambda} 1_V)$.
   (c) $\mathrm{im}(L) = \mathrm{im}(L^*)$.
   (d) $(\ker(L))^\perp = \mathrm{im}(L)$.

4. Assume that $L : V \to V$ is normal. Without using the spectral theorem, show:

   (a) $\ker(L) = \ker(L^k)$ for any $k \geq 1$. Hint: Use the self-adjoint operator $K = L^*L$.
   (b) $\mathrm{im}(L) = \mathrm{im}(L^k)$ for any $k \geq 1$.
   (c) $\ker(L - \lambda 1_V) = \ker\left((L - \lambda 1_V)^k\right)$ for any $k \geq 1$.
   (d) Show that the minimal polynomial of L has no multiple roots.

5. *(Characterization of Normal Operators)* Let $L : V \to V$ be a linear operator on a finite-dimensional inner product space. Show that L is normal if and only if $(L \circ E | L \circ E) = (L^* \circ E | L^* \circ E)$ for all orthogonal projections $E : V \to V$. Hint: Use the formula

$$(L_1 | L_2) = \sum_{i=1}^{n} (L_1(e_i) | L_2(e_i))$$

   from Exercise 4 in Sect. 3.5 for suitable choices of orthonormal bases $e_1, \ldots, e_n$ for V.

6. *(Reducibility of Normal Operators)* Let $L : V \to V$ be an operator on a finite-dimensional inner product space. Assume that $M \subset V$ is an L-invariant subspace and let $E : V \to V$ be the orthogonal projection onto M.

(a) Justify all of the steps in the calculation:

$$
\left(L^* \circ E \mid L^* \circ E\right) = \left(E^{\perp} \circ L^* \circ E \mid E^{\perp} \circ L^* \circ E\right) + \left(E \circ L^* \circ E \mid E \circ L^* \circ E\right)
$$
$$
= \left(E^{\perp} \circ L^* \circ E \mid E^{\perp} \circ L^* \circ E\right) + \left(E \circ L \circ E \mid E \circ L \circ E\right)
$$
$$
= \left(E^{\perp} \circ L^* \circ E \mid E^{\perp} \circ L^* \circ E\right) + \left(L \circ E \mid L \circ E\right).
$$

Hint: Use the result that $E^* = E$ from Sect. 3.6 and that $L(M) \subset M$ implies $E \circ L \circ E = L \circ E$ and Exercise 4 in Sect. 3.5 .

(b) If L is normal, use the previous exercise to conclude that M is L^*-invariant and $M^{\perp}$ is L-invariant, i.e., normal operators are completely reducible.

7. *(Characterization of Normal Operators)* Let $L : V \to V$ be a linear map on a finite-dimensional inner product space. Assume that L has the property that all L-invariant subspaces are also L^*-invariant.

   (a) Show that L is completely reducible (see Proposition 4.1.7).
   (b) Show that the matrix representation with respect to an orthonormal basis is diagonalizable when viewed as complex matrix.
   (c) Show that L is normal.

8. Assume that $L : V \to V$ satisfies $L^* L = \lambda 1_V$, for some $\lambda \in \mathbb{C}$. Show that L is normal.

9. Show that if a projection is normal, then it is an orthogonal projection.

10. Show that if $L : V \to V$ is normal and $p \in \mathbb{F}[t]$, then $p(L)$ is also normal. Moreover, if $\mathbb{F} = \mathbb{C}$, then the spectral resolution is given by

$$
p(L) = p(\lambda_1) \operatorname{proj}_{\ker(L - \lambda_1 1_V)} + \cdots + p(\lambda_k) \operatorname{proj}_{\ker(L - \lambda_k 1_V)}.
$$

11. Let $L, K : V \to V$ be normal. Show by example that neither $L + K$ nor LK need be normal.

12. Let A be an upper triangular matrix. Show that A is normal if and only if it is diagonal. Hint: Compute and compare the diagonal entries in AA^* and A^*A.

13. *(Characterization of Normal Operators)* Let $L : V \to V$ be an operator on a finite-dimensional complex inner product space. Show that L is normal if and only if $L^* = p(L)$ for some polynomial p.

14. *(Characterization of Normal Operators)* Let $L : V \to V$ be an operator on a finite-dimensional complex inner product space. Show that L is normal if and only if $L^* = LU$ for some unitary operator $U : V \to V$.

15. Let $L : V \to V$ be normal on a finite-dimensional complex inner product space. Show that $L = K^2$ for some normal operator K.

16. Give the canonical form for the linear operators that are both self-adjoint and unitary.

17. Give the canonical form for the linear operators that are both skew-adjoint and unitary.

4.5   Unitary Equivalence

In the special case where $V = \mathbb{F}^n$, the spectral theorem can be rephrased in terms of change of basis. Recall from Sect. 1.9 that if we pick a different basis $x_1, \ldots, x_n$ for $\mathbb{F}^n$, then the matrix representations for a linear map represented by A in the standard basis and B in the new basis are related by

$$A = \begin{bmatrix} x_1 \cdots x_n \end{bmatrix} B \begin{bmatrix} x_1 \cdots x_n \end{bmatrix}^{-1}.$$

In case $x_1, \ldots, x_n$ is an orthonormal basis, this simplifies to

$$A = \begin{bmatrix} x_1 \cdots x_n \end{bmatrix} B \begin{bmatrix} x_1 \cdots x_n \end{bmatrix}^*,$$

where $\begin{bmatrix} x_1 \cdots x_n \end{bmatrix}$ is a unitary or orthogonal operator.

Definition 4.5.1. Two $n \times n$ matrices A and B are said to be *unitarily equivalent* if $A = UBU^*$, where $U \in U_n$, i.e., U is an $n \times n$ matrix such that $U^*U = UU^* = 1_{\mathbb{F}^n}$. In case $U \in O_n \subset U_n$, we also say that the matrices are *orthogonally equivalent*.

The results from the previous two sections can now be paraphrased in the following way.

Corollary 4.5.2. *Let $A \in \mathrm{Mat}_{n \times n}(\mathbb{C})$*

(1) If A is normal, then A is unitarily equivalent to a diagonal matrix.
(2) If A is self-adjoint, then A is unitarily (or orthogonally) equivalent to a real diagonal matrix.
(3) If A is skew-adjoint, then A is unitarily equivalent to a diagonal matrix whose entries are purely imaginary.
(4) If A is unitary, then A is unitarily equivalent to a diagonal matrix whose diagonal entries are unit scalars.

Using the group properties of unitary matrices, one can easily show the next two results.

Proposition 4.5.3. *If A and B are unitarily equivalent, then*

(1) A is normal if and only if B is normal.
(2) A is self-adjoint if and only if B is self-adjoint.
(3) A is skew-adjoint if and only if B is skew-adjoint.
(4) A is unitary if and only if B is unitary.

Corollary 4.5.4. *Two normal operators are unitarily equivalent if and only if they have the same eigenvalues (counted with multiplicities).*

Example 4.5.5. The Pauli matrices are defined by

$$\begin{bmatrix} 0 & 1 \\ 1 & 0 \end{bmatrix}, \begin{bmatrix} 1 & 0 \\ 0 & -1 \end{bmatrix}, \begin{bmatrix} 0 & -i \\ i & 0 \end{bmatrix}.$$

They are all self-adjoint and unitary. Moreover, all have eigenvalues ± 1, so they are all unitarily equivalent.

If we multiply the Pauli matrices by i, we get three skew-adjoint and unitary matrices with eigenvalues $\pm i$:

$$\begin{bmatrix} 0 & 1 \\ -1 & 0 \end{bmatrix}, \begin{bmatrix} i & 0 \\ 0 & -i \end{bmatrix}, \begin{bmatrix} 0 & i \\ i & 0 \end{bmatrix}$$

that are also all unitarily equivalent. The eight matrices

$$\pm \begin{bmatrix} 1 & 0 \\ 0 & 1 \end{bmatrix}, \pm \begin{bmatrix} i & 0 \\ 0 & -i \end{bmatrix}, \pm \begin{bmatrix} 0 & 1 \\ -1 & 0 \end{bmatrix}, \pm \begin{bmatrix} 0 & i \\ i & 0 \end{bmatrix}$$

form a group that corresponds to the quaternions $\pm 1, \pm i, \pm j, \pm k$.

Example 4.5.6. The matrices

$$\begin{bmatrix} 1 & 1 \\ 0 & 2 \end{bmatrix}, \begin{bmatrix} 1 & 0 \\ 0 & 2 \end{bmatrix}$$

are not unitarily equivalent as the first is not normal while the second is normal. Note, however, that both are diagonalizable with the same eigenvalues.

Exercises

1. Decide which of the following matrices are unitarily equivalent:

$$A = \begin{bmatrix} 1 & 1 \\ 1 & 1 \end{bmatrix},$$

$$B = \begin{bmatrix} 2 & 2 \\ 0 & 0 \end{bmatrix},$$

$$C = \begin{bmatrix} 2 & 0 \\ 0 & 0 \end{bmatrix},$$

$$D = \begin{bmatrix} 1 & -i \\ i & 1 \end{bmatrix}.$$

2. Decide which of the following matrices are unitarily equivalent:

$$A = \begin{bmatrix} i & 0 & 0 \\ 0 & 1 & 0 \\ 0 & 0 & 1 \end{bmatrix},$$

$$B = \begin{bmatrix} 1 & -1 & 0 \\ i & i & 1 \\ 0 & 1 & 1 \end{bmatrix},$$

$$C = \begin{bmatrix} 1 & 0 & 0 \\ 1 & i & 1 \\ 0 & 0 & 1 \end{bmatrix},$$

$$D = \begin{bmatrix} 1+i & -\frac{1}{\sqrt{2}} - i\frac{1}{\sqrt{2}} & 0 \\ \frac{1}{\sqrt{2}} + i\frac{1}{\sqrt{2}} & 0 & 0 \\ 0 & 0 & 1 \end{bmatrix}.$$

3. Assume that $A, B \in \mathrm{Mat}_{n \times n}(\mathbb{C})$ are unitarily equivalent. Show that if A has a square root, i.e., $A = C^2$ for some $C \in \mathrm{Mat}_{n \times n}(\mathbb{C})$, then B also has a square root.
4. Assume that $A, B \in \mathrm{Mat}_{n \times n}(\mathbb{C})$ are unitarily equivalent. Show that if A is positive, i.e., A is self-adjoint and has positive eigenvalues, then B is also positive.
5. Assume that $A \in \mathrm{Mat}_{n \times n}(\mathbb{C})$ is normal. Show that A is unitarily equivalent to A^* if and only if A is self-adjoint.

4.6 Real Forms

In this section, we are going to explain the canonical forms for normal real linear operators that are not necessarily diagonalizable.

The idea is to follow the proof of the spectral theorem for complex normal operators. Thus, we use induction on dimension to obtain the desired canonical forms. To get the induction going, we decompose $L = A + B$, where $AB = BA$, $A = \frac{1}{2}(L + L^*)$ is symmetric and $B = \frac{1}{2}(L - L^*)$ is skew-symmetric. The spectral theorem can be applied to A so that V has an orthonormal basis of eigenvectors and the eigenspaces for A are B-invariant, since $AB = BA$. If $A \neq \alpha 1_V$, then we can find a nontrivial orthogonal decomposition of V that reduces L. In the case when $A = \alpha 1_V$, all subspaces of V are A-invariant. Thus, we use B to find invariant subspaces for L. To find such subspaces, observe that B^2 is symmetric and select an eigenvector/value pair $B^2(x) = \lambda x$. Since B maps x to $B(x)$ and $B(x)$ to $B^2(x) = \lambda x$ the subspace span $\{x, B(x)\}$ is B-invariant. If this subspace is one dimensional, then x is also an eigenvector for B, otherwise the subspace is two dimensional. As these subspaces are contained in the eigenspaces for A, we only need to figure out how B acts on them. In the one-dimensional case, it is spanned by an eigenvector of B. So the only case left to study is when $B : M \to M$ is skew-symmetric and M is two dimensional with no nontrivial invariant subspaces. In this case, we just select a unit vector $x \in M$ and note that $B(x) \neq 0$ as x would otherwise span a one-dimensional invariant subspace. In addition, for all $z \in V$, we have that z and $B(z)$ are always perpendicular as

$$(B(z)|z) = -(z|B(z))$$
$$= -(B(z)|z).$$

In particular, x and $B(x) / \|B(x)\|$ form an orthonormal basis for M. In this basis, the matrix representation for B is

$$\left[B(x) \; B\left(\tfrac{B(x)}{\|B(x)\|}\right) \right] = \left[x \; \tfrac{B(x)}{\|B(x)\|} \right] \begin{bmatrix} 0 & \gamma \\ \|B(x)\| & 0 \end{bmatrix}$$

as $B\left(\tfrac{B(x)}{\|B(x)\|}\right)$ is perpendicular to $B(x)$ and hence a multiple of x. Finally, we get that $\gamma = -\|B(x)\|$ since the matrix has to be skew-symmetric.

To complete the analysis, we use Proposition 4.1.7 to observe that the orthogonal complement of span $\{x, B(x)\}$ in $\ker(A - \alpha 1_V)$ is also B-invariant. All in all, this shows that V can be decomposed into one- and/or two-dimensional subspaces that are invariant under both A and B.

This shows what the canonical form for a real normal operator looks like.

Theorem 4.6.1. (The Canonical Form for Real Normal Operators) *Let $L : V \to V$ be a normal operator on a finite-dimensional real inner product space; then, we can find an orthonormal basis $e_1, \ldots, e_k, x_1, y_1, \ldots, x_l, y_l$ where $k + 2l = n$ and*

$$L(e_i) = \lambda_i e_i,$$
$$L(x_j) = \alpha_j x_j + \beta_j y_j,$$
$$L(y_j) = -\beta_j x_j + \alpha_j y_j,$$

and $\lambda_i, \alpha_j, \beta_j \in \mathbb{R}$. Thus, L has the matrix representation

$$
\begin{bmatrix}
\lambda_1 \cdots & 0 & 0 & 0 & \cdots & \cdots & 0 & 0 \\
\vdots & \ddots & \vdots & \vdots & \vdots & & & \\
0 & \cdots & \lambda_k & 0 & 0 & \cdots & & \\
0 & \cdots & 0 & \alpha_1 & -\beta_1 & 0 & \cdots & \vdots \\
0 & \cdots & 0 & \beta_1 & \alpha_1 & 0 & \cdots & \\
& & & 0 & 0 & \ddots & & \\
\vdots & & & & & \ddots & 0 & 0 \\
& & & & & & 0 & \alpha_l & -\beta_l \\
0 & & \cdots & & & & 0 & \beta_l & \alpha_l
\end{bmatrix}
$$

with respect to the basis $e_1, \ldots, e_k, x_1, y_1, \ldots, x_l, y_l$.

This yields two corollaries for skew-symmetric and orthogonal operators.

Corollary 4.6.2. (The Canonical Form for Real Skew-Adjoint Operators) *Let $L : V \to V$ be a skew-symmetric operator on a finite-dimensional real inner product space, then we can find an orthonormal basis $e_1, \ldots, e_k, x_1, y_1, \ldots, x_l, y_l$ where $k + 2l = n$ and*

$$
L(e_i) = 0,
$$
$$
L(x_j) = \beta_j y_j,
$$
$$
L(y_j) = -\beta_j x_j,
$$

and $\beta_j \in \mathbb{R}$. Thus, L has the matrix representation

$$
\begin{bmatrix}
0 \cdots & 0 & 0 & 0 & \cdots & \cdots & 0 & 0 \\
\vdots & \ddots & \vdots & \vdots & \vdots & & & \\
0 \cdots & 0 & 0 & 0 & \cdots & & & \\
0 \cdots & 0 & 0 & -\beta_1 & 0 & \cdots & \vdots & \\
0 \cdots & 0 & \beta_1 & 0 & 0 & \cdots & & \\
& & 0 & 0 & \ddots & & \\
\vdots & & & & \ddots & 0 & 0 \\
& & & & & 0 & 0 & -\beta_l \\
0 & & \cdots & & & 0 & \beta_l & 0
\end{bmatrix}
$$

with respect to the basis $e_1, \ldots, e_k, x_1, y_1, \ldots, x_l, y_l$.

Corollary 4.6.3. (The Canonical Form for Orthogonal Operators) *Let $O : V \to V$ be an orthogonal operator, then we can find an orthonormal basis $e_1, \ldots, e_k, x_1, y_1, \ldots, x_l, y_l$ where $k + 2l = n$ and*

$$O(e_i) = \pm e_i,$$
$$O(x_j) = \cos(\theta_j) x_j + \sin(\theta_j) y_j,$$
$$O(y_j) = -\sin(\theta_j) x_j + \cos(\theta_j) y_j,$$

and $\lambda_i, \alpha_j, \beta_j \in \mathbb{R}$. *Thus, L has the matrix representation*

$$\begin{bmatrix} \pm 1 & \cdots & 0 & 0 & 0 & \cdots & \cdots & 0 & 0 \\ \vdots & \ddots & \vdots & \vdots & \vdots & & & & \\ 0 & \cdots & \pm 1 & 0 & 0 & \cdots & & & \\ 0 & \cdots & 0 & \cos(\theta_1) & -\sin(\theta_1) & 0 & \cdots & & \vdots \\ 0 & \cdots & 0 & \sin(\theta_1) & \cos(\theta_1) & 0 & \cdots & & \\ & & & 0 & 0 & \ddots & & & \\ \vdots & & & & & \ddots & 0 & 0 \\ & & & & & & 0 & \cos(\theta_l) & -\sin(\theta_l) \\ 0 & & \cdots & & & & 0 & \sin(\theta_l) & \cos(\theta_l) \end{bmatrix}$$

with respect to the basis $e_1, \ldots, e_k, x_1, y_1, \ldots, x_l, y_l$.

Proof. We just need to justify the specific form of the eigenvalues. We know that as a unitary operator, all the eigenvalues look like $e^{i\theta}$. If they are real, they must therefore be ± 1. Otherwise, we use Euler's formula $e^{i\theta} = \cos\theta + i\sin\theta$ to get the desired form since matrices of the form

$$\begin{bmatrix} \alpha & -\beta \\ \beta & \alpha \end{bmatrix}$$

have eigenvalues $\alpha \pm i\beta$ by Example 4.4.9. $\qquad\square$

Note that we can artificially group some of the real eigenvalues in the decomposition of the orthogonal operators by using

$$\begin{bmatrix} 1 & 0 \\ 0 & 1 \end{bmatrix} = \begin{bmatrix} \cos 0 & -\sin 0 \\ \sin 0 & \cos 0 \end{bmatrix},$$

$$\begin{bmatrix} -1 & 0 \\ 0 & -1 \end{bmatrix} = \begin{bmatrix} \cos\pi & -\sin\pi \\ \sin\pi & \cos\pi \end{bmatrix}$$

By paring off as many eigenvectors for ± 1 as possible, we obtain

Corollary 4.6.4. *Let* $O : \mathbb{R}^{2n} \to \mathbb{R}^{2n}$ *be an orthogonal operator, then we can find an orthonormal basis where L has one of the following two types of the matrix representations:*

Type I:

$$
\begin{bmatrix}
\cos(\theta_1) & -\sin(\theta_1) & 0 & \cdots & 0 & 0 \\
\sin(\theta_1) & \cos(\theta_1) & 0 & \cdots & 0 & 0 \\
0 & 0 & \ddots & & & \\
\vdots & \vdots & & \ddots & 0 & 0 \\
0 & 0 & & 0 & \cos(\theta_n) & -\sin(\theta_n) \\
0 & 0 & & 0 & \sin(\theta_n) & \cos(\theta_n)
\end{bmatrix}
$$

Type II:

$$
\begin{bmatrix}
-1 & 0 & 0 & 0 & \cdots & & 0 & 0 \\
0 & 1 & 0 & 0 & \cdots & & 0 & 0 \\
0 & 0 & \cos(\theta_1) & -\sin(\theta_1) & 0 & \cdots & & \vdots \\
0 & 0 & \sin(\theta_1) & \cos(\theta_1) & 0 & \cdots & & \\
& & 0 & & 0 & \ddots & & \\
\vdots & \vdots & & & & \ddots & 0 & 0 \\
0 & 0 & & & & 0 & \cos(\theta_{n-1}) & -\sin(\theta_{n-1}) \\
0 & 0 & \cdots & & & 0 & \sin(\theta_{n-1}) & \cos(\theta_{n-1})
\end{bmatrix}
$$

Corollary 4.6.5. *Let* $O : \mathbb{R}^{2n+1} \to \mathbb{R}^{2n+1}$ *be an orthogonal operator, then we can find an orthonormal basis where L has one of the following two the matrix representations:*

Type I:

$$
\begin{bmatrix}
1 & 0 & 0 & 0 & \cdots & 0 & 0 \\
0 & \cos(\theta_1) & -\sin(\theta_1) & 0 & \cdots & & \vdots \\
0 & \sin(\theta_1) & \cos(\theta_1) & 0 & \cdots & & \\
0 & 0 & 0 & \ddots & & & \\
\vdots & & & & \ddots & 0 & 0 \\
0 & & & & 0 & \cos(\theta_n) & -\sin(\theta_n) \\
0 & \cdots & & & 0 & \sin(\theta_n) & \cos(\theta_n)
\end{bmatrix}
$$

Type II:

$$
\begin{bmatrix}
-1 & 0 & 0 & 0 & \cdots & 0 & 0 \\
0 & \cos(\theta_1) & -\sin(\theta_1) & 0 & \cdots & & \vdots \\
0 & \sin(\theta_1) & \cos(\theta_1) & 0 & \cdots & & \\
0 & 0 & 0 & \ddots & & & \\
\vdots & & & & \ddots & 0 & 0 \\
0 & & & & & \cos(\theta_n) & -\sin(\theta_n) \\
0 & \cdots & & & & \sin(\theta_n) & \cos(\theta_n)
\end{bmatrix}.
$$

Like with unitary equivalence (see Sect. 4.5), we also have the concept of orthogonal equivalence. One can with the appropriate modifications prove similar results about when matrices are orthogonally equivalent. The above results apparently give us the simplest types of matrices that real normal, skew-symmetric, and orthogonal operators are orthogonally equivalent to.

Note that type I operators have the property that -1 has even multiplicity, while for type II, -1 has odd multiplicity. In particular, we note that type I is the same as saying that the determinant is 1 while type II means that the determinant is -1. The collection of orthogonal transformations of type I is denoted SO_n. This set is a *subgroup* of O_n, i.e., if $A, B \in SO_n$, then $AB \in SO_n$. This is not obvious given what we know now, but the proof is quite simple using determinants.

Exercises

1. Explain what the canonical form is for real linear maps that are both orthogonal and skew-symmetric.
2. Let $L : V \to V$ be orthogonal on a finite-dimensional real inner product space and assume that $\dim(\ker(L + 1_V))$ is even. Show that $L = K^2$ for some orthogonal K.
3. Use the canonical forms to show

   (a) If $U \in U_n$, then $U = \exp(A)$ where A is skew-adjoint.
   (b) If $O \in O_n$ is of type I, then $O = \exp(A)$ where A is skew-symmetric.

4. Let $L : V \to V$ be skew-symmetric on a real inner product space. Show that $L = K^2$ for some K. Can you solve this using a skew-symmetric K?
5. Let $A \in O_n$. Show that the following conditions are equivalent:

   (a) A has type I.
   (b) The product of the real eigenvalues is 1.
   (c) The product of all real and complex eigenvalues is 1.
   (d) $\dim(\ker(L + 1_{\mathbb{R}^n}))$ is even.
   (e) $\chi_A(t) = t^n + \cdots + \alpha_1 t + (-1)^n$, i.e., the constant term is $(-1)^n$.

6. Assume that $A \in \mathrm{Mat}_{n \times n}(\mathbb{R})$ satisfies $AO = OA$ for all $O \in SO_n$. Show that

   (a) If $n = 2$, then

$$A = \begin{bmatrix} \alpha & -\beta \\ \beta & \alpha \end{bmatrix}.$$

   (b) If $n \geq 3$, then $A = \lambda 1_{\mathbb{R}^n}$.

7. Let $L : \mathbb{R}^3 \to \mathbb{R}^3$ be skew-symmetric.

   (a) Show that there is a unique vector $w \in \mathbb{R}^3$ such that $L(x) = w \times x$. w is known as the *Darboux vector* for L.
   (b) Show that the assignment $L \to w$ gives a linear isomorphism between skew-symmetric 3×3 matrices and vectors in $\mathbb{R}^3$.
   (c) Show that if $L_1(x) = w_1 \times x$ and $L_2(x) = w_2 \times x$, then the commutator

$$[L_1, L_2] = L_1 \circ L_2 - L_2 \circ L_1$$

   satisfies

$$[L_1, L_2](x) = (w_1 \times w_2) \times x$$

   Hint: This is equivalent to proving the so-called *Jacobi identity*:

$$(x \times y) \times z + (z \times x) \times y + (y \times z) \times x = 0.$$

   (d) Show that

$$L(x) = w_2(w_1|x) - w_1(w_2|x)$$

   is skew-symmetric and that

$$(w_1 \times w_2) \times x = w_2(w_1|x) - w_1(w_2|x).$$

   (e) Conclude that all skew-symmetric $L : \mathbb{R}^3 \to \mathbb{R}^3$ are of the form

$$L(x) = w_2(w_1|x) - w_1(w_2|x).$$

8. For $u_1, u_2 \in \mathbb{R}^n$:

   (a) Show that

$$L(x) = (u_1 \wedge u_2)(x) = (u_1|x) u_2 - (u_2|x) u_1$$

   defines a skew-symmetric operator.
   (b) Show that

$$u_1 \wedge u_2 = -u_2 \wedge u_1$$

$$(\alpha u_1 + \beta v_1) \wedge u_2 = \alpha(u_1 \wedge u_2) + \beta(v_1 \wedge u_2)$$

   (c) Show *Bianchi's identity*: For all $x, y, z \in \mathbb{R}^n$, we have

$$(x \wedge y)(z) + (z \wedge x)(y) + (y \wedge z)(x) = 0.$$

(d) When $n \geq 4$, show that not all skew-symmetric $L : \mathbb{R}^n \to \mathbb{R}^n$ are of the form
$L(x) = u_1 \wedge u_2$. Hint: Let $u_1, \ldots, u_4$ be linearly independent and consider

$$L = u_1 \wedge u_2 + u_3 \wedge u_4.$$

(e) Show that the skew-symmetric operators $e_i \wedge e_j$, where $i < j$, form a basis
for the space of skew-symmetric operators.

4.7   Orthogonal Transformations*

In this section, we are going to try to get a better grasp on orthogonal
transformations.

We start by specializing the above canonical forms for orthogonal transformations to the two situations where things can be visualized, namely, in dimensions
two and three.

Corollary 4.7.1. *Any orthogonal operator $O : \mathbb{R}^2 \to \mathbb{R}^2$ has one of the following
two forms in the standard basis:*
Either it is a rotation by θ and is of the form

Type I:

$$\begin{bmatrix} \cos(\theta) & -\sin(\theta) \\ \sin(\theta) & \cos(\theta) \end{bmatrix},$$

or it is a reflection in the line spanned by $(\cos \alpha, \sin \alpha)$ and has the form

Type II:

$$\begin{bmatrix} \cos(2\alpha) & \sin(2\alpha) \\ \sin(2\alpha) & -\cos(2\alpha) \end{bmatrix}.$$

Moreover, O is a rotation if $\chi_O(t) = t^2 - (2\cos\theta)t + 1$, and θ is given by $\cos\theta = \frac{1}{2}\mathrm{tr}O$, while O is a reflection if $\mathrm{tr}O = 0$ and $\chi_O(t) = t^2 - 1$.

Proof. We know that there is an orthonormal basis x_1, x_2 that puts O into one of the
two forms

$$\begin{bmatrix} \cos(\theta) & -\sin(\theta) \\ \sin(\theta) & \cos(\theta) \end{bmatrix}, \begin{bmatrix} 1 & 0 \\ 0 & -1 \end{bmatrix}.$$

We can write

$$x_1 = \begin{bmatrix} \cos(\alpha) \\ \sin(\alpha) \end{bmatrix}, x_2 = \pm \begin{bmatrix} -\sin(\alpha) \\ \cos(\alpha) \end{bmatrix}.$$

The sign on x_2 can have an effect on the matrix representation as we shall see. In the case of the rotation, it means a sign change in the angle; in the reflection case, it does not change the form at all.

To find the form of the matrix in the usual basis, we use the change of basis formula for matrix representations. Before doing this, let us note that the law of exponents

$$\exp(i(\theta + \alpha)) = \exp(i\theta)\exp(i\alpha)$$

tells us that the corresponding real 2×2 matrices satisfy

$$\begin{bmatrix} \cos(\alpha) & -\sin(\alpha) \\ \sin(\alpha) & \cos(\alpha) \end{bmatrix} \begin{bmatrix} \cos(\theta) & -\sin(\theta) \\ \sin(\theta) & \cos(\theta) \end{bmatrix}$$

$$= \begin{bmatrix} \cos(\alpha + \theta) & -\sin(\alpha + \theta) \\ \sin(\alpha + \theta) & \cos(\alpha + \theta) \end{bmatrix}.$$

Thus,

$$O = \begin{bmatrix} \cos(\alpha) & -\sin(\alpha) \\ \sin(\alpha) & \cos(\alpha) \end{bmatrix} \begin{bmatrix} \cos(\theta) & -\sin(\theta) \\ \sin(\theta) & \cos(\theta) \end{bmatrix} \begin{bmatrix} \cos(\alpha) & \sin(\alpha) \\ -\sin(\alpha) & \cos(\alpha) \end{bmatrix}$$

$$= \begin{bmatrix} \cos(\alpha) & -\sin(\alpha) \\ \sin(\alpha) & \cos(\alpha) \end{bmatrix} \begin{bmatrix} \cos(\theta) & -\sin(\theta) \\ \sin(\theta) & \cos(\theta) \end{bmatrix} \begin{bmatrix} \cos(-\alpha) & -\sin(-\alpha) \\ \sin(-\alpha) & \cos(-\alpha) \end{bmatrix}$$

$$= \begin{bmatrix} \cos(\theta) & -\sin(\theta) \\ \sin(\theta) & \cos(\theta) \end{bmatrix}$$

as expected. If x_2 is changed to $-x_2$, we have

$$O = \begin{bmatrix} \cos(\alpha) & \sin(\alpha) \\ \sin(\alpha) & -\cos(\alpha) \end{bmatrix} \begin{bmatrix} \cos(\theta) & -\sin(\theta) \\ \sin(\theta) & \cos(\theta) \end{bmatrix} \begin{bmatrix} \cos(\alpha) & \sin(\alpha) \\ \sin(\alpha) & -\cos(\alpha) \end{bmatrix}$$

$$= \begin{bmatrix} \cos(\alpha) & \sin(\alpha) \\ \sin(\alpha) & -\cos(\alpha) \end{bmatrix} \begin{bmatrix} \cos(-\theta) & \sin(-\theta) \\ -\sin(-\theta) & \cos(-\theta) \end{bmatrix} \begin{bmatrix} \cos(\alpha) & \sin(\alpha) \\ \sin(\alpha) & -\cos(\alpha) \end{bmatrix}$$

$$= \begin{bmatrix} \cos(\alpha - \theta) & \sin(\alpha - \theta) \\ \sin(\alpha - \theta) & -\cos(\alpha - \theta) \end{bmatrix} \begin{bmatrix} \cos(-\alpha) & -\sin(-\alpha) \\ -\sin(-\alpha) & -\cos(-\alpha) \end{bmatrix}$$

$$= \begin{bmatrix} \cos(-\theta) & -\sin(-\theta) \\ \sin(-\theta) & \cos(-\theta) \end{bmatrix}.$$

Finally, the reflection has the form

$$O = \begin{bmatrix} \cos(\alpha) & -\sin(\alpha) \\ \sin(\alpha) & \cos(\alpha) \end{bmatrix} \begin{bmatrix} 1 & 0 \\ 0 & -1 \end{bmatrix} \begin{bmatrix} \cos(\alpha) & \sin(\alpha) \\ -\sin(\alpha) & \cos(\alpha) \end{bmatrix}$$

$$= \begin{bmatrix} \cos(\alpha) & \sin(\alpha) \\ \sin(\alpha) & -\cos(\alpha) \end{bmatrix} \begin{bmatrix} \cos(\alpha) & \sin(\alpha) \\ -\sin(\alpha) & \cos(\alpha) \end{bmatrix}$$

$$= \begin{bmatrix} \cos(2\alpha) & \sin(2\alpha) \\ \sin(2\alpha) & -\cos(2\alpha) \end{bmatrix}.$$

$\square$

Note that there is clearly an ambiguity in what it should mean to be a rotation by θ as either of the two matrices

$$\begin{bmatrix} \cos(\pm\theta) & -\sin(\pm\theta) \\ \sin(\pm\theta) & \cos(\pm\theta) \end{bmatrix}$$

describe such a rotation. What is more, the same orthogonal transformation can have different canonical forms depending on what basis we choose as we just saw in the proof of the above theorem. Unfortunately, it is not possible to sort this out without being very careful about the choice of basis, specifically one needs the additional concept of orientation which in turn uses determinants.

We now turn to the three-dimensional situation.

Corollary 4.7.2. *Any orthogonal operator* $O : \mathbb{R}^3 \to \mathbb{R}^3$ *is either*

Type I: *It is a rotation in the plane that is perpendicular to the line representing the* $+1$ *eigenspace.*

Type II: *It is a rotation in the plane that is perpendicular to the* -1 *eigenspace followed by a reflection in that plane, corresponding to multiplying by* -1 *in the* -1 *eigenspace.*

As in the two-dimensional situation, we can also discover which case we are in by calculating the characteristic polynomial. For a rotation O in an axis, we have

$$\chi_O(t) = (t-1)\left(t^2 - (2\cos\theta)t + 1\right)$$

$$= t^3 - (1 + 2\cos\theta)t^2 + (1 + 2\cos\theta)t - 1$$

$$= t^3 - (\operatorname{tr}O)t^2 + (\operatorname{tr}O)t - 1,$$

while the case involving a reflection

$$\chi_O(t) = (t+1)\left(t^2 - (2\cos\theta)t + 1\right)$$

$$= t^3 - (-1 + 2\cos\theta)t^2 - (-1 + 2\cos\theta)t + 1$$

$$= t^3 - (\operatorname{tr}O)t^2 - (\operatorname{tr}O)t + 1.$$

Example 4.7.3. Imagine a cube that is centered at the origin and so that the edges and sides are parallel to coordinate axes and planes. We note that all of the orthogonal transformations that either reflect in a coordinate plane or form $90°, 180°,$ and $270°$ rotations around the coordinate axes are symmetries of the cube. Thus, the cube is mapped to itself via each of these isometries. In fact, the collection of all isometries that preserve the cube in this fashion is a (finite) group. It is evidently a subgroup of O_3. There are more symmetries than those already mentioned, namely, if we pick two antipodal vertices, then we can rotate the cube into itself by $120°$ and $240°$ rotations around the line going through these two points. What is even more surprising perhaps is that these rotations can be obtained by composing the already mentioned $90°$ rotations. To see this, let

$$O_x = \begin{bmatrix} 1 & 0 & 0 \\ 0 & 0 & -1 \\ 0 & 1 & 0 \end{bmatrix}, O_y = \begin{bmatrix} 0 & 0 & -1 \\ 0 & 1 & 0 \\ 1 & 0 & 0 \end{bmatrix}$$

be $90°$ rotations around the x- and y-axes, respectively. Then,

$$O_x O_y = \begin{bmatrix} 1 & 0 & 0 \\ 0 & 0 & -1 \\ 0 & 1 & 0 \end{bmatrix} \begin{bmatrix} 0 & 0 & -1 \\ 0 & 1 & 0 \\ 1 & 0 & 0 \end{bmatrix} = \begin{bmatrix} 0 & 0 & -1 \\ -1 & 0 & 0 \\ 0 & 1 & 0 \end{bmatrix},$$

$$O_y O_x = \begin{bmatrix} 0 & 0 & -1 \\ 0 & 1 & 0 \\ 1 & 0 & 0 \end{bmatrix} \begin{bmatrix} 1 & 0 & 0 \\ 0 & 0 & -1 \\ 0 & 1 & 0 \end{bmatrix} = \begin{bmatrix} 0 & -1 & 0 \\ 0 & 0 & -1 \\ 1 & 0 & 0 \end{bmatrix},$$

so we see that these two rotations do not commute. We now compute the (complex) eigenvalues via the characteristic polynomials in order to figure out what these new isometries look like. Since both matrices have zero trace, they have characteristic polynomial

$$\chi(t) = t^3 - 1.$$

Thus, they describe rotations where

$$\text{tr}(O) = 1 + 2\cos(\theta) = 0 \text{ or}$$

$$\theta = \pm \frac{2\pi}{3}$$

around the axis that corresponds to the 1 eigenvector. For $O_x O_y$, we have that $(1, -1, -1)$ is an eigenvector for 1, while for $O_y O_x$, we have $(-1, 1, -1)$. These two eigenvectors describe the directions for two different diagonals in the cube. Completing, say, $(1, -1, -1)$ to an orthonormal basis for $\mathbb{R}^3$, then tells us that

$$O_x O_y = \begin{bmatrix} \frac{1}{\sqrt{3}} & \frac{1}{\sqrt{2}} & \frac{1}{\sqrt{6}} \\ \frac{-1}{\sqrt{3}} & \frac{1}{\sqrt{2}} & \frac{-1}{\sqrt{6}} \\ \frac{-1}{\sqrt{3}} & 0 & \frac{2}{\sqrt{6}} \end{bmatrix} \begin{bmatrix} 1 & 0 & 0 \\ 0 & \cos\left(\pm\frac{2\pi}{3}\right) & -\sin\left(\pm\frac{2\pi}{3}\right) \\ 0 & \sin\left(\pm\frac{2\pi}{3}\right) & \cos\left(\pm\frac{2\pi}{3}\right) \end{bmatrix} \begin{bmatrix} \frac{1}{\sqrt{3}} & \frac{-1}{\sqrt{3}} & \frac{-1}{\sqrt{3}} \\ \frac{1}{\sqrt{2}} & \frac{1}{\sqrt{2}} & 0 \\ \frac{1}{\sqrt{6}} & \frac{-1}{\sqrt{6}} & \frac{2}{\sqrt{6}} \end{bmatrix}$$

$$= \begin{bmatrix} \frac{1}{\sqrt{3}} & \frac{1}{\sqrt{2}} & \frac{1}{\sqrt{6}} \\ \frac{-1}{\sqrt{3}} & \frac{1}{\sqrt{2}} & \frac{-1}{\sqrt{6}} \\ \frac{-1}{\sqrt{3}} & 0 & \frac{2}{\sqrt{6}} \end{bmatrix} \begin{bmatrix} 1 & 0 & 0 \\ 0 & -\frac{1}{2} & \mp\frac{\sqrt{3}}{2} \\ 0 & \pm\frac{\sqrt{3}}{2} & -\frac{1}{2} \end{bmatrix} \begin{bmatrix} \frac{1}{\sqrt{3}} & \frac{-1}{\sqrt{3}} & \frac{-1}{\sqrt{3}} \\ \frac{1}{\sqrt{2}} & \frac{1}{\sqrt{2}} & 0 \\ \frac{1}{\sqrt{6}} & \frac{-1}{\sqrt{6}} & \frac{2}{\sqrt{6}} \end{bmatrix}$$

$$= \begin{bmatrix} \frac{1}{\sqrt{3}} & \frac{1}{\sqrt{2}} & \frac{1}{\sqrt{6}} \\ \frac{-1}{\sqrt{3}} & \frac{1}{\sqrt{2}} & \frac{-1}{\sqrt{6}} \\ \frac{-1}{\sqrt{3}} & 0 & \frac{2}{\sqrt{6}} \end{bmatrix} \begin{bmatrix} 1 & 0 & 0 \\ 0 & -\frac{1}{2} & -\frac{\sqrt{3}}{2} \\ 0 & \frac{\sqrt{3}}{2} & -\frac{1}{2} \end{bmatrix} \begin{bmatrix} \frac{1}{\sqrt{3}} & \frac{-1}{\sqrt{3}} & \frac{-1}{\sqrt{3}} \\ \frac{1}{\sqrt{2}} & \frac{1}{\sqrt{2}} & 0 \\ \frac{1}{\sqrt{6}} & \frac{-1}{\sqrt{6}} & \frac{2}{\sqrt{6}} \end{bmatrix}.$$

The fact that we pick $+$ rather than $-$ depends on our orthonormal basis as we can see by changing the basis by a sign in the last column:

$$O_x O_y = \begin{bmatrix} \frac{1}{\sqrt{3}} & \frac{1}{\sqrt{2}} & \frac{-1}{\sqrt{6}} \\ \frac{-1}{\sqrt{3}} & \frac{1}{\sqrt{2}} & \frac{1}{\sqrt{6}} \\ \frac{-1}{\sqrt{3}} & 0 & \frac{-2}{\sqrt{6}} \end{bmatrix} \begin{bmatrix} 1 & 0 & 0 \\ 0 & -\frac{1}{2} & \frac{\sqrt{3}}{2} \\ 0 & -\frac{\sqrt{3}}{2} & -\frac{1}{2} \end{bmatrix} \begin{bmatrix} \frac{1}{\sqrt{3}} & \frac{-1}{\sqrt{3}} & \frac{-1}{\sqrt{3}} \\ \frac{1}{\sqrt{2}} & \frac{1}{\sqrt{2}} & 0 \\ \frac{-1}{\sqrt{6}} & \frac{1}{\sqrt{6}} & \frac{-2}{\sqrt{6}} \end{bmatrix}.$$

We are now ready to discuss how the two types of orthogonal transformations interact with each other when multiplied. Let us start with the two-dimensional situation. One can directly verify that

$$\begin{bmatrix} \cos\theta_1 & -\sin\theta_1 \\ \sin\theta_1 & \cos\theta_1 \end{bmatrix} \begin{bmatrix} \cos\theta_2 & -\sin\theta_2 \\ \sin\theta_2 & \cos\theta_2 \end{bmatrix}$$
$$= \begin{bmatrix} \cos(\theta_1 + \theta_2) & -\sin(\theta_1 + \theta_2) \\ \sin(\theta_1 + \theta_2) & \cos(\theta_1 + \theta_2) \end{bmatrix},$$

$$\begin{bmatrix} \cos\theta & -\sin\theta \\ \sin\theta & \cos\theta \end{bmatrix} \begin{bmatrix} \cos\alpha & \sin\alpha \\ \sin\alpha & -\cos\alpha \end{bmatrix} = \begin{bmatrix} \cos(\theta + \alpha) & \sin(\theta + \alpha) \\ \sin(\theta + \alpha) & -\cos(\theta + \alpha) \end{bmatrix},$$

$$\begin{bmatrix} \cos\alpha & \sin\alpha \\ \sin\alpha & -\cos\alpha \end{bmatrix} \begin{bmatrix} \cos\theta & -\sin\theta \\ \sin\theta & \cos\theta \end{bmatrix} = \begin{bmatrix} \cos(\alpha - \theta) & \sin(\alpha - \theta) \\ \sin(\alpha - \theta) & -\cos(\alpha - \theta) \end{bmatrix},$$

$$\begin{bmatrix} \cos\alpha_1 & \sin\alpha_1 \\ \sin\alpha_1 & -\cos\alpha_1 \end{bmatrix} \begin{bmatrix} \cos\alpha_2 & \sin\alpha_2 \\ \sin\alpha_2 & -\cos\alpha_2 \end{bmatrix}$$
$$= \begin{bmatrix} \cos(\alpha_1 - \alpha_2) & -\sin(\alpha_1 - \alpha_2) \\ \sin(\alpha_1 - \alpha_2) & \cos(\alpha_1 - \alpha_2) \end{bmatrix}.$$

Thus, we see that if the transformations are of the same type, their product has type I, while if they have different type, their product has type II. This is analogous to multiplying positive and negative numbers. This result actually holds in all dimensions and has a very simple proof using determinants. Euler was the first to observe this phenomenon in the three-dimensional case. What we are going to look into here is the observation that any rotation (type I) in O_2 is a product of two reflections. More specifically, if $\theta = \alpha_1 - \alpha_2$, then the above calculation shows that

$$\begin{bmatrix} \cos\theta & -\sin\theta \\ \sin\theta & \cos\theta \end{bmatrix} = \begin{bmatrix} \cos\alpha_1 & \sin\alpha_1 \\ \sin\alpha_1 & -\cos\alpha_1 \end{bmatrix} \begin{bmatrix} \cos\alpha_2 & \sin\alpha_2 \\ \sin\alpha_2 & -\cos\alpha_2 \end{bmatrix}.$$

Definition 4.7.4. To pave the way for a higher dimensional analogue of this, we define $A \in O_n$ to be a *reflection* if it has the canonical form

$$A = O \begin{bmatrix} -1 & 0 & & 0 \\ 0 & 1 & & \\ & & \ddots & \\ 0 & & & 1 \end{bmatrix} O^*.$$

This implies that BAB^* is also a reflection for all $B \in O_n$. To get a better picture of what A does, we note that the -1 eigenvector gives the reflection in the hyperplane spanned by the $(n-1)$-dimensional $+1$ eigenspace. If z is a unit eigenvector for -1, then we can write A in the following way:

$$A(x) = R_z(x) = x - 2(x|z)z.$$

To see why this is true, first note that if x is an eigenvector for $+1$, then it is perpendicular to z, and hence,

$$x - 2(x|z)z = x.$$

In case $x = z$, we have

$$z - 2(z|z)z = z - 2z$$

$$= -z$$

as desired. We can now prove an interesting and important lemma.

Lemma 4.7.5. (E. Cartan) *Let $A \in O_n$. If A has type I, then A is a product of an even number of reflections, while if A has type II, then it is a product of an odd number of reflections.*

Proof. A very simple alternate proof can be found in the exercises.

The canonical form for A can be expressed as follows:

$$A = OI_{\pm} R_1 \cdots R_l O^*,$$

where O is the orthogonal change of basis matrix, each R_i corresponds to a rotation on a two-dimensional subspace M_i, and

$$I_{\pm} = \begin{bmatrix} \pm 1 & 0 & & 0 \\ 0 & 1 & & \\ & & \ddots & \\ 0 & & & 1 \end{bmatrix},$$

where $+$ is used for type I and $-$ is used for type II. The above two-dimensional construction shows that each rotation is a product of two reflections on M_i. If we extend these two-dimensional reflections to be the identity on $M_i^{\perp}$, then they become reflections on the whole space. Thus, we have

$$A = OI_{\pm} (A_1 B_1) \cdots (A_l B_l) O^*,$$

where $I_{\pm}$ is either the identity or a reflection and $A_1, B_1, \ldots, A_l, B_l$ are all reflections. Finally,

$$A = OI_{\pm} (A_1 B_1) \cdots (A_l B_l) O^*$$
$$= (OI_{\pm} O^*) (OA_1 O^*) (OB_1 O^*) \cdots (OA_l O^*) (OB_l O^*).$$

This proves the claim. $\square$

Remark 4.7.6. The converse to this lemma is also true, namely, that any even number of reflection compose to a type I orthogonal transformation, while an odd number yields one of type II. This proof of this fact is very simple if one uses determinants.

Exercises

1. Decide the type and what the rotation and/or line of reflection is for each the matrices

$$\begin{bmatrix} \frac{1}{2} & \frac{\sqrt{3}}{2} \\ -\frac{\sqrt{3}}{2} & \frac{1}{2} \end{bmatrix},$$

$$\begin{bmatrix} \frac{1}{2} & \frac{\sqrt{3}}{2} \\ \frac{\sqrt{3}}{2} & -\frac{1}{2} \end{bmatrix}.$$

2. Decide on the type, ± 1 eigenvector and possible rotation angles on the orthogonal complement for the ± 1 eigenvector for the matrices:

$$
\begin{bmatrix}
-\dfrac{1}{3} & -\dfrac{2}{3} & -\dfrac{2}{3} \\[4pt]
\dfrac{2}{3} & \dfrac{1}{3} & \dfrac{2}{3} \\[4pt]
-\dfrac{2}{3} & -\dfrac{2}{3} & -\dfrac{1}{3}
\end{bmatrix},
$$

$$
\begin{bmatrix}
0 & 0 & 1 \\
0 & -1 & 0 \\
1 & 0 & 0
\end{bmatrix},
$$

$$
\begin{bmatrix}
\dfrac{2}{3} & -\dfrac{2}{3} & \dfrac{1}{3} \\[4pt]
\dfrac{2}{3} & \dfrac{1}{3} & -\dfrac{2}{3} \\[4pt]
-\dfrac{1}{3} & -\dfrac{2}{3} & -\dfrac{2}{3}
\end{bmatrix},
$$

$$
\begin{bmatrix}
\dfrac{1}{3} & \dfrac{2}{3} & \dfrac{2}{3} \\[4pt]
\dfrac{2}{3} & -\dfrac{2}{3} & \dfrac{1}{3} \\[4pt]
\dfrac{2}{3} & \dfrac{1}{3} & -\dfrac{2}{3}
\end{bmatrix}.
$$

3. Write the matrices from Exercises 1 and 2 as products of reflections.

4. Let $O \in O_3$ and assume we have a Darboux vector $u \in \mathbb{R}^3$ such that for all $x \in \mathbb{R}^3$,

$$
\frac{1}{2}\left(O - O^t\right)(x) = u \times x.
$$

(See also Exercise 7 in Sect. 4.6).

(a) Show that u determines the axis of rotation by showing that $O(u) = \pm u$.
(b) Show that the rotation is determined by $|\sin \theta| = |u|$.
(c) Show that for any $O \in O_3$, we can find a Darboux vector $u \in \mathbb{R}^3$ such that the above formula holds.

5. *(Euler)* Define the rotations around the three coordinate axes in $\mathbb{R}^3$ by

$$
O_x(\alpha) = \begin{bmatrix}
1 & 0 & 0 \\
0 & \cos\alpha & -\sin\alpha \\
0 & \sin\alpha & \cos\alpha
\end{bmatrix},
$$

$$
O_y(\beta) = \begin{bmatrix}
\cos\beta & 0 & -\sin\beta \\
0 & 1 & 0 \\
\sin\beta & 0 & \cos\beta
\end{bmatrix},
$$

$$O_z(\gamma) = \begin{bmatrix} \cos\gamma & -\sin\gamma & 0 \\ \sin\gamma & \cos\gamma & 0 \\ 0 & 0 & 1 \end{bmatrix}.$$

(a) Show that any $O \in SO\,(3)$ is of the form $O = O_x\,(\alpha)\,O_y\,(\beta)\,O_z\,(\gamma)$. The angles α, β, γ are called the *Euler angles* for O. Hint:

$$O_x\,(\alpha)\,O_y\,(\beta)\,O_z\,(\gamma) = \begin{bmatrix} \cos\beta\cos\gamma & -\cos\beta\sin\gamma & -\sin\beta \\ & & -\sin\alpha\cos\beta \\ & & \cos\alpha\sin\beta \end{bmatrix}$$

(b) Show that $O_x\,(\alpha)\,O_y\,(\beta)\,O_z\,(\gamma) \in SO\,(3)$ for all α, β, γ.
(c) Show that if $O_1, O_2 \in SO\,(3)$, then also $O_1 O_2 \in SO\,(3)$.

6. Find the matrix representations with respect to the canonical basis for $\mathbb{R}^3$ for all of the orthogonal matrices that describe a rotation by θ in span $\{(1,1,0),(1,2,1)\}$.
7. Show, without using canonical forms or Cartan's lemma, that if $O \in O_n$, then O is a composition of at most n reflections. Hint: For $x \in \mathbb{R}^n$, select a reflection R that takes x to Ox. Then, show that RO fixes x and conclude that RO also fixes the orthogonal complement.
8. Let $z \in \mathbb{R}^n$ be a unit vector and

$$R_z\,(x) = x - 2\,(x|z)\,z$$

the reflection in the hyperplane perpendicular to z.

(a) Show that

$$R_z = R_{-z},$$
$$(R_z)^{-1} = R_z.$$

(b) If $y, z \in \mathbb{R}^n$ are linearly independent unit vectors, then show that $R_y R_z \in O_n$ is a rotation on $M = \text{span}\,\{y, z\}$ and the identity on $M^\perp$.
(c) Show that the angle θ of rotation is given by the relationship

$$\cos\theta = -1 + 2\,|(y|z)|^2$$
$$= \cos\,(2\psi),$$

where $(y|z) = \cos\,(\psi)$.

9. Let S_n denote the group of permutations. These are the bijective maps from $\{1, 2, \ldots, n\}$ to itself. The group product is composition, and inverses are the inverse maps. Show that the map defined by sending $\sigma \in S_n$ to the permutation matrix O_σ defined by $O_\sigma\,(e_i) = e_{\sigma(i)}$ is a group homomorphism

$$S_n \to O_n,$$

i.e., show $O_\sigma \in O_n$ and $O_{\sigma\circ\tau} = O_\sigma \circ O_\tau$. (See also Example 1.7.7.)

10. Let $A \in O_4$.

    (a) Show that we can find a two-dimensional subspace $M \subset \mathbb{R}^4$ such that M
        and $M^{\perp}$ are both invariant under A.

    (b) Show that we can choose M so that $A|_{M^{\perp}}$ is rotation and $A|_M$ is a rotation
        precisely when A is type I while $A|_M$ is a reflection when A has type II.

    (c) Show that if A has type I, then

$$\chi_A(t) = t^4 - 2\left(\cos(\theta_1) + \cos(\theta_2)\right) t^3$$
$$+ \left(2 + 4\cos(\theta_1)\cos(\theta_2)\right) t^2 - 2\left(\cos(\theta_1) + \cos(\theta_2)\right) t + 1$$
$$= t^4 - (\operatorname{tr}(A)) t^3 + (2 + \operatorname{tr}(A|_M)\operatorname{tr}(A|_{M^{\perp}})) t^2 - (\operatorname{tr}(A)) t + 1,$$

      where $\operatorname{tr}(A) = \operatorname{tr}(A|_M) + \operatorname{tr}(A|_{M^{\perp}})$.

    (d) Show that if A has type II, then

$$\chi_A(t) = t^4 - (2\cos(\theta)) t^3 + (2\cos\theta) t - 1$$
$$= t^4 - (\operatorname{tr}(A)) t^3 + (\operatorname{tr}(A)) t - 1$$
$$= t^4 - (\operatorname{tr}(A|_{M^{\perp}})) t^3 + (\operatorname{tr}(A|_{M^{\perp}})) t - 1.$$

4.8 Triangulability*

There is a result that gives a simple form for general complex linear maps in an orthonormal basis. This is a sort of consolation prize for operators without any special properties relating to the inner product structure. In Sects. 4.9 and 4.10 on the singular value decomposition and the polar composition we shall encounter some other simplified forms for general linear maps between inner product spaces.

Theorem 4.8.1. (Schur's Theorem) *Let* $L : V \to V$ *be a linear operator on a finite-dimensional complex inner product space. It is possible to find an orthonormal basis* $e_1, \ldots, e_n$ *such that the matrix representation* $[L]$ *is upper triangular in this basis, i.e.,*

$$L = \begin{bmatrix} e_1 \cdots e_n \end{bmatrix} [L] \begin{bmatrix} e_1 \cdots e_n \end{bmatrix}^*$$

$$= \begin{bmatrix} e_1 \cdots e_n \end{bmatrix} \begin{bmatrix} \alpha_{11} & \alpha_{12} & \cdots & \alpha_{1n} \\ 0 & \alpha_{22} & \cdots & \alpha_{2n} \\ \vdots & \vdots & \ddots & \vdots \\ 0 & 0 & \cdots & \alpha_{nn} \end{bmatrix} \begin{bmatrix} e_1 \cdots e_n \end{bmatrix}^*.$$

Before discussing how to prove this result, let us consider a few examples.

Example 4.8.2. Note that

$$\begin{bmatrix} 1 & 1 \\ 0 & 2 \end{bmatrix}, \begin{bmatrix} 0 & 1 \\ 0 & 0 \end{bmatrix}$$

are both in the desired form. The former matrix is diagonalizable but not with respect to an orthonormal basis. So within that framework, we cannot improve its canonical form. The latter matrix is not diagonalizable so there is nothing else to discuss.

Example 4.8.3. Any 2×2 matrix A can be put into upper triangular form by finding an eigenvector e_1 and then selecting e_2 to be orthogonal to e_1. Specifically,

$$\begin{bmatrix} Ae_1 & Ae_2 \end{bmatrix} = \begin{bmatrix} e_1 & e_2 \end{bmatrix} \begin{bmatrix} \lambda & \beta \\ 0 & \gamma \end{bmatrix}.$$

Proof. (of Schur's theorem) Note that if we have the desired form

$$\begin{bmatrix} L(e_1) & \cdots & L(e_n) \end{bmatrix} = \begin{bmatrix} e_1 & \cdots & e_n \end{bmatrix} \begin{bmatrix} \alpha_{11} & \alpha_{12} & \cdots & \alpha_{1n} \\ 0 & \alpha_{22} & \cdots & \alpha_{2n} \\ \vdots & \vdots & \ddots & \vdots \\ 0 & 0 & \cdots & \alpha_{nn} \end{bmatrix},$$

then we can construct a *flag* of invariant subspaces

$$\{0\} \subset V_1 \subset V_2 \subset \cdots \subset V_{n-1} \subset V,$$

where $\dim V_k = k$ and $L(V_k) \subset V_k$, defined by $V_k = \operatorname{span}\{e_1, \ldots, e_k\}$. Conversely, given such a flag of subspaces, we can find the orthonormal basis by selecting unit vectors $e_k \in V_k \cap V_{k-1}^{\perp}$.

In order to exhibit such a flag, we use an induction argument along the lines of what we did when proving the spectral theorems for self-adjoint and normal operators (Theorems 4.3.4 and 4.4.6). In this case, the proof of Schur's theorem is reduced to showing that any complex linear map has an invariant subspace of dimension $\dim V - 1$. To see why this is true, consider the adjoint $L^* : V \to V$ and select an eigenvalue/vector pair $L^*(y) = \mu y$ (note that in order to find an eigenvalue, we must invoke the Fundamental Theorem of Algebra 2.1.8). Then, define $V_{n-1} = y^{\perp} = \{x \in V : (x|y) = 0\}$ and note that for $x \in V_{n-1}$, we have

$$(L(x)|y) = (x|L^*(y))$$
$$= (x|\mu y)$$
$$= \mu(x|y)$$
$$= 0.$$

Thus, V_{n-1} is L-invariant. □

Example 4.8.4. Let

$$A = \begin{bmatrix} 0 & 0 & 1 \\ 1 & 0 & 0 \\ 1 & 1 & 0 \end{bmatrix}.$$

To find the basis that puts A into upper triangular form, we can always use an eigenvalue e_1 for A as the first vector. To use the induction, we need one for A^* as well. Note, however, that if $Ax = \lambda x$ and $A^* y = \mu y$, then

$$\lambda (x|y) = (\lambda x|y)$$
$$= (Ax|y)$$
$$= (x|A^* y)$$
$$= (x|\mu y)$$
$$= \bar{\mu} (x|y).$$

So x and y are perpendicular as long as $\lambda \neq \bar{\mu}$. Having selected e_1, we should then select e_3 as an eigenvector for A^* where the eigenvalue is not conjugate to the one for e_1. Next, we note that $e_3^\perp$ is invariant and contains e_1. Thus, we can easily find $e_2 \in e_3^\perp$ as a vector perpendicular to e_1. This then gives the desired basis for A.

Now, let us implement this on the original matrix. First, note that 0 is not an eigenvalue for either matrix as $\ker (A) = \{0\} = \ker (A^*)$. This is a little unlucky of course. Thus, we must find λ such that $(A - \lambda 1_{\mathbb{C}^3}) x = 0$ has a nontrivial solution. This means that we should study the augmented system

$$\begin{bmatrix} -\lambda & 0 & 1 & 0 \\ 1 & -\lambda & 0 & 0 \\ 1 & 1 & -\lambda & 0 \end{bmatrix}$$

$$\begin{bmatrix} 1 & 1 & -\lambda & 0 \\ 1 & -\lambda & 0 & 0 \\ -\lambda & 0 & 1 & 0 \end{bmatrix}$$

$$\begin{bmatrix} 1 & 1 & -\lambda & 0 \\ 0 & -\lambda - 1 & \lambda & 0 \\ 0 & \lambda & 1 - \lambda^2 & 0 \end{bmatrix}$$

$$\begin{bmatrix} 1 & 1 & -\lambda & 0 \\ 0 & \lambda & 1 - \lambda^2 & 0 \\ 0 & \lambda + 1 & -\lambda & 0 \end{bmatrix}$$

$$\begin{bmatrix} 1 & 1 & -\lambda & 0 \\ 0 & \lambda & 1 - \lambda^2 & 0 \\ 0 & 0 & -\lambda - \frac{\lambda+1}{\lambda}\left(1 - \lambda^2\right) & 0 \end{bmatrix}.$$

In order to find a nontrivial solution to the last equation, the characteristic equation

$$\lambda \left(-\lambda - \frac{\lambda + 1}{\lambda} \left(1 - \lambda^2 \right) \right) = \lambda^3 - \lambda - 1$$

must vanish. This is not a pretty equation to solve but we do know that it has a solution which is real. We run into the same equation when considering A^* and we know that we can find yet another solution that is either complex or a different real number. Thus, we can conclude that we can put this matrix into upper triangular form. Despite the simple nature of the matrix, the upper triangular form is not very pretty.

Schur's theorem evidently does not depend on our earlier theorems such as the spectral theorem. In fact, all of those results can be reproved using the Schur's theorem. The spectral theorem itself can, for instance, be proved by simply observing that the matrix representation for a normal operator must be normal if the basis is orthonormal. But an upper triangular matrix can only be normal if it is diagonal.

One of the nice uses of Schur's theorem is to linear differential equations. Assume that we have a system $L(x) = \dot{x} - Ax = b$, where $A \in \mathrm{Mat}_{n \times n} (\mathbb{C})$, $b \in \mathbb{C}^n$. Then, find a basis arranged as a matrix U so that $U^* A U$ is upper triangular. If we let $x = Uy$, then the system can be rewritten as $U\dot{y} - AUy = b$, which is equivalent to solving

$$K(y) = \dot{y} - U^* A U y = U^* b.$$

Since $U^* A U$ is upper triangular, it will look like

$$\begin{bmatrix} \dot{y}_1 \\ \vdots \\ \dot{y}_{n-1} \\ \dot{y}_n \end{bmatrix} - \begin{bmatrix} \beta_{11} & \cdots & \beta_{1,n-1} & \beta_{1,n} \\ \vdots & \ddots & \vdots & \vdots \\ 0 & \cdots & \beta_{n-1,n-1} & \beta_{n-1,n} \\ 0 & \cdots & 0 & \beta_{nn} \end{bmatrix} \begin{bmatrix} y_1 \\ \vdots \\ y_{n-1} \\ y_n \end{bmatrix} = \begin{bmatrix} \gamma_1 \\ \vdots \\ \gamma_{n-1} \\ \gamma_n \end{bmatrix}.$$

Now, start by solving the last equation $\dot{y}_n - \beta_{nn} y_n = \gamma_n$ and then successively solve backwards; using that, we know how to solve linear equations of the form $\dot{z} - \alpha z = f(t)$. Finally, translate back to $x = U^* y$ to find x. Note that this also solves any particular initial value problem $x(t_0) = x_0$ as we know how to solve each of the systems with a fixed initial value at t_0. Specifically, $\dot{z} - \alpha z = f(t)$, $z(t_0) = z_0$ has the unique solution

$$z(t) = z_0 \exp \left(\alpha (t - t_0) \right) \int_{t_0}^{t} \exp \left(-\alpha (s - t_0) \right) f(s) \, ds$$

$$= z_0 \exp (\alpha t) \int_{t_0}^{t} \exp (-\alpha s) f(s) \, ds.$$

Note that the procedure only uses that A is a matrix whose entries are complex numbers. The constant b can in fact be allowed to have smooth functions as entries without changing a single step in the construction.

We could, of course, have used the Jordan canonical form (Theorem 2.8.3) as an upper triangular representative for A as well. The advantage of Schur's theorem is that the transition matrix is unitary and therefore easy to invert.

Exercises

1. Show that for any linear map $L : V \to V$ on an n-dimensional vector space, where the field of scalars $\mathbb{F} \subset \mathbb{C}$, we have $\operatorname{tr} L = \lambda_1 + \cdots + \lambda_n$, where $\lambda_1, \ldots, \lambda_n$ are the complex roots of $\chi_L(t)$ counted with multiplicities. Hint: First go to a matrix representation $[L]$, then consider this as a linear map on $\mathbb{C}^n$ and triangularize it.

2. Let $L : V \to V$, where V is a real finite-dimensional inner product space, and assume that $\chi_L(t)$ splits, i.e., all roots are real. Show that there is an orthonormal basis in which the matrix representation for L is upper triangular.

3. Use Schur's theorem to prove that if $A \in \operatorname{Mat}_{n \times n}(\mathbb{C})$ and $\varepsilon > 0$, then we can find $A_\varepsilon \in \operatorname{Mat}_{n \times n}(\mathbb{C})$ such that $\|A - A_\varepsilon\| \le \varepsilon$ and the n eigenvalues for A_ε are distinct. Conclude that any complex linear operator on a finite-dimensional inner product space can be approximated by diagonalizable operators.

4. Let $L : V \to V$ be a linear operator on a finite-dimensional complex inner product space and let $p \in \mathbb{C}[t]$. Show that μ is an eigenvalue for $p(L)$ if and only if $\mu = p(\lambda)$ where λ is an eigenvalue for L.

5. Show that a linear operator $L : V \to V$ on an n-dimensional inner product space is normal if and only if

$$\operatorname{tr}\left(L^*L\right) = |\lambda_1|^2 + \cdots + |\lambda_n|^2,$$

where $\lambda_1, \ldots, \lambda_n$ are the complex roots of the characteristic polynomial $\chi_L(t)$.

6. Let $L : V \to V$ be an invertible linear operator on an n-dimensional complex inner product space. If $\lambda_1, \ldots, \lambda_n$ are the eigenvalues for L counted with multiplicities, then

$$\|L^{-1}\| \le C_n \frac{\|L\|^{n-1}}{|\lambda_1| \cdots |\lambda_n|}$$

for some constant C_n that depends only on n. Hint: If $Ax = b$ and A is upper triangular, show that there are constants

$$1 = C_{n,n} \le C_{n,n-1} \le \cdots \le C_{n,1}$$

such that

$$|\xi_k| \le C_{n,k} \frac{\|b\| \|A\|^{n-k}}{|\alpha_{nn} \cdots \alpha_{kk}|},$$

$$A = \begin{bmatrix} \alpha_{11} & \alpha_{12} & \cdots & \alpha_{1n} \\ 0 & \alpha_{22} & \cdots & \alpha_{2n} \\ \vdots & \vdots & \ddots & \vdots \\ 0 & 0 & \cdots & \alpha_{nn} \end{bmatrix},$$

$$x = \begin{bmatrix} \xi_1 \\ \vdots \\ \xi_n \end{bmatrix}.$$

Then, bound $L^{-1}(e_i)$ using that $L\left(L^{-1}(e_i)\right) = e_i$.

7. Let $A \in \mathrm{Mat}_{n \times n}(\mathbb{C})$ and $\lambda \in \mathbb{C}$ be given and assume that there is a unit vector x such that

$$\|Ax - \lambda x\| < \frac{\varepsilon^n}{C_n \|A - \lambda 1_V\|^{n-1}}.$$

Show that there is an eigenvalue λ' for A such that

$$|\lambda - \lambda'| < \varepsilon.$$

Hint: Use the above exercise to conclude that if

$$(A - \lambda 1_V)(x) = b,$$

$$\|b\| < \frac{\varepsilon^n}{C_n \|A - \lambda 1_V\|^{n-1}}$$

and all eigenvalues for $A - \lambda 1_V$ have absolute value $\ge \varepsilon$, then $\|x\| < 1$.

8. Let $A \in \mathrm{Mat}_{n \times n}(\mathbb{C})$ be given and assume that $\|A - B\| < \delta$ for some small δ.

(a) Show that all eigenvalues for A and B lie in the compact set $K = \{z : |z| \le \|A\| + 1\}$.

(b) Show that if $\lambda \in K$ is no closer than ε to any eigenvalue for A, then

$$\left\|(\lambda 1_V - A)^{-1}\right\| < C_n \frac{(2 \|A\| + 2)^{n-1}}{\varepsilon^n}.$$

(c) Using

$$\delta = \frac{\varepsilon^n}{C_n (2 \|A\| + 2)^{n-1}},$$

show that any eigenvalue for B is within ε of some eigenvalue for A.

(d) Show that

$$\left\|(\lambda 1_V - B)^{-1}\right\| \le C_n \frac{(2\,\|A\| + 2)^{n-1}}{\varepsilon^n}$$

and that any eigenvalue for A is within ε of an eigenvalue for B.

9. Show directly that the solution to $\dot{z} - \alpha z = f(t)$, $z(t_0) = z_0$ is unique. Conclude that the initial value problems for systems of differential equations with constant coefficients have unique solutions.

10. Find the general solution to the system $\dot{x} - Ax = b$, where

(a) $A = \begin{bmatrix} 0 & 1 \\ 1 & 2 \end{bmatrix}$.

(b) $A = \begin{bmatrix} 1 & 1 \\ 1 & 2 \end{bmatrix}$.

(c) $A = \begin{bmatrix} -\frac{1}{2} & \frac{1}{2} \\ -\frac{1}{2} & \frac{1}{2} \end{bmatrix}$.

4.9   The Singular Value Decomposition*

Using the results we have developed so far, it is possible to obtain some very nice decompositions for general linear maps as well. First, we treat the so-called singular value decomposition. Note that general linear maps $L : V \to W$ do not have eigenvalues. The *singular values* of L that we define below are a good substitute for eigenvalues when we have a map between inner product spaces.

Theorem 4.9.1 (The Singular Value Decomposition). *Let $L : V \to W$ be a linear map between finite-dimensional inner product spaces. There is an orthonormal basis $e_1, \ldots, e_m$ for V such that $(L(e_i)|L(e_j)) = 0$ if $i \ne j$. Moreover, we can find orthonormal bases $e_1, \ldots, e_m$ for V and $f_1, \ldots, f_n$ for W so that*

$$L(e_1) = \sigma_1 f_1, \ldots, L(e_k) = \sigma_k f_k,$$
$$L(e_{k+1}) = \cdots = L(e_m) = 0$$

for some $k \le m$. In particular,

$$L = \begin{bmatrix} f_1 & \cdots & f_n \end{bmatrix} [L] \begin{bmatrix} e_1 & \cdots & e_m \end{bmatrix}^*$$

$$= \begin{bmatrix} f_1 & \cdots & f_n \end{bmatrix} \begin{bmatrix} \sigma_1 & 0 & \cdots & & \\ 0 & \ddots & 0 & & \\ \vdots & 0 & \sigma_k & 0 & \\ & & 0 & 0 & \\ & & & & \ddots \end{bmatrix} \begin{bmatrix} e_1 & \cdots & e_m \end{bmatrix}^*.$$

Proof. Use the spectral theorem (Theorem 4.3.4) on $L^*L : V \to V$ to find an orthonormal basis $e_1, \ldots, e_m$ for V such that $L^*L\,(e_i) = \lambda_i e_i$. Then,

$$\left(L\,(e_i)\,|\,L\,(e_j)\right) = \left(L^*L\,(e_i)\,|\,e_j\right) = \left(\lambda_i e_i\,|\,e_j\right) = \lambda_i \delta_{ij}.$$

Next, reorder if necessary so that $\lambda_1, \ldots, \lambda_k \neq 0$ and define

$$f_i = \frac{L\,(e_i)}{\|L\,(e_i)\|}, \quad i = 1, \ldots, k.$$

Finally, select $f_{k+1}, \ldots, f_n$ so that we get an orthonormal basis for W.
In this way, we see that $\sigma_i = \|L\,(e_i)\|$. Finally, note that

$$L\,(e_{k+1}) = \cdots = L\,(e_m) = 0$$

since $\|L\,(e_i)\|^2 = \lambda_i$ for all i.   □

The values $\sigma = \sqrt{\lambda}$ where λ is an eigenvalue for L^*L are called the *singular values* of L. We often write the decomposition of L as follows:

$$L = U \Sigma \tilde{U}^*,$$

$$U = \left[\, f_1 \cdots f_n \,\right],$$

$$\tilde{U} = \left[\, e_1 \cdots e_m \,\right],$$

$$\Sigma = \begin{bmatrix} \sigma_1 & 0 & \cdots & & \\ 0 & \ddots & 0 & & \\ \vdots & 0 & \sigma_k & 0 & \\ & & 0 & 0 & \\ & & & & \ddots \end{bmatrix}$$

and we generally order the singular values $\sigma_1 \geq \cdots \geq \sigma_k$.

The singular value decomposition gives us a nice way of studying systems $Lx = b$, when L is not necessarily invertible. In this case, L has a partial or generalized inverse called the *Moore-Penrose inverse*. The construction is quite simple. Take a linear map $L : V \to W$, then use Theorems 3.5.4 and 1.11.7 to conclude that $L|_{(\ker(L))^\perp} : (\ker(L))^\perp \to \operatorname{im}(L)$ is an isomorphism. Thus, we can define the generalized inverse $L^\dagger : W \to V$ in such a way that

$$\ker\left(L^\dagger\right) = (\operatorname{im}(L))^\perp,$$

$$\operatorname{im}\left(L^\dagger\right) = (\ker(L))^\perp,$$

$$L^\dagger|_{\operatorname{im}(L)} = \left(L|_{(\ker(L))^\perp} : (\ker(L))^\perp \to \operatorname{im}(L)\right)^{-1}.$$

If we have picked orthonormal bases that yield the singular value decomposition, then

$$L^\dagger(f_1) = \sigma_1^{-1} f_1, \ldots, L^\dagger(f_k) = \sigma_k^{-1} f_k,$$
$$L^\dagger(f_{k+1}) = \cdots = L^\dagger(f_n) = 0.$$

Or in matrix form using $L = U\Sigma\tilde{U}^*$, we have

$$L^\dagger = \tilde{U}\Sigma^\dagger U^*,$$

where

$$\Sigma^\dagger = \begin{bmatrix} \sigma_1^{-1} & 0 & \cdots & & \\ 0 & \ddots & 0 & & \\ \vdots & 0 & \sigma_k^{-1} & 0 & \\ & & 0 & 0 & \\ & & & & \ddots \end{bmatrix}.$$

This generalized inverse can now be used to try to solve $Lx = b$ for given $b \in W$. Before explaining how that works, we list some of the important properties of the generalized inverse.

Proposition 4.9.2. *Let* $L : V \to W$ *be a linear map between finite-dimensional inner product spaces and* $L^\dagger$ *the Moore-Penrose inverse. Then:*

(1) $(\lambda L)^\dagger = \lambda^{-1} L^\dagger$ *if* $\lambda \neq 0$.
(2) $(L^\dagger)^\dagger = L$.
(3) $(L^*)^\dagger = (L^\dagger)^*$.
(4) $LL^\dagger$ *is an orthogonal projection with* $\text{im}(LL^\dagger) = \text{im}(L)$ *and* $\ker(LL^\dagger) = \ker(L^*) = \ker(L^\dagger)$.
(5) $L^\dagger L$ *is an orthogonal projection with* $\text{im}(L^\dagger L) = \text{im}(L^*) = \text{im}(L^\dagger)$ *and* $\ker(L^\dagger L) = \ker(L)$.
(6) $L^\dagger LL^\dagger = L^\dagger$.
(7) $LL^\dagger L = L$.

Proof. All of these properties can be proven using the abstract definition. Instead, we shall see how the matrix representation coming from the singular value decomposition can also be used to prove the results. Conditions (1)–(3) are straightforward to prove using that the singular value decomposition of L yields singular value decompositions of both $L^\dagger$ and L^*.

To prove (4) and (5), we use the matrix representation to see that

$$L^\dagger L = \tilde{U}\,\Sigma^\dagger U^* U \Sigma \tilde{U}^*$$

$$= \tilde{U} \begin{bmatrix} 1 & 0 & \cdots & & & \\ 0 & \ddots & 0 & & & \\ \vdots & 0 & 1 & 0 & & \\ & & 0 & 0 & & \\ & & & & \ddots \end{bmatrix} \tilde{U}^*$$

and similarly

$$LL^\dagger = U \begin{bmatrix} 1 & 0 & \cdots & & & \\ 0 & \ddots & 0 & & & \\ \vdots & 0 & 1 & 0 & & \\ & & 0 & 0 & & \\ & & & & \ddots \end{bmatrix} U^*.$$

This proves that these maps are orthogonal projections as the bases are orthonormal. It also yields the desired properties for kernels and images.

Finally, (6) and (7) now follow via a similar calculation using the matrix representations.   □

To solve $Lx = b$ for given $b \in W$, we can now use

Corollary 4.9.3. $Lx = b$ *has a solution if and only if* $b = LL^\dagger b$. *Moreover, if b is a solution, then all solutions are given by*

$$x = L^\dagger b + \left(1_V - L^\dagger L\right) z,$$

where $z \in V$. The smallest solution is given by

$$x_0 = L^\dagger b.$$

In case $b \neq LL^\dagger b$, the best approximate solutions are given by

$$x = L^\dagger b + \left(1_V - L^\dagger L\right) z, z \in V$$

again with

$$x_0 = L^\dagger b$$

being the smallest.

Proof. Since $LL^\dagger$ is the orthogonal projection onto $\mathrm{im}\,(L)$, we see that $b \in \mathrm{im}\,(L)$ if and only if $b = LL^\dagger b$. This means that $b = L\left(L^\dagger b\right)$ so that $x_0 = L^\dagger b$ is a solution to the system. Next, we note that $\left(1_V - L^\dagger L\right)$ is the orthogonal projection onto $(\text{im}\,(L^*))^{\perp} = \ker\,(L)$. Thus, all solutions are of the desired form. Finally, as $L^{\dagger}b \in \text{im}\,(L^*)$, the Pythagorean theorem implies that

$$\left\| L^{\dagger}b + \left(1_V - L^{\dagger}L\right) z \right\|^2 = \left\| L^{\dagger}b \right\|^2 + \left\| \left(1_V - L^{\dagger}L\right) z \right\|^2$$

showing that

$$\left\| L^{\dagger}b \right\|^2 \leq \left\| L^{\dagger}b + \left(1_V - L^{\dagger}L\right) z \right\|^2$$

for all z.

The last statement is a consequence of the fact that $LL^{\dagger}b$ is the element in $\text{im}\,(L)$ that is closest to b since $LL^{\dagger}$ is an orthogonal projection.   □

Exercises

1. Show that the singular value decomposition of a self-adjoint operator L with nonnegative eigenvalues looks like $U\Sigma U^*$ where the diagonal entries of Σ are the eigenvalues of L.

2. Find the singular value decompositions of

$$\begin{bmatrix} 0 & 1 \\ 0 & 1 \\ 1 & 0 \end{bmatrix} \text{ and } \begin{bmatrix} 0 & 0 & 1 \\ 1 & 1 & 0 \end{bmatrix}.$$

3. Find the generalized inverses to

$$\begin{bmatrix} 0 & 1 \\ 0 & 0 \end{bmatrix} \text{ and } \begin{bmatrix} 0 & 0 & 0 \\ 1 & 0 & 0 \\ 0 & 1 & 1 \end{bmatrix}.$$

4. Let $L : V \to W$ be a linear operator between finite-dimensional inner product spaces and $\sigma_1 \geq \cdots \geq \sigma_k$ the singular values of L. Show that the results of the section can be rephrased as follows: There exist orthonormal bases $e_1, \ldots, e_m$ for V and $f_1, \ldots, f_n$ for W such that

$$L\,(x) = \sigma_1\,(x|e_1)\,f_1 + \cdots + \sigma_k\,(x|e_k)\,f_k,$$

$$L^*\,(y) = \sigma_1\,(y|f_1)\,e_1 + \cdots + \sigma_k\,(y|f_k)\,e_k,$$

$$L^{\dagger}\,(y) = \sigma_1^{-1}\,(y|f_1)\,e_1 + \cdots + \sigma_k^{-1}\,(y|f_k)\,e_k.$$

5. Let $L : V \to V$ be a linear operator on a finite-dimensional inner product space. Show that L is an isometry if and only if $\ker\,(L) = \{0\}$ and all singular values are 1.

6. Let $L : V \rightarrow W$ be a linear operator between finite-dimensional inner product spaces. Show that

$$\|L\| = \sigma_1,$$

   where σ_1 is the largest singular value of L (see Theorem 3.3.8 for the definition of $\|L\|$).

7. Let $L : V \rightarrow W$ be a linear operator between finite-dimensional inner product spaces. Show that if there are orthonormal bases $e_1, \ldots, e_m$ for V and $f_1, \ldots, f_n$ for W such that $L(e_i) = \tau_i f_i$, $i \leq k$ and $L(e_i) = 0$, $i > k$, then the τ_is are the singular values of L.

8. Let $L : V \rightarrow W$ be a nontrivial linear operator between finite-dimensional inner product spaces.

   (a) If $e_1, \ldots, e_m$ is an orthonormal basis for V, show that

   $$\text{tr}\left(L^*L\right) = \|L(e_1)\|^2 + \cdots + \|L(e_m)\|^2.$$

   (b) If $\sigma_1 \geq \cdots \geq \sigma_k$ are the singular values for L, show that

   $$\text{tr}\left(L^*L\right) = \sigma_1^2 + \cdots + \sigma_k^2.$$

4.10   The Polar Decomposition*

In this section, we are going to study general linear operators $L : V \rightarrow V$ on a finite-dimensional inner product space. These can be decomposed in a manner similar to the polar coordinate decomposition of complex numbers: $z = e^{i\theta} |z|$.

Theorem 4.10.1 (The Polar Decomposition). *; Let $L : V \rightarrow V$ be a linear operator on a finite-dimensional inner product space; then, $L = WS$, where W is unitary (or orthogonal) and S is self-adjoint with nonnegative eigenvalues. Moreover, if L is invertible, then W and S are uniquely determined by L.*

Proof. The proof is similar to the construction of the singular value decomposition (Theorem 4.9.1). In fact, we can use the singular value decomposition to prove the polar decomposition:

$$L = U \Sigma \tilde{U}^*$$
$$= U \tilde{U}^* \tilde{U} \Sigma \tilde{U}^*$$
$$= \left(U \tilde{U}^*\right)\left(\tilde{U} \Sigma \tilde{U}^*\right).$$

Thus, we define

$$W = U \tilde{U}^*,$$
$$S = \tilde{U} \Sigma \tilde{U}^*.$$

Clearly, W is unitary as it is a composition of two isometries. And S is certainly self-adjoint with nonnegative eigenvalues as we have diagonalized it with an orthonormal basis, and Σ has nonnegative diagonal entries.

Finally, assume that L is invertible and

$$L = WS = \tilde{W}T,$$

where $W, \tilde{W}$ are unitary and S, T are self-adjoint with positive eigenvalues. Then, S and T must also be invertible and

$$ST^{-1} = \tilde{W}W^{-1}$$
$$= \tilde{W}W^*.$$

This implies that ST^{-1} is unitary. Thus,

$$\left(ST^{-1}\right)^{-1} = \left(ST^{-1}\right)^*$$
$$= \left(T^*\right)^{-1}S^*$$
$$= T^{-1}S,$$

and therefore,

$$1_V = T^{-1}SST^{-1}$$
$$= T^{-1}S^2T^{-1}.$$

This means that $S^2 = T^2$. Since both operators are self-adjoint and have nonnegative eigenvalues, this implies that $S = T$ (see Exercise 8 in Sect. 4.3) and hence $\tilde{W} = W$ as desired. □

There is also a decomposition $L = SW$, where $S = U\Sigma U^*$ and $W = U\tilde{U}^*$. From this, it is clear that S and W need not be the same in the two decompositions unless $U = \tilde{U}$ in the singular value decomposition. This is equivalent to L being normal (see also exercises).

Recall from Sect. 1.13 that we have the *general linear group* $Gl_n\,(\mathbb{F}) \subset \text{Mat}_{n \times n}\,(\mathbb{F})$ of invertible $n \times n$ matrices. Further, define $PS_n\,(\mathbb{F}) \subset \text{Mat}_{n \times n}\,(\mathbb{F})$ as being the self-adjoint positive matrices, i.e., the eigenvalues are positive. The polar decomposition says that we have bijective (nonlinear) maps (i.e., one-to-one and onto maps)

$$Gl_n\,(\mathbb{C}) \approx U_n \times PS_n\,(\mathbb{C})\,,$$
$$Gl_n\,(\mathbb{R}) \approx O_n \times PS_n\,(\mathbb{R})\,,$$

given by $A = WS \longleftrightarrow (W, S)$. These maps are in fact *homeomorphisms*, i.e., both $(W, S) \mapsto WS$ and $A = WS \mapsto (W, S)$ are continuous. The first map only involves matrix multiplication, so it is obviously continuous. That $A = WS \to (W, S)$ is continuous takes a little more work. Assume that $A_k = W_k S_k$ and that $A_k \to A = WS \in Gl_n$. Then, we need to show that $W_k \to W$ and $S_k \to S$. The space of unitary or orthogonal operators is compact. So any subsequence of W_k has a convergent subsequence. Now, assume that $W_{k_l} \to \bar{W}$, then also $S_{k_l} = \left(W_{k_l}^*\right) A_{k_l} \to \bar{W}^* A$. Thus, $A = \bar{W}\left(\bar{W}^* A\right)$, which implies by the uniqueness of the polar decomposition that $\bar{W} = W$ and $S_{k_l} \to S$. This means that convergent subsequences of W_k always converge to W; this in turn implies that $W_k \to W$. We then conclude that also $S_k \to S$.

Next, we note that PS_n is a *convex cone*. This means that if $A, B \in PS_n$, then also $sA + tB \in PS_n$ for all $t, s > 0$. It is obvious that $sA + tB$ is self-adjoint. To see that all eigenvalues are positive, we use that $(Ax|x), (Bx|x) > 0$ for all $x \neq 0$ to see that

$$((sA + tB)(x)|x) = s(Ax|x) + t(Bx|x) > 0.$$

The importance of this last observation is that we can deform any matrix $A = WS$ via

$$A_t = W(tI + (1-t)A) \in Gl_n$$

into a unitary or orthogonal matrix. This means that many *topological* properties of Gl_n can be investigated by studying the compact groups U_n and O_n.

The simplest example of this is that $Gl_n(\mathbb{C})$ is path connected, i.e., for any two matrices $A, B \in Gl_n(\mathbb{C})$, there is a continuous path $C : [0, \alpha] \to Gl_n(\mathbb{C})$ such that $C(0) = A$ and $C(\alpha) = B$. By way of contrast, $Gl_n(\mathbb{R})$ has two path connected components. We can see these two facts directly when $n = 1$ as $Gl_1(\mathbb{C}) = \{\alpha \in \mathbb{C} : \alpha \neq 0\}$ is connected, while $Gl_1(\mathbb{R}) = \{\alpha \in \mathbb{R} : \alpha \neq 0\}$ consists of the two components corresponding the positive and negative numbers. For general n, we can prove the claim by using the canonical form for unitary and orthogonal matrices. In the unitary situation, we have that any $U \in U_n$ looks like

$$U = BDB^*$$

$$= B \begin{bmatrix} \exp(i\theta_1) & & 0 \\ & \ddots & \\ 0 & & \exp(i\theta_n) \end{bmatrix} B^*,$$

where $B \in U_n$. Then, define

$$D(t) = \begin{bmatrix} \exp(it\theta_1) & & 0 \\ & \ddots & \\ 0 & & \exp(it\theta_n) \end{bmatrix}.$$

Hence, $D(t) \in U_n$ and $U(t) = BD(t)B^* \in U_n$ defines a path that at $t = 0$ is I_n and at $t = 1$ is U. Thus, any unitary transformation can be joined to the identity matrix inside U_n.

In the orthogonal case, we see using the real canonical form that a similar deformation using

$$\begin{bmatrix} \cos(t\theta_i) & -\sin(t\theta_i) \\ \sin(t\theta_i) & \cos(t\theta_i) \end{bmatrix}$$

will deform any orthogonal transformation to one of the following two matrices:

$$\begin{bmatrix} 1 & 0 & & 0 \\ 0 & 1 & & 0 \\ & & \ddots & \\ 0 & 0 & & 1 \end{bmatrix} \text{ or } O \begin{bmatrix} -1 & 0 & & 0 \\ 0 & 1 & & 0 \\ & & \ddots & \\ 0 & 0 & & 1 \end{bmatrix} O^t.$$

Here

$$O \begin{bmatrix} -1 & 0 & & 0 \\ 0 & 1 & & 0 \\ & & \ddots & \\ 0 & 0 & & 1 \end{bmatrix} O^t$$

is the same as the reflection R_x where x is the first column vector in O (-1 eigenvector). We then have to show that $1_{\mathbb{R}^n}$ and R_x cannot be joined to each other inside O_n. This is done by contradiction. Thus, assume that $A(t)$ is a continuous path with

$$A(0) = \begin{bmatrix} 1 & 0 & & 0 \\ 0 & 1 & & 0 \\ & & \ddots & \\ 0 & 0 & & 1 \end{bmatrix},$$

$$A(1) = O \begin{bmatrix} -1 & 0 & & 0 \\ 0 & 1 & & 0 \\ & & \ddots & \\ 0 & 0 & & 1 \end{bmatrix} O^t,$$

$$A(t) \in O_n, \text{ for all } t \in [0, 1].$$

The characteristic polynomial

$$\chi_{A(t)}(\lambda) = \lambda^n + \cdots + \alpha_0(t)$$

has coefficients that vary continuously with t (the proof of this uses determinants). However, $\alpha_0(0) = (-1)^n$, while $\alpha_0(1) = (-1)^{n-1}$. Thus, the intermediate value theorem tells us that $\alpha_0(t_0) = 0$ for some $t_0 \in (0, 1)$. But this implies that $\lambda = 0$ is a root of $A(t_0)$, thus contradicting that $A(t_0) \in O_n \subset Gl_n$.

Exercises

1. Find the polar decompositions for

$$\begin{bmatrix} \alpha & -\beta \\ \beta & \alpha \end{bmatrix}, \begin{bmatrix} \alpha & \beta \\ -\beta & \alpha \end{bmatrix}, \text{ and } \begin{bmatrix} \alpha & 1 \\ 0 & \alpha \end{bmatrix}.$$

2. Find the polar decompositions for

$$\begin{bmatrix} 0 & \beta & 0 \\ \alpha & 0 & 0 \\ 0 & 0 & \gamma \end{bmatrix} \text{ and } \begin{bmatrix} 1 & -1 & 0 \\ 0 & 0 & 2 \\ 1 & 1 & 0 \end{bmatrix}.$$

3. If $L : V \to V$ is a linear operator on a finite-dimensional inner product space, define the *Cayley transform* of L as $(L + 1_V)(L - 1_V)^{-1}$.

   (a) If L is skew-adjoint, show that $(L + 1)(L - 1)^{-1}$ is an isometry that does not have -1 as an eigenvalue.
   (b) Show that $U \to (U - 1_V)(U + 1_V)^{-1}$ takes isometries that do not have -1 as an eigenvalue to skew-adjoint operators and is an inverse to the Cayley transform.

4. Let $L : V \to V$ be a linear operator on a finite-dimensional inner product space. Show that $L = SW$, where W is unitary (or orthogonal) and S is self-adjoint with nonnegative eigenvalues. Moreover, if L is invertible, then W and S are unique. Show by example that the operators in this polar decomposition do not have to be the same as in the $L = WS$ decomposition.

5. Let $L = WS$ be the unique polar decomposition of an invertible operator $L : V \to V$ on a finite-dimensional inner product space V. Show that L is normal if and only if $WS = SW$.

6. The purpose of this exercise is to check some properties of the exponential map $\exp : \text{Mat}_{n \times n}(\mathbb{F}) \to Gl_n(\mathbb{F})$. You may want to consult Sect. 3.7 for the definition and various elementary properties.

   (a) Show that exp maps normal operators to normal operators.
   (b) Show that exp maps self-adjoint operators to positive self-adjoint operators and that it is a homeomorphism, i.e., it is one-to-one, onto, continuous, and the inverse is also continuous.
   (c) Show that exp maps skew-adjoint operators to isometries but is not one-to-one. In the complex case, show that it is onto.

7. Let $L : V \to V$ be normal and $L = S + A$, where S is self-adjoint and A skew-adjoint. Recall that since L is normal S and A commute.

   (a) Show that $\exp(S)\exp(A) = \exp(A)\exp(S)$ is the polar decomposition of $\exp(L)$.
   (b) Show that any invertible normal transformation can be written as $\exp(L)$ for some normal L.

4.11   Quadratic Forms*

Conic sections are those figures we obtain by intersecting a cone with a plane. Analytically, this is the problem of determining all of the intersections of a cone given by $z = x^2 + y^2$ with a plane $z = ax + by + c$.

We can picture what these intersections look like by shining a flash light on a wall. The light emanating from the flash light describes a cone which is then intersected by the wall. The figures we get are circles, ellipses, parabolae, and hyperbolae, depending on how we hold the flash light.

These questions naturally lead to the more general question of determining the figures described by the equation

$$ax^2 + bxy + cy^2 + dx + ey + f = 0.$$

We shall see below that it is possible to make a linear change of coordinates, that depends only on the quadratic quantities, such that the equation is transformed into an equation of the simpler form:

$$a'\left(x'\right)^2 + c'\left(y'\right)^2 + d'x' + e'y' + f' = 0.$$

It is now easy to see that the solutions to such an equation consist of a circle, ellipse, parabola, hyperbola, or the degenerate cases of two lines, a point or nothing. Moreover a, b, c together determine the type of the figure as long as it is not degenerate.

Aside from the aesthetic virtues of this problem, it also comes up naturally when solving the two-body problem from physics, a rather remarkable coincidence between beauty and the real world. Another application is to the problem of deciding when a function in two variables has a maximum, minimum, or neither at a critical point.

The goal here is to study this problem in the more general case with n variables and show how the spectral theorem can be brought in to help our investigations. We shall also explain the use in multivariable calculus.

A *quadratic form* Q in n real variables $x = (x_1, \ldots, x_n)$ is a function of the form

$$Q(x) = \sum_{1 \le i \le j \le n} a_{ij} x_i x_j.$$

The term $x_i x_j$ only appears once in this sum. We can artificially have it appear twice so that the sum is more symmetric

$$Q(x) = \sum_{i,j=1}^{n} a'_{ij} x_i x_j,$$

where $a'_{ii} = a_{ii}$ and $a'_{ij} = a'_{ji} = a_{ij}/2$. If we define A as the matrix whose entries are a'_{ij} and use the inner product on $\mathbb{R}^n$, then the quadratic form can be written in the more abstract and condensed form

$$Q(x) = (Ax|x).$$

The important observation is that A is a symmetric real matrix and hence self-adjoint. This means that we can find a new orthonormal basis for $\mathbb{R}^n$ that diagonalizes A. If this basis is given by the matrix B, then

$$A = BDB^{-1}$$

$$= B \begin{bmatrix} \sigma_1 & & 0 \\ & \ddots & \\ 0 & & \sigma_n \end{bmatrix} B^{-1}$$

$$= B \begin{bmatrix} \sigma_1 & & 0 \\ & \ddots & \\ 0 & & \sigma_n \end{bmatrix} B^t.$$

If we define new coordinates by

$$\begin{bmatrix} y_1 \\ \vdots \\ y_n \end{bmatrix} = B^{-1} \begin{bmatrix} x_1 \\ \vdots \\ x_n \end{bmatrix}, \quad \text{or}$$

$$x = By,$$

then

$$Q(x) = (Ax|x)$$
$$= (ABy|By)$$
$$= (B^t ABy|y)$$
$$= Q'(y).$$

Since B is an orthogonal matrix, we have that $B^{-1} = B^t$ and hence $B^t AB = B^{-1}AB = D$. Thus,

$$Q'(y) = \sigma_1 y_1^2 + \cdots + \sigma_n y_n^2$$

in the new coordinates.

The general classification of the types of quadratic forms is given by

(1) If all of $\sigma_1, \ldots, \sigma_n$ are positive or negative, then it is said to be *elliptic*.
(2) If all of $\sigma_1, \ldots, \sigma_n$ are nonzero and there are both negative and positive values, then it said to be *hyperbolic*.
(3) If at least one of $\sigma_1, \ldots, \sigma_n$ is zero, then it is called *parabolic*.

In the case of two variables, this makes perfect sense as $x^2 + y^2 = r^2$ is a circle (special ellipse), $x^2 - y^2 = f$ two branches of a hyperbola, and $x^2 = f$ a parabola. The first two cases occur when $\sigma_1 \cdots \sigma_n \neq 0$. In this case, the quadratic form is said to be *nondegenerate*. In the parabolic case, $\sigma_1 \cdots \sigma_n = 0$ and we say that the quadratic form is degenerate.

Having obtained this simple classification it would be nice to find a way of characterizing these types directly from the characteristic polynomial of A without having to find the roots. This is actually not too hard to accomplish.

Lemma 4.11.1 (Descartes' Rule of Signs). *Let*

$$p(t) = t^n + a_{n-1}t^{n-1} + \cdots + a_1 t + a_0 = (t - \lambda_1) \cdots (t - \lambda_n),$$

where $a_0, \ldots, a_{n-1}, \lambda_1, \ldots, \lambda_n \in \mathbb{R}$.

(1) 0 is a root of $p(t)$ if and only if $a_0 = 0$.
(2) All roots of $p(t)$ are negative if and only if $a_{n-1}, \ldots, a_0 > 0$.
(3) If n is odd, then all roots of $p(t)$ are positive if and only if $a_{n-1} < 0$, $a_{n-2} > 0, \ldots, a_1 > 0$, $a_0 < 0$.
(4) If n is even, then all roots of $p(t)$ are positive if and only if $a_{n-1} < 0$, $a_{n-2} > 0, \ldots, a_1 < 0$, $a_0 > 0$.

Proof. Descartes' rule is actually more general as it relates the number of positive roots to the number of times the coefficients change sign. The simpler version, however, suffices for our purposes.

Part 1 is obvious as $p(0) = a_0$.

The relationship

$$t^n + a_{n-1}t^{n-1} + \cdots + a_1 t + a_0 = (t - \lambda_1) \cdots (t - \lambda_n)$$

clearly shows that $a_{n-1}, \ldots, a_0 > 0$ if $\lambda_1, \ldots, \lambda_n < 0$. Conversely, if $a_{n-1}, \ldots, a_0 > 0$, then it is obvious that $p(t) > 0$ for all $t \geq 0$.

For the other two properties, consider $q(t) = p(-t)$ and use (2). $\square$

This lemma gives us a very quick way of deciding whether a given quadratic form is parabolic or elliptic. If it is not one of these two types, then we know it has to be hyperbolic.

Example 4.11.2. The matrix

$$\begin{bmatrix} 2 & 3 & 0 \\ 3 & -2 & 4 \\ 0 & 4 & -2 \end{bmatrix}$$

has characteristic polynomial $t^3 + 2t^2 - 29t + 6$. The coefficients do not conform to the patterns that guarantee that the roots are all positive or negative so we conclude that the corresponding quadratic form is hyperbolic.

Example 4.11.3. Let Q be a quadratic form corresponding to the matrix

$$A = \begin{bmatrix} 6 & 1 & 2 & 3 \\ 1 & 5 & 0 & 4 \\ 2 & 0 & 2 & 0 \\ 3 & 4 & 0 & 7 \end{bmatrix}.$$

The characteristic polynomial is given by $t^4 - 20t^3 + 113t^2 - 200t + 96$. In this case, the coefficients tell us that the roots must be positive.

Exercises

1. A bilinear form on a vector space V is a function $B : V \times V \to \mathbb{F}$ such that $x \to B(x, y)$ and $y \to B(x, y)$ are both linear. Show that a quadratic form Q always looks like $Q(x) = B(x, x)$, where B is a bilinear form.
2. A bilinear form is said to be symmetric, respectively skew-symmetric, if $B(x, y) = B(y, x)$, respectively $B(x, y) = -B(y, x)$ for all x, y.

   (a) Show that a quadratic form looks like $Q(x) = B(x, x)$ where B is symmetric.
   (b) Show that $B(x, x) = 0$ for all $x \in V$ if and only if B is skew-symmetric.

3. Let B be a bilinear form on $\mathbb{R}^n$ or $\mathbb{C}^n$.

   (a) Show that $B(x, y) = (Ax|y)$ for some matrix A.
   (b) Show that B is symmetric if and only if A is symmetric.
   (c) Show that B is skew-symmetric if and only if A is skew-symmetric.
   (d) If $x = Cx'$ is a change of basis, show that if B corresponds to A in the standard basis, then it corresponds to $C^t A C$ in the new basis.

4. Let $Q(x)$ be a quadratic form on $\mathbb{R}^n$. Show that there is an orthogonal basis where

   $$Q(z) = -z_1^2 - \cdots - z_k^2 + z_{k+1}^2 + \cdots + z_l^2,$$

   where $0 \le k \le l \le n$. Hint: Use the orthonormal basis that diagonalized Q and adjust the lengths of the basis vectors.

5. Let $B\,(x, y)$ be a skew-symmetric form on $\mathbb{R}^n$.

   (a) If $B\,(x, y) = (Ax|y)$ where $A = \begin{bmatrix} 0 & -\beta \\ \beta & 0 \end{bmatrix}$, $\beta \in \mathbb{R}$, then show that there is

   a basis for $\mathbb{R}^2$ where $B\,(x', y')$ corresponds to $A' = \begin{bmatrix} 0 & -1 \\ 1 & 0 \end{bmatrix}$.

   (b) If $B\,(x, y)$ is a skew-symmetric bilinear form on $\mathbb{R}^n$, then there is a basis
   where $B\,(x', y')$ corresponds to a matrix of the type

$$A' = \begin{bmatrix}
0 & -1 & \cdots & 0 & 0 & 0 & \cdots & 0 \\
1 & 0 & & 0 & 0 & 0 & \vdots & 0 \\
\vdots & \vdots & \ddots & 0 & 0 & 0 & \vdots & 0 \\
0 & 0 & & 0 & 0 & -1 & 0 & 0 \\
0 & 0 & & 0 & 1 & 0 & 0 & 0 \\
0 & 0 & & 0 & 0 & 0 & \cdots & 0 \\
0 & 0 & \cdots & 0 & 0 & \vdots & \ddots & \vdots \\
0 & 0 & \cdots & 0 & 0 & 0 & \cdots & 0
\end{bmatrix}$$

6. Show that for a quadratic form $Q\,(z)$ on $\mathbb{C}^n$, we can always change coordinates
   to make it look like
$$Q'\,(z') = (z_1')^2 + \cdots + (z_n')^2.$$

7. Show that $Q\,(x, y) = ax^2 + 2bxy + cy^2$ is elliptic when $ac - b^2 > 0$, hyperbolic
   when $ac - b^2 < 0$, and parabolic when $ac - b^2 = 0$.

8. If A is a symmetric real matrix, then show that $tI + A$ defines an elliptic quadratic
   form when $|t|$ is sufficiently large.

9. Decide for each of the following matrices whether or not the corresponding
   quadratic form is elliptic, hyperbolic, or parabolic:

   (a)
$$\begin{bmatrix}
-7 & -2 & -3 & 0 \\
-2 & -6 & -4 & 0 \\
-3 & -4 & -5 & 2 \\
0 & 0 & 2 & -3
\end{bmatrix}.$$

   (b)
$$\begin{bmatrix}
7 & 3 & -3 & 4 \\
3 & 2 & -1 & 0 \\
-3 & -1 & 5 & -2 \\
4 & 0 & -2 & 10
\end{bmatrix}.$$

(c)

$$\begin{bmatrix} -8 & -3 & 0 & -2 \\ -3 & -1 & -1 & 0 \\ 0 & -1 & 1 & 3 \\ -2 & 0 & 3 & -3 \end{bmatrix}.$$

(d)

$$\begin{bmatrix} 15 & 2 & 3 & 4 \\ 2 & 4 & 2 & 0 \\ 3 & 2 & 3 & -2 \\ 4 & 0 & -2 & 5 \end{bmatrix}.$$

Chapter 5
Determinants

5.1 Geometric Approach

Before plunging in to the theory of determinants, we are going to make an attempt at defining them in a more geometric fashion. This works well in low dimensions and will serve to motivate our more algebraic constructions in subsequent sections.

From a geometric point of view, the determinant of a linear operator $L : V \to V$ is a scalar $\det(L)$ that measures how L changes the volume of solids in V. To understand how this works, we obviously need to figure out how volumes are computed in V. In this section, we will study this problem in dimensions 1 and 2. In subsequent sections, we take a more axiomatic and algebraic approach, but the ideas come from what we present here.

Let V be one-dimensional and assume that the scalar field is $\mathbb{R}$ so as to keep things as geometric as possible. We already know that $L : V \to V$ must be of the form $L(x) = \lambda x$ for some $\lambda \in \mathbb{R}$. This λ clearly describes how L changes the length of vectors as $\|L(x)\| = |\lambda| \|x\|$. The important and surprising thing to note is that while we need an inner product to compute the length of vectors, it is not necessary to know the norm in order to compute how L changes the length of vectors.

Let now V be two-dimensional. If we have a real inner product, then we can talk about areas of simple geometric configurations. We shall work with parallelograms as they are easy to define, one can easily find their area, and linear operators map parallelograms to parallelograms. Given $x, y \in V$, the parallelogram $\pi(x, y)$ with sides x and y is defined by

$$\pi(x, y) = \{sx + ty : s, t \in [0, 1]\}.$$

The area of $\pi(x, y)$ can be computed by the usual formula where one multiplies the base length with the height. If we take x to be the base, then the height is the projection of y onto to orthogonal complement of x. Thus, we get the formula (see also Fig. 5.1)

P. Petersen, *Linear Algebra*, Undergraduate Texts in Mathematics,
DOI 10.1007/978-1-4614-3612-6_5, © Springer Science+Business Media New York 2012

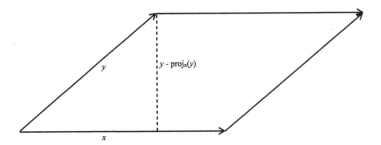

Fig. 5.1 Area of a parallelogram

$$\text{area}\,(\pi\,(x,y)) = \|x\|\,\|y - \text{proj}_x\,(y)\|$$

$$= \|x\|\left\|y - \frac{(y|x)\,x}{\|x\|^2}\right\|.$$

This expression does not appear to be symmetric in x and y, but if we square it, we get

$$(\text{area}\,(\pi\,(x,y)))^2 = (x|x)\,(y - \text{proj}_x\,(y)\,|y - \text{proj}_x\,(y))$$

$$= (x|x)\,((y|y) - 2\,(y|\text{proj}_x\,(y)) + (\text{proj}_x\,(y)\,|\text{proj}_x\,(y)))$$

$$= (x|x)\left((y|y) - 2\left(y\left|\frac{(y|x)\,x}{\|x\|^2}\right.\right) + \left(\frac{(y|x)\,x}{\|x\|^2}\left|\frac{(y|x)\,x}{\|x\|^2}\right.\right)\right)$$

$$= (x|x)\,(y|y) - (x|y)^2,$$

which is symmetric in x and y. Now assume that

$$x' = \alpha x + \beta y$$
$$y' = \gamma x + \delta y$$

or

$$[x'\ y'] = [x\ y]\begin{bmatrix}\alpha & \gamma \\ \beta & \delta\end{bmatrix};$$

then, we see that

$$(\text{area}\,(\pi\,(x',y')))^2$$

$$= (x'|x')\,(y'|y') - (x'|y')^2$$

$$= (\alpha x + \beta y|\alpha x + \beta y)\,(\gamma x + \delta y|\gamma x + \delta y) - (\alpha x + \beta y|\gamma x + \delta y)^2$$

$$= \left(\alpha^2 \left(x|x\right) + 2\alpha\beta \left(x|y\right) + \beta^2 \left(y|y\right)\right) \left(\gamma^2 \left(x|x\right) + 2\gamma\delta \left(x|y\right) + \delta^2 \left(y|y\right)\right)$$

$$- \left(\alpha\gamma \left(x|x\right) + \left(\alpha\delta + \beta\gamma\right) \left(x|y\right) + \beta\delta \left(y|y\right)\right)^2$$

$$= \left(\alpha^2\delta^2 + \beta^2\gamma^2 - 2\alpha\beta\gamma\delta\right) \left(\left(x|x\right) \left(y|y\right) - \left(x|y\right)^2\right)$$

$$= \left(\alpha\delta - \beta\gamma\right)^2 \left(\text{area} \left(\pi \left(x, y\right)\right)\right)^2 .$$

This tells us several things. First, if we know how to compute the area of just one parallelogram, then we can use linear algebra to compute the area of any other parallelogram by simply expanding the base vectors for the new parallelogram in terms of the base vectors of the given parallelogram. This has the surprising consequence that the ratio of the areas of two parallelograms does not depend upon the inner product! With this in mind, we can then define the determinant of a linear operator $L : V \to V$ so that

$$\left(\det \left(L\right)\right)^2 = \frac{\left(\text{area} \left(\pi \left(L \left(x\right), L \left(y\right)\right)\right)\right)^2}{\left(\text{area} \left(\pi \left(x, y\right)\right)\right)^2}.$$

To see that this does not depend on x and y, we chose x' and y' as above and note that

$$\left[L \left(x'\right) \ L \left(y'\right) \right] = \left[L \left(x\right) \ L \left(y\right) \right] \begin{bmatrix} \alpha & \gamma \\ \beta & \delta \end{bmatrix}$$

and

$$\frac{\left(\text{area} \left(\pi \left(L \left(x'\right), L \left(y'\right)\right)\right)\right)^2}{\left(\text{area} \left(\pi \left(x', y'\right)\right)\right)^2} = \frac{\left(\alpha\delta - \beta\gamma\right)^2 \left(\text{area} \left(\pi \left(L \left(x\right), L \left(y\right)\right)\right)\right)^2}{\left(\alpha\delta - \beta\gamma\right)^2 \left(\text{area} \left(\pi \left(x, y\right)\right)\right)^2}$$

$$= \frac{\left(\text{area} \left(\pi \left(L \left(x\right), L \left(y\right)\right)\right)\right)^2}{\left(\text{area} \left(\pi \left(x, y\right)\right)\right)^2}.$$

Thus, $\left(\det \left(L\right)\right)^2$ depends neither on the inner product that is used to compute the area nor on the vectors x and y. Finally, we can refine the definition so that

$$\det \left(L\right) = \begin{vmatrix} \alpha & \gamma \\ \beta & \delta \end{vmatrix} = \alpha\delta - \beta\gamma, \text{ where}$$

$$\left[L \left(x\right) \ L \left(y\right) \right] = \left[x \ y \right] \begin{bmatrix} \alpha & \gamma \\ \beta & \delta \end{bmatrix}.$$

This introduces a sign in the definition which one can also easily check does not depend on the choice of x and y.

This approach generalizes to higher dimensions, but it also runs into a little trouble. The keen observer might have noticed that the formula for the area is in fact a determinant:

$$
(\text{area} \, (\pi \, (x, y)))^2 = (x|x) \, (y|y) - (x|y)^2
$$

$$
= \begin{vmatrix} (x|x) \ (x|y) \\ (x|y) \ (y|y) \end{vmatrix}.
$$

When passing to higher dimensions, it will become increasingly harder to justify how the volume of a parallelepiped depends on the base vectors without using a determinant. Thus, we encounter a bit of a vicious circle when trying to define determinants in this fashion.

The other problem is that we used only real scalars. One can modify the approach to also work for complex numbers, but beyond that, there is not much hope. The approach we take below is mirrored on the constructions here but works for general scalar fields.

5.2   Algebraic Approach

As was done in the previous section, we are going to separate the idea of volumes and determinants, the latter being exclusively for linear operators and a quantity which is independent of other structures on the vector space. Since volumes are used to define determinants, we start by defining what a volume form is. Unlike the more motivational approach we took in the previous section, we will take a more strictly axiomatic approach.

Let V be an n-dimensional vector space over $\mathbb{F}$.

Definition 5.2.1. A *volume form*

$$
\text{vol} : \overbrace{V \times \cdots \times V}^{n \text{ times}} \to \mathbb{F}
$$

is a multi-linear map, i.e., it is linear in each variable if the others are fixed, which is also alternating. More precisely, if $x_1, \ldots, x_{i-1}, x_{i+1}, \ldots, x_n \in V$, then

$$
x \to \text{vol} \, (x_1, \ldots, x_{i-1}, x, x_{i+1}, \ldots, x_n)
$$

is linear, and for $i < j$, we have the *alternating property* when x_i and x_j are transposed:

$$
\text{vol} \, (\ldots, x_i, \ldots, x_j, \ldots) = -\text{vol} \, (\ldots, x_j, \ldots, x_i, \ldots).
$$

In Sect. 5.4 below, we shall show that such volume forms always exist. In this section, we are going to establish some important properties and also give some methods for computing volumes.

Proposition 5.2.2. *Let* vol : $V \times \cdots \times V \to \mathbb{F}$ *be a volume form on an n-dimensional vector space over* $\mathbb{F}$*. Then,*

(1) vol $(\ldots, x, \ldots, x, \ldots) = 0$.
(2) vol $(x_1, \ldots, x_{i-1}, x_i + y, x_{i+1}, \ldots, x_n) =$ vol $(x_1, \ldots, x_{i-1}, x_i, x_{i+1}, \ldots, x_n)$
    if $y = \sum_{k \neq i} \alpha_k x_k$ *is a linear combination of* $x_1, \ldots, x_{i-1}, x_{i+1}, \ldots x_n$.
(3) vol $(x_1, \ldots, x_n) = 0$ *if* $x_1, \ldots, x_n$ *are linearly dependent.*
(4) *If* vol $(x_1, \ldots, x_n) \neq 0$*, then* $x_1, \ldots, x_n$ *form a basis for V.*

Proof. (1) The alternating property tells us that

$$\text{vol} (\ldots, x, \ldots, x, \ldots) = -\text{vol} (\ldots, x, \ldots, x, \ldots)$$

if we switch x and x. Thus, vol $(\ldots, x, \ldots, x, \ldots) = 0$.
(2) Let $y = \sum_{k \neq i} \alpha_k x_k$ and use linearity to conclude

$$\text{vol} (x_1, \ldots, x_{i-1}, x_i + y, x_{i+1}, \ldots, x_n)$$
$$= \text{vol} (x_1, \ldots, x_{i-1}, x_i, x_{i+1}, \ldots, x_n)$$
$$+ \sum_{k \neq i} \alpha_k \text{vol} (x_1, \ldots, x_{i-1}, x_k, x_{i+1}, \ldots, x_n) .$$

Since x_k is always equal to one of $x_1, \ldots, x_{i-1}, x_{i+1}, \ldots, x_n$, we see that

$$\alpha_k \text{vol} (x_1, \ldots, x_{i-1}, x_k, x_{i+1}, \ldots, x_n) = 0.$$

This implies the claim.
(3) If $x_1 = 0$, we are finished. Otherwise, Lemma 1.12.3 shows that there is $k \geq 1$ such that $x_k = \sum_{i=1}^{k-1} \alpha_i x_i$. Then, (2) implies that

$$\text{vol} (x_1, \ldots, 0 + x_k, \ldots, x_n) = \text{vol} (x_1, \ldots, 0, \ldots, x_n)$$
$$= 0.$$

(4) From (3), we have that $x_1, \ldots, x_n$ are linearly independent. Since V has dimension n, they must also form a basis. $\qquad\square$

Note that in the above proof, we had to use that $1 \neq -1$ in the scalar field. This is certainly true for the fields we work with. When working with more general fields such as $\mathbb{F} = \{0, 1\}$, we need to modify the alternating property. Instead, we assume that the volume form vol $(x_1, \ldots, x_n)$ satisfies vol $(x_1, \ldots, x_n) = 0$ whenever $x_i = x_j$. This in turn implies the alternating property. To prove this note that if $x = x_i + x_j$, then

$$0 = \mathrm{vol}\left(\ldots, \underset{x}{\overset{i\,\text{th place}}{\underline{}}}, \ldots, \underset{x}{\overset{j\,\text{th place}}{\underline{}}}, \ldots\right)$$

$$= \mathrm{vol}\left(\ldots, \underset{x_i + x_j}{\overset{i\,\text{th place}}{\underline{}}}, \ldots, \underset{x_i + x_j}{\overset{j\,\text{th place}}{\underline{}}}, \ldots\right)$$

$$= \mathrm{vol}\left(\ldots, \underset{x_i}{\overset{i\,\text{th place}}{\underline{}}}, \ldots, \underset{x_i}{\overset{j\,\text{th place}}{\underline{}}}, \ldots\right)$$

$$+ \mathrm{vol}\left(\ldots, \underset{x_j}{\overset{i\,\text{th place}}{\underline{}}}, \ldots, \underset{x_i}{\overset{j\,\text{th place}}{\underline{}}}, \ldots\right)$$

$$+ \mathrm{vol}\left(\ldots, \underset{x_i}{\overset{i\,\text{th place}}{\underline{}}}, \ldots, \underset{x_j}{\overset{j\,\text{th place}}{\underline{}}}, \ldots\right)$$

$$+ \mathrm{vol}\left(\ldots, \underset{x_j}{\overset{i\,\text{th place}}{\underline{}}}, \ldots, \underset{x_j}{\overset{j\,\text{th place}}{\underline{}}}, \ldots\right)$$

$$= \mathrm{vol}\left(\ldots, \underset{x_j}{\overset{i\,\text{th place}}{\underline{}}}, \ldots, \underset{x_i}{\overset{j\,\text{th place}}{\underline{}}}, \ldots\right)$$

$$+ \mathrm{vol}\left(\ldots, \underset{x_i}{\overset{i\,\text{th place}}{\underline{}}}, \ldots, \underset{x_j}{\overset{j\,\text{th place}}{\underline{}}}, \ldots\right),$$

which shows that the form is alternating.

Theorem 5.2.3. (Uniqueness of Volume Forms) *Let* $\mathrm{vol}_1, \mathrm{vol}_2 : V \times \cdots \times V \to \mathbb{F}$ *be two volume forms on an n-dimensional vector space over* $\mathbb{F}$. *If* vol_2 *is nontrivial, then* $\mathrm{vol}_1 = \lambda \mathrm{vol}_2$ *for some* $\lambda \in \mathbb{F}$.

Proof. If we assume that vol_2 is nontrivial, then we can find $x_1, \ldots, x_n \in V$ so that $\mathrm{vol}_2(x_1, \ldots, x_n) \neq 0$. Then, define λ so that

$$\mathrm{vol}_1(x_1, \ldots, x_n) = \lambda \mathrm{vol}_2(x_1, \ldots, x_n).$$

If $z_1, \ldots, z_n \in V$, then we can write

$$\begin{bmatrix} z_1 \cdots z_n \end{bmatrix} = \begin{bmatrix} x_1 \cdots x_n \end{bmatrix} A$$

$$= \begin{bmatrix} x_1 \cdots x_n \end{bmatrix} \begin{bmatrix} \alpha_{11} & \cdots & \alpha_{1n} \\ \vdots & \ddots & \vdots \\ \alpha_{n1} & \cdots & \alpha_{nn} \end{bmatrix}.$$

For any volume form vol, we have

$$\mathrm{vol}(z_1, \ldots, z_n) = \mathrm{vol}\left(\sum_{i_1=1}^{n} x_{i_1} \alpha_{i_1 1}, \ldots, \sum_{i_n=1}^{n} x_{i_n} \alpha_{i_n n}\right)$$

$$= \sum_{i_1=1}^{n} \alpha_{i_1 1} \mathrm{vol} \left(x_{i_1}, \ldots, \sum_{i_n=1}^{n} \alpha_{i_n n} x_{i_n} \right)$$

$$\vdots$$

$$= \sum_{i_1,\ldots,i_n=1}^{n} \alpha_{i_1 1} \cdots \alpha_{i_n n} \mathrm{vol} \left(x_{i_1}, \ldots, x_{i_n} \right).$$

The first thing we should note is that $\mathrm{vol}\,(x_{i_1}, \ldots, x_{i_n}) = 0$ if any two of the indices $i_1, \ldots, i_n$ are equal. When doing the sum

$$\sum_{i_1,\ldots,i_n=1}^{n} \alpha_{i_1 1} \cdots \alpha_{i_n n} \mathrm{vol}\,(x_{i_1}, \ldots, x_{i_n}),$$

we can therefore assume that all of the indices $i_1, \ldots, i_n$ are different. This means that by switching indices around, we have

$$\mathrm{vol}\,(x_{i_1}, \ldots, x_{i_n}) = \pm \mathrm{vol}\,(x_1, \ldots, x_n),$$

where the sign $\pm$ depends on the number of switches we have to make in order to rearrange $i_1, \ldots, i_n$ to get back to the standard ordering $1, \ldots, n$. Since this number of switches does not depend on vol but only on the indices, we obtain the desired result:

$$\mathrm{vol}_1\,(z_1, \ldots, z_n) = \sum_{i_1,\ldots,i_n=1}^{n} \pm \alpha_{i_1 1} \cdots \alpha_{i_n n} \mathrm{vol}_1\,(x_1, \ldots, x_n)$$

$$= \sum_{i_1,\ldots,i_n=1}^{n} \pm \alpha_{i_1 1} \cdots \alpha_{i_n n} \lambda \mathrm{vol}_2\,(x_1, \ldots, x_n)$$

$$= \lambda \sum_{i_1,\ldots,i_n=1}^{n} \pm \alpha_{i_1 1} \cdots \alpha_{i_n n} \mathrm{vol}_2\,(x_1, \ldots, x_n)$$

$$= \lambda \mathrm{vol}_2\,(z_1, \ldots, z_n).$$

$\square$

From the proof of this theorem, we also obtain one of the crucial results about volumes that we mentioned in the previous section.

Corollary 5.2.4. *If $x_1, \ldots, x_n \in V$ is a basis for V, then any volume form vol is completely determined by its value $\mathrm{vol}\,(x_1, \ldots, x_n)$.*

This corollary could be used to create volume forms by simply defining

$$\mathrm{vol}\,(z_1, \ldots, z_n) = \sum_{i_1,\ldots,i_n} \pm \alpha_{i_1 1} \cdots \alpha_{i_n n} \mathrm{vol}\,(x_1, \ldots, x_n),$$

where $\{i_1, \ldots, i_n\} = \{1, \ldots, n\}$. For that to work, we would have to show that the sign $\pm$ is well defined in the sense that it does not depend on the particular way in which we reorder $i_1, \ldots, i_n$ to get $1, \ldots, n$. While this is certainly true, we shall not prove this combinatorial fact here. Instead, we observe that if we have a volume form that is nonzero on $x_1, \ldots, x_n$, then the fact that $\text{vol}\,(x_{i_1}, \ldots, x_{i_n})$ is a multiple of $\text{vol}\,(x_1, \ldots, x_n)$ tells us that this sign is well defined and so does not depend on the way in which $1, \ldots, n$ was rearranged to get $i_1, \ldots, i_n$. We use the notation sign $(i_1, \ldots, i_n)$ for the sign we get from

$$\text{vol}\,(x_{i_1}, \ldots, x_{i_n}) = \text{sign}\,(i_1, \ldots, i_n)\,\text{vol}\,(x_1, \ldots, x_n).$$

Finally, we need to check what happens when we restrict it to subspaces. To this end, let vol be a nontrivial volume form on V and $M \subset V$ a k-dimensional subspace of V. If we fix vectors $y_1, \ldots, y_{n-k} \in V$, then we can define a form on M by

$$\text{vol}_M\,(x_1, \ldots, x_k) = \text{vol}\,(x_1, \ldots, x_k, y_1, \ldots, y_{n-k}),$$

where $x_1, \ldots, x_k \in M$. It is clear that vol_M is linear in each variable and also alternating as vol has those properties. Moreover, if $y_1, \ldots, y_{n-k}$ form a basis for a complement to M in V, then $x_1, \ldots, x_k, y_1, \ldots, y_{n-k}$ will be a basis for V as long as $x_1, \ldots, x_k$ is a basis for M. In this case, vol_M becomes a nontrivial volume form as well. If, however, some linear combination of $y_1, \ldots, y_{n-k}$ lies in M, then it follows that $\text{vol}_M = 0$.

Exercises

1. Let V be a three-dimensional real inner product space and vol a volume form so that $\text{vol}\,(e_1, e_2, e_3) = 1$ for some orthonormal basis. For $x, y \in V$, define $x \times y$ as the unique vector such that

$$\text{vol}\,(x, y, z) = \text{vol}\,(z, x, y) = (z|x \times y).$$

   (a) Show that $x \times y = -y \times x$ and that $x \to x \times y$ is linear.
   (b) Show that

$$(x_1 \times y_1|x_2 \times y_2) = (x_1|x_2)\,(y_1|y_2) - (x_1|y_2)\,(x_2|y_1).$$

   (c) Show that

$$\|x \times y\| = \|x\|\,\|y\|\,|\sin\theta|,$$

   where

$$\cos\theta = \frac{(x, y)}{\|x\|\,\|y\|}.$$

(d) Show that

$$x \times (y \times z) = (x|z)\, y - (x|y)\, z.$$

(e) Show that the Jacobi identity holds

$$x \times (y \times z) + z \times (x \times y) + y \times (z \times x) = 0.$$

2. Let $x_1, \ldots, x_n \in \mathbb{R}^n$ and do a Gram–Schmidt procedure so as to obtain a QR decomposition

$$\begin{bmatrix} x_1 \cdots x_n \end{bmatrix} = \begin{bmatrix} e_1 \cdots e_n \end{bmatrix} \begin{bmatrix} r_{11} & \cdots & r_{1n} \\ & \ddots & \vdots \\ 0 & & r_{nn} \end{bmatrix}$$

Show that

$$\mathrm{vol}\,(x_1, \ldots, x_n) = r_{11} \cdots r_{nn}\, \mathrm{vol}\,(e_1, \ldots, e_n),$$

where

$$r_{11} = \|x_1\|,$$
$$r_{22} = \|x_2 - \mathrm{proj}_{x_1}(x_2)\|,$$
$$\vdots$$
$$r_{nn} = \|x_n - \mathrm{proj}_{M_{n-1}}(x_n)\|.$$

and explain why $r_{11} \cdots r_{nn}$ gives the geometrically defined volume that comes from the formula where one multiplies height and base "area" and in turn uses that same principle to compute base "area".

3. Show that

$$\mathrm{vol}\left(\begin{bmatrix} \alpha \\ \beta \end{bmatrix}, \begin{bmatrix} \gamma \\ \delta \end{bmatrix} \right) = \alpha\delta - \gamma\beta$$

defines a volume form on $\mathbb{F}^2$ such that $\mathrm{vol}\,(e_1, e_2) = 1$.

4. Show that we can define a volume form on $\mathbb{F}^3$ by

$$\mathrm{vol}\left(\begin{bmatrix} \alpha_{11} \\ \alpha_{21} \\ \alpha_{31} \end{bmatrix}, \begin{bmatrix} \alpha_{12} \\ \alpha_{22} \\ \alpha_{32} \end{bmatrix}, \begin{bmatrix} \alpha_{13} \\ \alpha_{23} \\ \alpha_{33} \end{bmatrix} \right)$$

$$= \alpha_{11} \mathrm{vol}\left(\begin{bmatrix} \alpha_{22} \\ \alpha_{32} \end{bmatrix}, \begin{bmatrix} \alpha_{23} \\ \alpha_{33} \end{bmatrix} \right)$$

$$- \alpha_{12} \mathrm{vol}\left(\begin{bmatrix} \alpha_{21} \\ \alpha_{31} \end{bmatrix}, \begin{bmatrix} \alpha_{23} \\ \alpha_{33} \end{bmatrix} \right)$$

$$+ \alpha_{13} \text{vol} \left(\begin{bmatrix} \alpha_{21} \\ \alpha_{31} \end{bmatrix}, \begin{bmatrix} \alpha_{22} \\ \alpha_{32} \end{bmatrix} \right)$$

$$= \alpha_{11}\alpha_{22}\alpha_{33} + \alpha_{12}\alpha_{23}\alpha_{31} + \alpha_{13}\alpha_{32}\alpha_{21}$$

$$- \alpha_{11}\alpha_{23}\alpha_{32} - \alpha_{33}\alpha_{12}\alpha_{21} - \alpha_{22}\alpha_{13}\alpha_{31}.$$

5. Assume that $\text{vol}(e_1, \ldots, e_4) = 1$ for the standard basis in $\mathbb{R}^4$. Using the permutation formula for the volume form, determine with a minimum of calculations the sign for the volume of the columns in each of the matrices:

(a)

$$\begin{bmatrix} 1000 & -1 & 2 & -1 \\ 1 & 1000 & 1 & 2 \\ 3 & -2 & 1 & 1000 \\ 2 & -1 & 1000 & 2 \end{bmatrix}$$

(b)

$$\begin{bmatrix} 2 & 1000 & 2 & -1 \\ 1 & -1 & 1000 & 2 \\ 3 & -2 & 1 & 1000 \\ 1000 & -1 & 1 & 2 \end{bmatrix}$$

(c)

$$\begin{bmatrix} 2 & -2 & 2 & 1000 \\ 1 & -1 & 1000 & 2 \\ 3 & 1000 & 1 & -1 \\ 1000 & -1 & 1 & 2 \end{bmatrix}$$

(d)

$$\begin{bmatrix} 2 & -2 & 1000 & -1 \\ 1 & 1000 & 2 & 2 \\ 3 & -1 & 1 & 1000 \\ 1000 & -1 & 1 & 2 \end{bmatrix}$$

5.3   How to Calculate Volumes

Before establishing the existence of the volume form, we shall try to use what we learned in the previous section in a more concrete fashion to calculate $\text{vol}(z_1, \ldots, z_n)$. Assume that $\text{vol}(z_1, \ldots, z_n)$ is a volume form on V and that there is a basis $x_1, \ldots, x_n$ for V where $\text{vol}(x_1, \ldots, x_n)$ is known. First, observe that when

$$\begin{bmatrix} z_1 & \cdots & z_n \end{bmatrix} = \begin{bmatrix} x_1 & \cdots & x_n \end{bmatrix} A$$

and $A = [\alpha_{ij}]$ is an upper triangular matrix, then $\alpha_{i_11} \cdots \alpha_{i_nn} = 0$ unless $i_1 \leq 1, \ldots, i_n \leq n$. Since we also need all the indices $i_1, \ldots, i_n$ to be distinct, this implies that $i_1 = 1, \ldots, i_n = n$. Thus, we obtain the simple relationship

$$\text{vol}(z_1, \ldots, z_n) = \alpha_{11} \cdots \alpha_{nn} \text{vol}(x_1, \ldots, x_n).$$

While we cannot expect this to happen too often, it is possible to change $z_1, \ldots, z_n$ to vectors $y_1, \ldots, y_n$ in such a way that

$$\text{vol}(z_1, \ldots, z_n) = \pm \text{vol}(y_1, \ldots, y_n)$$

and

$$[y_1 \cdots y_n] = [x_1 \cdots x_n] A,$$

where A is upper triangular.

To construct the y_is, we simply use *elementary column operations* (see also Sect. 1.13 and Exercise 6 in that section). This works in almost the same way as Gauss elimination but with the twist that we are multiplying by matrices on the right. The allowable operations are

(1) Interchanging vectors z_k and z_l.
(2) Multiplying z_l by $\alpha \in \mathbb{F}$ and adding it to z_k.

The first operation changes the volume by a sign, while the second leaves the volume unchanged. So if $[y_1 \cdots y_n]$ is obtained from $[z_1 \cdots z_n]$ through these operations, then we have

$$\text{vol}(z_1, \ldots, z_n) = \pm \text{vol}(y_1, \ldots, y_n).$$

The minus sign occurs exactly when we have done an odd number of interchanges. We now need to explain why we can obtain $[y_1 \cdots y_n]$ such that

$$[y_1 \cdots y_n] = [x_1 \cdots x_n] \begin{bmatrix} \alpha_{11} & \alpha_{12} & \cdots & \alpha_{1n} \\ 0 & \alpha_{22} & & \alpha_{2n} \\ \vdots & & \ddots & \vdots \\ 0 & 0 & \cdots & \alpha_{nn} \end{bmatrix}.$$

One issue to note is that the process might break down if $z_1, \ldots, z_n$ are linearly dependent. In that case, we have vol $= 0$.

Instead of describing the procedure abstractly, let us see how it works in practice. In the case of $\mathbb{F}^n$, we assume that we are using a volume form such that $\text{vol}(e_1, \ldots, e_n) = 1$ for the canonical basis. Since that uniquely defines the volume form, we introduce some special notation for it:

$$|A| = |x_1 \cdots x_n| = \text{vol}(x_1, \ldots, x_n),$$

where $A \in \text{Mat}_{n \times n}(\mathbb{F})$ is the matrix such that

$$[x_1 \cdots x_n] = [e_1 \cdots e_n] A$$

Example 5.3.1. Let

$$\left[z_1\ z_2\ z_3 \right] = \begin{bmatrix} 0 & 1 & 0 \\ 0 & 0 & 3 \\ -2 & 0 & 0 \end{bmatrix}.$$

We can rearrange this into

$$\left[z_2\ z_3\ z_1 \right] = \begin{bmatrix} 1 & 0 & 0 \\ 0 & 3 & 0 \\ 0 & 0 & -2 \end{bmatrix}.$$

This takes two transpositions. Thus,

$$\begin{aligned}
\mathrm{vol}\,(z_1, z_2, z_3) &= \mathrm{vol}\,(z_2, z_3, z_1) \\
&= 1 \cdot 3 \cdot (-2)\,\mathrm{vol}\,(e_1, e_2, e_3) \\
&= -6\mathrm{vol}\,(e_1, e_2, e_3).
\end{aligned}$$

Example 5.3.2. Let

$$\left[z_1\ z_2\ z_3\ z_4 \right] = \begin{bmatrix} 3 & 0 & 1 & 3 \\ 1 & -1 & 2 & 0 \\ -1 & 1 & 0 & -2 \\ -3 & 1 & 1 & -3 \end{bmatrix}.$$

$$\begin{vmatrix} 3 & 0 & 1 & 3 \\ 1 & -1 & 2 & 0 \\ -1 & 1 & 0 & -2 \\ -3 & 1 & 1 & -3 \end{vmatrix}$$

$$= \begin{vmatrix} 0 & 1 & 2 & 3 \\ 1 & -1 & 2 & 0 \\ 1 & \frac{1}{3} & -\frac{2}{3} & -2 \\ 0 & 0 & 0 & -3 \end{vmatrix} \quad \text{after eliminating entries in row 4,}$$

$$= \begin{vmatrix} 3 & 2 & 2 & 3 \\ 4 & 0 & 2 & 0 \\ 0 & 0 & -\frac{2}{3} & -2 \\ 0 & 0 & 0 & -3 \end{vmatrix} \quad \text{after eliminating entries in row 3,}$$

$$= \begin{vmatrix} 2 & 3 & 2 & 3 \\ 0 & 4 & 2 & 0 \\ 0 & 0 & -\frac{2}{3} & -2 \\ 0 & 0 & 0 & -3 \end{vmatrix} \quad \text{after switching column 1 and 2.}$$

Thus, we get

$$\text{vol}\,(z_1,\ldots,z_4) = -2\cdot 4\cdot\left(-\frac{2}{3}\right)\cdot(-3)\,\text{vol}\,(e_1,\ldots,e_4)$$

$$= -16\text{vol}\,(e_1,\ldots,e_4)\,.$$

Example 5.3.3. Let us try to find

$$\begin{vmatrix} 1 & 1 & 1 & \cdots & 1 \\ 1 & 2 & 2 & \cdots & 2 \\ 1 & 2 & 3 & \cdots & 3 \\ \vdots & \vdots & \vdots & \ddots & \vdots \\ 1 & 2 & 3 & \cdots & n \end{vmatrix}$$

Instead of starting with the last column vector, we are going to start with the first. This will lead us to a lower triangular matrix, but otherwise, we are using the same principles.

$$\begin{vmatrix} 1 & 1 & 1 & \cdots & 1 \\ 1 & 2 & 2 & \cdots & 2 \\ 1 & 2 & 3 & \cdots & 3 \\ \vdots & \vdots & \vdots & \ddots & \vdots \\ 1 & 2 & 3 & \cdots & n \end{vmatrix} = \begin{vmatrix} 1 & 0 & 0 & \cdots & 0 \\ 1 & 1 & 1 & \cdots & 1 \\ 1 & 1 & 2 & \cdots & 2 \\ \vdots & \vdots & \vdots & \ddots & \vdots \\ 1 & 1 & 2 & \cdots & n-1 \end{vmatrix}$$

$$= \begin{vmatrix} 1 & 0 & 0 & \cdots & 0 \\ 1 & 1 & 0 & \cdots & 0 \\ 1 & 1 & 1 & \cdots & 1 \\ \vdots & \vdots & \vdots & \ddots & \vdots \\ 1 & 1 & 1 & \cdots & n-2 \end{vmatrix}$$

$$\vdots$$

$$= \begin{vmatrix} 1 & 0 & 0 & \cdots & 0 \\ 1 & 1 & 0 & \cdots & 0 \\ 1 & 1 & 1 & \cdots & 0 \\ \vdots & \vdots & \vdots & \ddots & \vdots \\ 1 & 1 & 1 & \cdots & 1 \end{vmatrix}$$

$$= 1.$$

Exercises

1. The following problem was first considered by Leibniz and appears to be the first use of determinants. Let $A \in \mathrm{Mat}_{(n+1) \times n}(\mathbb{F})$ and $b \in \mathbb{F}^{n+1}$. Show that:

   (a) If there is a solution to the overdetermined system $Ax = b$, $x \in \mathbb{F}^n$, then the augmented matrix satisfies $|A|b| = 0$.
   (b) Conversely, if A has rank $(A) = n$ and $|A|b| = 0$, then there is a solution to $Ax = b$, $x \in \mathbb{F}^n$.

2. Compute

$$\begin{vmatrix} 1 & 1 & 1 & \cdots & 1 \\ 0 & 1 & 1 & \cdots & 1 \\ 1 & 0 & 1 & \cdots & 1 \\ \vdots & \vdots & \vdots & \ddots & \vdots \\ 1 & \cdots & 1 & 0 & 1 \end{vmatrix}$$

3. Let $x_1, \ldots, x_k \in \mathbb{R}^n$ and assume that $\mathrm{vol}\,(e_1, \ldots, e_n) = 1$. Show that

$$|G\,(x_1, \ldots, x_k)| \le \|x_1\|^2 \cdots \|x_k\|^2,$$

   where $G\,(x_1, \ldots, x_k)$ is the Gram matrix whose ij entries are the inner products $(x_j|x_i)$. Look at Exercise 4 in Sect. 3.5 for the definition of the Gram matrix and use Exercise 2 in Sect. 5.2.

4. Let $x_1, \ldots, x_k \in \mathbb{R}^n$ and assume that $\mathrm{vol}\,(e_1, \ldots, e_n) = 1$.

   (a) Show that
$$G\,(x_1, \ldots, x_n) = \begin{bmatrix} x_1 & \cdots & x_n \end{bmatrix}^* \begin{bmatrix} x_1 & \cdots & x_n \end{bmatrix}.$$

   (b) Show that
$$|G\,(x_1, \ldots, x_n)| = |\mathrm{vol}\,(x_1, \ldots, x_n)|^2.$$

   (c) Using the previous exercise, conclude that Hadamard's inequality holds
$$|\mathrm{vol}\,(x_1, \ldots, x_n)|^2 \le \|x_1\|^2 \cdots \|x_n\|^2.$$

   (d) When is
$$|\mathrm{vol}\,(x_1, \ldots, x_n)|^2 = \|x_1\|^2 \cdots \|x_n\|^2?$$

5. Assume that $\mathrm{vol}\,(e_1, \ldots, e_4) = 1$ for the standard basis in $\mathbb{R}^4$. Find the volumes:

   (a)
$$\begin{vmatrix} 0 & -1 & 2 & -1 \\ 1 & 0 & 1 & 2 \\ 3 & -2 & 1 & 0 \\ 2 & -1 & 0 & 2 \end{vmatrix}$$

(b)
$$\begin{vmatrix} 2 & 0 & 2 & -1 \\ 1 & -1 & 0 & 2 \\ 3 & -2 & 1 & 1 \\ 0 & -1 & 1 & 2 \end{vmatrix}$$

(c)
$$\begin{vmatrix} 2 & -2 & 2 & 0 \\ 1 & -1 & 1 & 2 \\ 3 & 0 & 1 & -1 \\ 1 & -1 & 1 & 2 \end{vmatrix}$$

(d)
$$\begin{vmatrix} 2 & -2 & 0 & -1 \\ 1 & 1 & 2 & 2 \\ 3 & -1 & 1 & 1 \\ 1 & -1 & 1 & 2 \end{vmatrix}$$

5.4 Existence of the Volume Form

The construction of $\mathrm{vol}\,(x_1,\ldots,x_n)$ proceeds by induction on the dimension of V. We start with a basis $e_1,\ldots,e_n \in V$ that is assumed to have unit volume. Next, we assume, by induction, that there is a volume form vol^{n-1} on $\mathrm{span}\,\{e_2,\ldots,e_n\}$ such that $e_2,\ldots,e_n$ has unit volume. Finally, let $E : V \to V$ be the projection onto $\mathrm{span}\,\{e_2,\ldots,e_n\}$ whose kernel is $\mathrm{span}\,\{e_1\}$. For a collection $x_1,\ldots,x_n \in V$, we decompose $x_i = \alpha_i e_1 + E\,(x_i)$. The volume form vol^n on V is then defined by

$$\mathrm{vol}^n\,(x_1,\ldots,x_n) = \sum_{k=1}^{n}(-1)^{k-1}\alpha_k\,\mathrm{vol}^{n-1}\left(E\,(x_1),\ldots,\widehat{E\,(x_k)},\ldots,E\,(x_n)\right).$$

(Recall that $\widehat{a}$ means that a has been eliminated). This is essentially like defining the volume via a Laplace expansion along the first row. As α_k, E, and vol^{n-1} are linear, it is obvious that the new vol^n form is linear in each variable. The alternating property follows if we can show that the form vanishes when $x_i = x_j$. This is done as follows:

$$\mathrm{vol}^n\,(\ldots,x_i,\ldots x_j,\ldots)$$
$$= \sum_{k\neq i,j}(-1)^{k-1}\alpha_k\,\mathrm{vol}^{n-1}\left(\ldots,E\,(x_i),\ldots,\widehat{E\,(x_k)},\ldots,E\,(x_j),\ldots\right)$$
$$+ (-1)^{i-1}\alpha_i\,\mathrm{vol}^{n-1}\left(\ldots,\widehat{E\,(x_i)},\ldots,E\,(x_j),\ldots\right)$$

$$+ (-1)^{j-1} \alpha_j \, \text{vol}^{n-1} \left(\ldots, E\left(x_i\right), \ldots, E\left(\widehat{x_j}\right), \ldots \right)$$

Using that $E\left(x_i\right) = E\left(x_j\right)$ and vol^{n-1} is alternating on span $\{e_2, \ldots, e_n\}$ shows

$$\text{vol}^{n-1} \left(\ldots, E\left(x_i\right), \ldots, \widehat{E\left(x_k\right)}, \ldots, E\left(x_j\right), \ldots \right) = 0$$

Hence,

$$\text{vol}^n \left(\ldots, x_i, \ldots x_j, \ldots \right)$$
$$= (-1)^{i-1} \alpha_i \, \text{vol}^{n-1} \left(\ldots, \widehat{E\left(x_i\right)}, \ldots, E\left(x_j\right), \ldots \right)$$
$$+ (-1)^{j-1} \alpha_j \, \text{vol}^{n-1} \left(\ldots, E(x_i), \ldots, \widehat{E(x_j)}, \ldots \right)$$
$$= (-1)^{i-1} (-1)^{j-1-i} \alpha_i \, \text{vol}^{n-1} \left(\ldots, E\left(x_{i-1}\right), \underset{E\left(x_j\right)}{\overset{i^{\text{th}} \text{ place}}{}}, E\left(x_{i+1}\right) \ldots \right)$$
$$+ (-1)^{j-1} \alpha_j \, \text{vol}^{n-1} \left(\ldots, E\left(x_i\right), \ldots, \widehat{E(x_j)}, \ldots \right),$$

where moving $E\left(x_j\right)$ to the ith-place in the expression

$$\text{vol}^{n-1} \left(\ldots, \widehat{E\left(x_i\right)}, \ldots, E\left(x_j\right), \ldots \right)$$

requires $j - 1 - i$ moves since $E\left(x_j\right)$ is in the $(j - 1)$-place. Using that $\alpha_i = \alpha_j$ and $E\left(x_i\right) = E\left(x_j\right)$, this shows

$$\text{vol}^n \left(\ldots, x_i, \ldots x_j, \ldots \right) = (-1)^{j-2} \alpha_i \, \text{vol}^{n-1} \left(\ldots, \underset{E\left(x_j\right)}{\overset{i^{\text{th}} \text{ place}}{}}, \ldots, \ldots \right)$$
$$+ (-1)^{j-1} \alpha_j \, \text{vol}^{n-1} \left(\ldots, E\left(x_i\right), \ldots, \widehat{E(x_j)}, \ldots \right)$$
$$= 0.$$

Aside from defining the volume form, we also get a method for calculating volumes using induction on dimension. In $\mathbb{F}$, we just define $\text{vol}\left(x\right) = x$. For $\mathbb{F}^2$, we have

$$\text{vol}\left(\begin{bmatrix} \alpha \\ \beta \end{bmatrix}, \begin{bmatrix} \gamma \\ \delta \end{bmatrix} \right) = \alpha\delta - \gamma\beta.$$

In $\mathbb{F}^3$, we get

$$\mathrm{vol}\left(\begin{bmatrix}\alpha_{11}\\\alpha_{21}\\\alpha_{31}\end{bmatrix},\begin{bmatrix}\alpha_{12}\\\alpha_{22}\\\alpha_{32}\end{bmatrix},\begin{bmatrix}\alpha_{13}\\\alpha_{23}\\\alpha_{33}\end{bmatrix}\right)$$

$$= \alpha_{11}\mathrm{vol}\left(\begin{bmatrix}\alpha_{22}\\\alpha_{32}\end{bmatrix},\begin{bmatrix}\alpha_{23}\\\alpha_{33}\end{bmatrix}\right)$$

$$- \alpha_{12}\mathrm{vol}\left(\begin{bmatrix}\alpha_{21}\\\alpha_{31}\end{bmatrix},\begin{bmatrix}\alpha_{23}\\\alpha_{33}\end{bmatrix}\right)$$

$$+ \alpha_{13}\mathrm{vol}\left(\begin{bmatrix}\alpha_{21}\\\alpha_{31}\end{bmatrix},\begin{bmatrix}\alpha_{22}\\\alpha_{32}\end{bmatrix}\right)$$

$$= \alpha_{11}\alpha_{22}\alpha_{33} + \alpha_{12}\alpha_{23}\alpha_{31} + \alpha_{13}\alpha_{21}\alpha_{32}$$

$$- \alpha_{11}\alpha_{32}\alpha_{23} - \alpha_{12}\alpha_{21}\alpha_{33} - \alpha_{13}\alpha_{31}\alpha_{22}$$

$$= \alpha_{11}\alpha_{22}\alpha_{33} + \alpha_{12}\alpha_{23}\alpha_{31} + \alpha_{13}\alpha_{32}\alpha_{21}$$

$$- \alpha_{11}\alpha_{23}\alpha_{32} - \alpha_{33}\alpha_{12}\alpha_{21} - \alpha_{22}\alpha_{13}\alpha_{31}.$$

In the above definition, there is, of course, nothing special about the choice of basis $e_1,\ldots,e_n$ or the ordering of the basis. Let us refer to the specific choice of volume form as vol_1 as we are expanding along the first row. If we switch e_1 and e_k, then we are apparently expanding along the kth row instead. This defines a volume form vol_k. By construction, we have

$$\mathrm{vol}_1(e_1,\ldots,e_n) = 1,$$

$$\mathrm{vol}_k\left(e_k, e_2, \ldots, \overset{k\text{th place}}{\underset{e_1}{}}, \ldots, e_n\right) = 1.$$

Thus,

$$\mathrm{vol}_1 = (-1)^{k-1}\,\mathrm{vol}_k$$

$$= (-1)^{k+1}\,\mathrm{vol}_k.$$

So if we wish to calculate vol_1 by an expansion along the kth row, we need to remember the extra sign $(-1)^{k+1}$. In the case of $\mathbb{F}^n$, we define the volume form vol to be vol_1 as constructed above. In this case, we shall often just write

$$\left|\, x_1 \cdots x_n \,\right| = \mathrm{vol}(x_1,\ldots,x_n)$$

as in the previous section.

Example 5.4.1. Let us try this with the example from the previous section:

$$\begin{bmatrix} z_1 & z_2 & z_3 & z_4 \end{bmatrix} = \begin{bmatrix} 3 & 0 & 1 & 3 \\ 1 & -1 & 2 & 0 \\ -1 & 1 & 0 & -2 \\ -3 & 1 & 1 & -3 \end{bmatrix}.$$

Expansion along the first row gives

$$\begin{vmatrix} z_1 & z_2 & z_3 & z_4 \end{vmatrix} = 3 \begin{vmatrix} -1 & 2 & 0 \\ 1 & 0 & -2 \\ 1 & 1 & -3 \end{vmatrix} - 0 \begin{vmatrix} 1 & 2 & 0 \\ -1 & 0 & -2 \\ -3 & 1 & -3 \end{vmatrix}$$

$$+ 1 \begin{vmatrix} 1 & -1 & 0 \\ -1 & 1 & -2 \\ -3 & 1 & -3 \end{vmatrix} - 3 \begin{vmatrix} 1 & -1 & 2 \\ -1 & 1 & 0 \\ -3 & 1 & 1 \end{vmatrix}$$

$$= 3 \cdot 0 - 0 + 1 \cdot (-4) - 3 \cdot 4$$

$$= -16.$$

Expansion along the second row gives

$$\begin{vmatrix} z_1 & z_2 & z_3 & z_4 \end{vmatrix} = -1 \begin{vmatrix} 0 & 1 & 3 \\ 1 & 0 & -2 \\ 1 & 1 & -3 \end{vmatrix} + (-1) \begin{vmatrix} 3 & 1 & 3 \\ -1 & 0 & -2 \\ -3 & 1 & -3 \end{vmatrix}$$

$$-2 \begin{vmatrix} 3 & 0 & 3 \\ -1 & 1 & -2 \\ -3 & 1 & -3 \end{vmatrix} + 0 \begin{vmatrix} 3 & 0 & 1 \\ -1 & 1 & 0 \\ -3 & 1 & 1 \end{vmatrix}$$

$$= -1 \cdot 4 - 1 \cdot 6 - 2 \cdot 3 + 0$$

$$= -16.$$

Definition 5.4.2. The general formula in $\mathbb{F}^n$ for expanding along the kth row in an $n \times n$ matrix $A = \begin{bmatrix} x_1 & \cdots & x_n \end{bmatrix}$ is called the *Laplace expansion* along the kth row and looks like

$$|A| = (-1)^{k+1} \alpha_{k1} |A_{k1}| + (-1)^{k+2} \alpha_{k2} |A_{k2}| + \cdots + (-1)^{k+n} \alpha_{kn} |A_{kn}|$$

$$= \sum_{i=1}^{n} (-1)^{k+i} \alpha_{ki} |A_{ki}|.$$

Here α_{ij} is the ij entry in A, i.e., the ith coordinate for x_j, and A_{ij} is the *companion* $(n-1) \times (n-1)$ matrix for α_{ij}. The matrix A_{ij} is constructed from A by eliminating the ith row and jth column. Note that the exponent for -1 is $i + j$ when we are at the ij entry α_{ij}.

Example 5.4.3. This expansion gives us a very intriguing formula for the determinant that looks like we have used the chain rule for differentiation in several variables. To explain this, let us think of $|A|$ as a function in the entries x_{ij}. The expansion along the kth row then looks like

$$|A| = (-1)^{k+1} x_{k1} |A_{k1}| + (-1)^{k+2} x_{k2} |A_{k2}| + \cdots + (-1)^{k+n} x_{kn} |A_{kn}|.$$

From the definition of $|A_{kj}|$, it follows that it does depend on the variables x_{ki}. Thus,

$$\frac{\partial |A|}{\partial x_{ki}} = (-1)^{k+1} \frac{\partial x_{k1}}{\partial x_{ki}} |A_{k1}| + (-1)^{k+2} \frac{\partial x_{k2}}{\partial x_{ki}} |A_{k2}| + \cdots + (-1)^{k+n} \frac{\partial x_{kn}}{\partial x_{ki}} |A_{kn}|$$

$$= (-1)^{k+i} |A_{ki}|.$$

Replacing $(-1)^{k+i} |A_{ki}|$ by the partial derivative then gives us the formula

$$|A| = x_{k1} \frac{\partial |A|}{\partial x_{k1}} + x_{k2} \frac{\partial |A|}{\partial x_{k2}} + \cdots + x_{kn} \frac{\partial |A|}{\partial x_{kn}}$$

$$= \sum_{i=1}^{n} x_{ki} \frac{\partial |A|}{\partial x_{ki}}.$$

Since we get the same answer for each k, this implies

$$n |A| = \sum_{i,j=1}^{n} x_{ij} \frac{\partial |A|}{\partial x_{ij}}.$$

Exercises

1. Find the determinant of the following $n \times n$ matrix where all entries are 1 except the entries just below the diagonal which are 0:

$$\begin{vmatrix} 1 & 1 & 1 & \cdots & 1 \\ 0 & 1 & 1 & \cdots & 1 \\ 1 & 0 & 1 & \cdots & \vdots \\ \vdots & 1 & \ddots & \ddots & 1 \\ 1 & \cdots & 1 & 0 & 1 \end{vmatrix}$$

2. Find the determinant of the following $n \times n$ matrix:

$$\begin{vmatrix} 1 & \cdots & 1 & 1 & 1 \\ 2 & \cdots & 2 & 2 & 1 \\ 3 & \cdots & 3 & 1 & \vdots \\ \vdots & & 1 & \cdots & 1 \\ n & 1 & \cdots & 1 & 1 \end{vmatrix}$$

3. (The Vandermonde Determinant)

   (a) Show that

   $$\begin{vmatrix} 1 & \cdots & 1 \\ \lambda_1 & \cdots & \lambda_n \\ \vdots & & \vdots \\ \lambda_1^{n-1} & \cdots & \lambda_n^{n-1} \end{vmatrix} = \prod_{i<j} (\lambda_i - \lambda_j).$$

   (b) When $\lambda_1, \ldots, \lambda_n$ are the complex roots of a polynomial $p(t) = t^n + a_{n-1}t^{n-1} + \cdots + a_1 t + a_0$, we define the discriminant of p as

   $$\Delta = D = \left(\prod_{i<j} (\lambda_i - \lambda_j) \right)^2.$$

   When $n = 2$, show that this conforms with the usual definition. In general, one can compute Δ from the coefficients of p. Show that Δ is real if p is real.

4. Let S_n be the group of permutations, i.e., bijective maps from $\{1, \ldots, n\}$ to itself. These are generally denoted by σ and correspond to a switching of indices, $\sigma(k) = i_k, k = 1, \ldots, n$. Consider the polynomial in n variables

   $$p(x_1, \ldots, x_n) = \prod_{i<j} (x_i - x_j).$$

   (a) Show that if $\sigma \in S_n$ is a permutation, then

   $$\text{sign}(\sigma) p(x_1, \ldots, x_n) = p\left(x_{\sigma(1)}, \ldots, x_{\sigma(n)}\right)$$

   for some sign sign $(\sigma) \in \{\pm 1\}$.

   (b) Show that the sign function $S_n \to \{\pm 1\}$ is a homomorphism, i.e., $\text{sign}(\sigma \tau) = \text{sign}(\sigma) \text{sign}(\tau)$.

   (c) Using the above characterization, show that sign (σ) can be determined by the number of inversions in the permutation. An inversion in σ is a pair of consecutive integers whose order is reversed, i.e., $\sigma(i) > \sigma(i+1)$.

5. Let $A_n = \left[\alpha_{ij}\right]$ be a real skew-symmetric $n \times n$ matrix, i.e., $\alpha_{ij} = -\alpha_{ji}$.

   (a) Show that $|A_2| = \alpha_{12}^2$.
   (b) Show that $|A_4| = (\alpha_{12}\alpha_{34} + \alpha_{14}\alpha_{23} - \alpha_{13}\alpha_{24})^2$.
   (c) Show that $|A_{2n}| \geq 0$.
   (d) Show that $|A_{2n+1}| = 0$.

6. Show that the $n \times n$ matrix satisfies

$$
\begin{vmatrix}
\alpha & \beta & \beta & \cdots & \beta \\
\beta & \alpha & \beta & \cdots & \beta \\
\beta & \beta & \alpha & \cdots & \beta \\
\vdots & \vdots & \vdots & \ddots & \vdots \\
\beta & \beta & \beta & \cdots & \alpha
\end{vmatrix}
= (\alpha + (n-1)\beta)(\alpha - \beta)^{n-1} .
$$

7. Show that the $n \times n$ matrix

$$
A_n =
\begin{bmatrix}
\alpha_1 & 1 & 0 & \cdots & 0 \\
-1 & \alpha_2 & 1 & \cdots & 0 \\
0 & -1 & \alpha_3 & \cdots & 0 \\
\vdots & \vdots & \vdots & \ddots & \vdots \\
0 & 0 & 0 & \cdots & \alpha_n
\end{bmatrix}
$$

   satisfies

$$
|A_1| = \alpha_1
$$
$$
|A_2| = 1 + \alpha_1\alpha_2,
$$
$$
|A_n| = \alpha_n |A_{n-1}| + |A_{n-2}| .
$$

8. Show that an $n \times m$ matrix has (column) rank $\geq k$ if and only there is a submatrix of size $k \times k$ with nonzero determinant. Use this to prove that row and column ranks are equal.

9. Here are some problems that discuss determinants and geometry.

   (a) Show that the area of the triangle whose vertices are

$$
\begin{bmatrix} \alpha_1 \\ \beta_1 \end{bmatrix}, \begin{bmatrix} \alpha_2 \\ \beta_2 \end{bmatrix}, \begin{bmatrix} \alpha_3 \\ \beta_3 \end{bmatrix} \in \mathbb{R}^2
$$

   is given by

$$
\frac{1}{2}
\begin{vmatrix}
1 & 1 & 1 \\
\alpha_1 & \alpha_2 & \alpha_3 \\
\beta_1 & \beta_2 & \beta_3
\end{vmatrix} .
$$

(b) Show that three vectors

$$\begin{bmatrix} \alpha_1 \\ \beta_1 \end{bmatrix}, \begin{bmatrix} \alpha_2 \\ \beta_2 \end{bmatrix}, \begin{bmatrix} \alpha_3 \\ \beta_3 \end{bmatrix} \in \mathbb{R}^2$$

satisfy

$$\begin{vmatrix} 1 & 1 & 1 \\ \alpha_1 & \alpha_2 & \alpha_3 \\ \beta_1 & \beta_2 & \beta_3 \end{vmatrix} = 0$$

if and only if they are collinear, i.e., lie on a line $l = \{at + b : t \in \mathbb{R}\}$, where $a, b \in \mathbb{R}^2$.

(c) Show that four vectors

$$\begin{bmatrix} \alpha_1 \\ \beta_1 \\ \gamma_1 \end{bmatrix}, \begin{bmatrix} \alpha_2 \\ \beta_2 \\ \gamma_2 \end{bmatrix}, \begin{bmatrix} \alpha_3 \\ \beta_3 \\ \gamma_3 \end{bmatrix}, \begin{bmatrix} \alpha_4 \\ \beta_4 \\ \gamma_4 \end{bmatrix} \in \mathbb{R}^3$$

satisfy

$$\begin{vmatrix} 1 & 1 & 1 & 1 \\ \alpha_1 & \alpha_2 & \alpha_3 & \alpha_4 \\ \beta_1 & \beta_2 & \beta_3 & \beta_4 \\ \gamma_1 & \gamma_2 & \gamma_3 & \gamma_4 \end{vmatrix} = 0$$

if and only if they are coplanar, i.e., lie in the same plane $\pi = \{x \in \mathbb{R}^3 : (a, x) = \alpha\}$.

10. Let

$$\begin{bmatrix} \alpha_1 \\ \beta_1 \end{bmatrix}, \begin{bmatrix} \alpha_2 \\ \beta_2 \end{bmatrix}, \begin{bmatrix} \alpha_3 \\ \beta_3 \end{bmatrix} \in \mathbb{R}^2$$

be three points in the plane.

(a) If $\alpha_1, \alpha_2, \alpha_3$ are distinct, then the equation for the parabola $y = ax^2 + bx + c$ passing through the three given points is given by

$$\frac{\begin{vmatrix} 1 & 1 & 1 & 1 \\ x & \alpha_1 & \alpha_2 & \alpha_3 \\ x^2 & \alpha_1^2 & \alpha_2^2 & \alpha_3^2 \\ y & \beta_1 & \beta_2 & \beta_3 \end{vmatrix}}{\begin{vmatrix} 1 & 1 & 1 \\ \alpha_1 & \alpha_2 & \alpha_3 \\ \alpha_1^2 & \alpha_2^2 & \alpha_3^2 \end{vmatrix}} = 0.$$

(b) If the points are not collinear, then the equation for the circle $x^2 + y^2 + ax + by + c = 0$ passing through the three given points is given by

$$\frac{\begin{vmatrix} 1 & 1 & 1 & 1 \\ x & \alpha_1 & \alpha_2 & \alpha_3 \\ y & \beta_1 & \beta_2 & \beta_3 \\ x^2 + y^2 & \alpha_1^2 + \beta_1^2 & \alpha_2^2 + \beta_2^2 & \alpha_3^2 + \beta_3^2 \end{vmatrix}}{\begin{vmatrix} 1 & 1 & 1 \\ \alpha_1 & \alpha_2 & \alpha_3 \\ \beta_1 & \beta_2 & \beta_3 \end{vmatrix}} = 0.$$

5.5 Determinants of Linear Operators

Definition 5.5.1. To define the *determinant* of a linear operator $L : V \to V$, we simply observe that $\mathrm{vol}\,(L\,(x_1), \ldots, L\,(x_n))$ defines an alternating n-form that is linear in each variable. Thus,

$$\mathrm{vol}\,(L\,(x_1), \ldots, L\,(x_n)) = \det(L)\,\mathrm{vol}\,(x_1, \ldots, x_n)$$

for some scalar $\det(L) \in \mathbb{F}$. This is the determinant of L.

We note that a different volume form $\mathrm{vol}_1\,(x_1, \ldots, x_n)$ gives the same definition of the determinant. To see this, we first use that $\mathrm{vol}_1 = \lambda \mathrm{vol}$ and then observe that

$$\mathrm{vol}_1\,(L\,(x_1), \ldots, L\,(x_n)) = \lambda \mathrm{vol}\,(L\,(x_1), \ldots, L\,(x_n))$$
$$= \det(L)\,\lambda \mathrm{vol}\,(x_1, \ldots, x_n)$$
$$= \det(L)\,\mathrm{vol}_1\,(x_1, \ldots, x_n).$$

If $e_1, \ldots, e_n$ is chosen so that $\mathrm{vol}\,(e_1, \ldots, e_n) = 1$, then we get the simpler formula

$$\mathrm{vol}\,(L\,(e_1), \ldots, L\,(e_n)) = \det(L).$$

This leads us to one of the standard formulas for the determinant of a matrix. From the properties of volume forms (see Proposition 5.2.2), we obtain

$$\det(L) = \mathrm{vol}\,(L\,(e_1), \ldots, L\,(e_n))$$
$$= \sum \alpha_{i_1 1} \cdots \alpha_{i_n n}\,\mathrm{vol}\,(e_{i_1}, \ldots, e_{i_n})$$
$$= \sum \pm \alpha_{i_1 1} \cdots \alpha_{i_n n}$$
$$= \sum \mathrm{sign}\,(i_1, \ldots, i_n)\,\alpha_{i_1 1} \cdots \alpha_{i_n n},$$

where $[\alpha_{ij}] = [L]$ is the matrix representation for L with respect to $e_1, \ldots, e_n$. This formula is often used as the definition of determinants. Note that it also shows that $\det(L) = \det([L])$ since

$$\left[L(e_1) \cdots L(e_n)\right] = \left[e_1 \cdots e_n\right][L]$$

$$= \left[e_1 \cdots e_n\right] \begin{bmatrix} \alpha_{11} & \cdots & \alpha_{1n} \\ \vdots & \ddots & \vdots \\ \alpha_{n1} & \cdots & \alpha_{nn} \end{bmatrix}.$$

The next proposition contains the fundamental properties for determinants.

Proposition 5.5.2. (Determinant Characterization of Invertibility) *Let V be an n-dimensional vector space.*

(1) If $L, K : V \to V$ are linear operators, then

$$\det(L \circ K) = \det(L)\det(K).$$

(2) $\det(\alpha 1_V) = \alpha^n$.
(3) If L is invertible, then

$$\det L^{-1} = \frac{1}{\det L}.$$

(4) If $\det(L) \neq 0$, then L is invertible.

Proof. For any $x_1, \ldots, x_n$, we have

$$\det(L \circ K)\,\mathrm{vol}(x_1, \ldots, x_n) = \mathrm{vol}(L \circ K(x_1), \ldots, L \circ K(x_n))$$
$$= \det(L)\,\mathrm{vol}(K(x_1), \ldots, L(x_n))$$
$$= \det(L)\det(K)\,\mathrm{vol}(x_1, \ldots, x_n).$$

The second property follows from

$$\mathrm{vol}(\alpha x_1, \ldots, \alpha x_n) = \alpha^n \mathrm{vol}(x_1, \ldots, x_n).$$

For the third, we simply use that $1_V = L \circ L^{-1}$ so

$$1 = \det(L)\det(L^{-1}).$$

For the last property, select a basis $x_1, \ldots, x_n$ for V. Then,

$$\mathrm{vol}(L(x_1), \ldots, L(x_n)) = \det(L)\,\mathrm{vol}(x_1, \ldots, x_n)$$
$$\neq 0.$$

Thus, $L(x_1), \ldots, L(x_n)$ is also a basis for V. This implies that L is invertible. $\square$

One can in fact show that any map $\Delta :$ Hom $(V, V) \to \mathbb{F}$ such that

$$\Delta (K \circ L) = \Delta (K) \Delta (L)$$
$$\Delta (1_V) = 1$$

depends only on the determinant of the operator (see also exercises).

We have some further useful and interesting results for determinants of matrices.

Proposition 5.5.3. *If $A \in$ Mat$_{n \times n}$ $(\mathbb{F})$ can be written in block form*

$$A = \begin{bmatrix} A_{11} & A_{12} \\ 0 & A_{22} \end{bmatrix},$$

where $A_{11} \in$ Mat$_{n_1 \times n_1}$ $(\mathbb{F})$, $A_{12} \in$ Mat$_{n_1 \times n_2}$ $(\mathbb{F})$, *and* $A_{22} \in$ Mat$_{n_2 \times n_2}$ $(\mathbb{F})$, $n_1 + n_2 = n$, *then*

$$\det A = \det A_{11} \det A_{22}.$$

Proof. Write the canonical basis for $\mathbb{F}^n$ as $e_1, \ldots, e_{n_1}, f_1, \ldots, f_{n_2}$ according to the block decomposition. Next, observe that A can be written as a composition in the following way:

$$A = \begin{bmatrix} A_{11} & A_{12} \\ 0 & A_{22} \end{bmatrix}$$

$$= \begin{bmatrix} 1 & A_{12} \\ 0 & A_{22} \end{bmatrix} \begin{bmatrix} A_{11} & 0 \\ 0 & 1 \end{bmatrix}$$

$$= BC$$

Thus, it suffices to show that

$$\det \begin{bmatrix} 1 & A_{12} \\ 0 & A_{22} \end{bmatrix} = \det B$$

$$= \det (A_{22})$$

and

$$\det \begin{bmatrix} A_{11} & 0 \\ 0 & 1 \end{bmatrix} = \det C$$

$$= \det (A_{11}).$$

To prove the last formula, note that for fixed $f_1, \ldots, f_{n_2}$ and

$$x_1, \ldots, x_{n_1} \in \text{span} \{e_1, \ldots, e_{n_1}\}$$

the volume form

$$\text{vol} (x_1, \ldots, x_{n_1}, f_1, \ldots, f_{n_2})$$

defines the usual volume form on span $\{e_1, \ldots, e_{n_1}\} = \mathbb{F}^{n_1}$. Thus,

$$\det C = \text{vol} \left(C\left(e_1\right), \ldots, C\left(e_{n_1}\right), C\left(f_1\right), \ldots, C\left(f_{n_2}\right) \right)$$
$$= \text{vol} \left(A_{11}\left(e_1\right), \ldots, A_{11}\left(e_{n_1}\right), f_1, \ldots, f_{n_2} \right)$$
$$= \det A_{11}.$$

For the first equation, we observe

$$\det B = \text{vol} \left(B\left(e_1\right), \ldots, B\left(e_{n_1}\right), B\left(f_1\right), \ldots, B\left(f_{n_2}\right) \right)$$
$$= \text{vol} \left(e_1, \ldots, e_{n_1}, A_{12}\left(f_1\right) + A_{22}\left(f_1\right), \ldots, A_{12}\left(f_{n_2}\right) + A_{22}\left(f_{n_2}\right) \right)$$
$$= \text{vol} \left(e_1, \ldots, e_{n_1}, A_{22}\left(f_1\right), \ldots, A_{22}\left(f_{n_2}\right) \right)$$

since $A_{12}\left(f_j\right) \in \text{span} \{e_1, \ldots, e_{n_1}\}$. Then, we get $\det B = \det A_{22}$ as before. $\square$

Proposition 5.5.4. *If* $A \in \text{Mat}_{n \times n}\left(\mathbb{F}\right)$, *then* $\det A = \det A^t$.

Proof. First note that the result is obvious if A is upper triangular. Using row operations, we can always find an invertible P such that PA is upper triangular and where P is a product of the elementary matrices of the types I_{ij} and $R_{ij}\left(\alpha\right)$. The row interchange matrices I_{ij} are symmetric, i.e., $I_{ij}^t = I_{ij}$ and have $\det I_{ij} = -1$. While $R_{ji}\left(\alpha\right)$ is upper or lower triangular with 1s on the diagonal. Hence $\left(R_{ij}\left(\alpha\right)\right)^t = R_{ji}\left(\alpha\right)$ and $\det R_{ij}\left(\alpha\right) = 1$. In particular, it follows that $\det P = \det P^t = \pm 1$. Thus,

$$\det A = \frac{\det \left(PA\right)}{\det P}$$
$$= \frac{\det \left(\left(PA\right)^t\right)}{\det \left(P\right)^t}$$
$$= \frac{\det \left(A^t P^t\right)}{\det \left(P\right)^t}$$
$$= \det \left(A^t\right)$$

$\square$

Remark 5.5.5. This last proposition tells us that the determinant map $A \rightarrow |A|$ defined on $\text{Mat}_{n \times n}\left(\mathbb{F}\right)$ is linear and alternating in both columns and rows. This can be extremely useful when calculating determinants. It also tells us that one can do Laplace expansions along columns as well as rows.

Exercises

1. Find the determinant of

$$L : \text{Mat}_{n \times n} (\mathbb{F}) \to \text{Mat}_{n \times n} (\mathbb{F})$$

$$L (X) = X^t.$$

2. Find the determinant of $L : P_n \to P_n$ where

(a) $L (p (t)) = p (-t)$
(b) $L (p (t)) = p (t) + p (-t)$
(c) $L (p) = Dp = p'$

3. Find the determinant of $L = p (D)$, for $p \in \mathbb{C} [t]$ when restricted to the spaces

(a) $V = P_n$
(b) $V = \text{span} \{\exp (\lambda_1 t) , \dots , \exp (\lambda_n t)\}$

4. Let $L : V \to V$ be an operator on a finite-dimensional inner product space. Show that

$$\overline{\det (L)} = \det (L^*) .$$

5. Let V be an n-dimensional inner product space and vol a volume form so that $\text{vol} (e_1, \dots , e_n) = 1$ for some orthonormal basis $e_1, \dots , e_n$.

(a) Show that if $L : V \to V$ is an isometry, then $|\det L| = 1$.
(b) Show that the set of isometries L with $\det L = 1$ is a group.

6. Show that $O \in O_n$ has type I if and only if $\det (O) = 1$. Conclude that SO_n is a group.

7. Given $A \in \text{Mat}_{n \times n} (\mathbb{F})$, consider the two linear operators $L_A (X) = AX$ and $R_A (X) = XA$ on $\text{Mat}_{n \times n} (\mathbb{F})$. Compute the determinant for these operators in terms of the determinant for A (see Example 1.7.6).

8. Show that if $L : V \to V$ is a linear operator and vol a volume form on V, then

$$\text{tr} (A) \text{vol} (x_1, \dots , x_n) = \text{vol} (L (x_1) , \dots , x_n)$$

$$+ \text{vol} (x_1, L (x_2) , \dots , x_n)$$

$$\vdots$$

$$+ \text{vol} (x_1, \dots , L (x_n)) .$$

9. Show that

$$p (t) = \det \begin{bmatrix} 1 & \cdots & 1 & 1 \\ \lambda_1 & \cdots & \lambda_n & t \\ \vdots & & \vdots & \vdots \\ \lambda_1^n & \cdots & \lambda_n^n & t^n \end{bmatrix}$$

defines a polynomial of degree n whose roots are $\lambda_1, \ldots, \lambda_n$. Compute k where

$$p\,(t) = k\,(t - \lambda_1) \cdots (t - \lambda_n)$$

by doing a Laplace expansion along the last column.

10. Assume that the $n \times n$ matrix A has a block decomposition

$$A = \begin{bmatrix} A_{11} & A_{12} \\ A_{21} & A_{22} \end{bmatrix},$$

where A_{11} is an invertible matrix. Show that

$$\det\,(A) = \det\,(A_{11}) \det\,\left(A_{22} - A_{21} A_{11}^{-1} A_{12}\right).$$

Hint: Select a suitable product decomposition of the form

$$\begin{bmatrix} A_{11} & A_{12} \\ A_{21} & A_{22} \end{bmatrix} = \begin{bmatrix} B_{11} & 0 \\ B_{21} & B_{22} \end{bmatrix} \begin{bmatrix} C_{11} & C_{12} \\ 0 & C_{22} \end{bmatrix}.$$

11. *(Jacobi's Theorem)* Let A be an invertible $n \times n$ matrix. Assume that A and A^{-1} have block decompositions

$$A = \begin{bmatrix} A_{11} & A_{12} \\ A_{21} & A_{22} \end{bmatrix},$$

$$A^{-1} = \begin{bmatrix} A'_{11} & A'_{12} \\ A'_{21} & A'_{22} \end{bmatrix}.$$

Show

$$\det\,(A) \det\,\left(A'_{22}\right) = \det\,(A_{11}).$$

Hint: Compute the matrix product

$$\begin{bmatrix} A_{11} & A_{12} \\ A_{21} & A_{22} \end{bmatrix} \begin{bmatrix} 1 & A'_{12} \\ 0 & A'_{22} \end{bmatrix}.$$

12. Let $A = \mathrm{Mat}_{n \times n}\,(\mathbb{F})$. We say that A has an *LU* decomposition if $A = LU$, where L is lower triangular with 1s on the diagonal and U is upper triangular. Show that A has an *LU* decomposition if all the *leading principal minors* have nonzero determinants. The leading principal $k \times k$ minor is the $k \times k$ submatrix gotten from A by eliminating the last $n - k$ rows and columns. (See also Exercise 4 in Sect. 1.13.)

13. *(Sylvester's Criterion)* Let A be a real and symmetric $n \times n$ matrix. Show that A has positive eigenvalues if and only if all leading principal minors have

positive determinant. Hint: As with the $A = LU$ decomposition in the previous exercise, show by induction on n that $A = U^*U$, where U is upper triangular. Such a decomposition is also called a *Cholesky factorization*.

14. *(Characterization of Determinant Functions)* Let $\Delta : \text{Mat}_{n \times n}(\mathbb{F}) \to \mathbb{F}$ be a function such that

$$\Delta(AB) = \Delta(A)\Delta(B),$$

$$\Delta(1_{\mathbb{F}^n}) = 1.$$

(a) Show that there is a function $f : \mathbb{F} \to \mathbb{F}$ satisfying

$$f(\alpha\beta) = f(\alpha)f(\beta)$$

such that $\Delta(A) = f(\det(A))$. Hint: Use Exercise 8 in Sect. 1.13 to show that

$$\Delta\left(I_{ij}\right) = \pm 1,$$

$$\Delta(M_i(\alpha)) = \Delta(M_1(\alpha)),$$

$$\Delta(R_{kl}(\alpha)) = \Delta(R_{kl}(1)) = \Delta(R_{12}(1)),$$

and define $f(\alpha) = \Delta(M_1(\alpha))$.

(b) If $\mathbb{F} = \mathbb{R}$ and n is even, show that $\Delta(A) = |\det(A)|$ defines a function such that

$$\Delta(AB) = \Delta(A)\Delta(B),$$

$$\Delta(\lambda 1_{\mathbb{R}^n}) = \lambda^n.$$

(c) If $\mathbb{F} = \mathbb{C}$ and in addition $\Delta(\lambda 1_{\mathbb{C}^n}) = \lambda^n$, then show that $\Delta(A) = \det(A)$.

(d) If $\mathbb{F} = \mathbb{R}$ and in addition $\Delta(\lambda 1_{\mathbb{R}^n}) = \lambda^n$, where n is odd, then show that $\Delta(A) = \det(A)$.

5.6   Linear Equations

Cramer's rule is a formula for the solution to n linear equations in n variables when we know that only one solution exists. We will generalize this construction a bit so as to see that it can be interpreted as an inverse to the isomorphism

$$\begin{bmatrix} x_1 & \cdots & x_n \end{bmatrix} : \mathbb{F}^n \to V$$

when $x_1, \ldots, x_n$ is a basis.

Theorem 5.6.1. *Let V be an n-dimensional vector space and* vol *a volume form. If $x_1, \ldots, x_n$ is a basis for V and $x = x_1\alpha_1 + \cdots + x_n\alpha_n$ is the expansion of $x \in V$ with respect to that basis, then*

$$\alpha_1 = \frac{\text{vol}(x, x_2, \ldots, x_n)}{\text{vol}(x_1, \ldots, x_n)},$$

$$\vdots \quad \vdots$$

$$\alpha_i = \frac{\text{vol}(x_1, \ldots, x_{i-1}, x, x_{i+1}, \ldots, x_n)}{\text{vol}(x_1, \ldots, x_n)},$$

$$\vdots \quad \vdots$$

$$\alpha_n = \frac{\text{vol}(x_1, \ldots, x_{n-1}, x)}{\text{vol}(x_1, \ldots, x_n)}.$$

Proof. First note that each

$$\frac{\text{vol}(x_1, \ldots, x_{i-1}, x, x_{i+1}, \ldots, x_n)}{\text{vol}(x_1, \ldots, x_n)}$$

is linear in x. Thus,

$$L(x) = \begin{bmatrix} \frac{\text{vol}(x, x_2, \ldots, x_n)}{\text{vol}(x_1, \ldots, x_n)} \\ \vdots \\ \frac{\text{vol}(x_1, \ldots, x_{i-1}, x, x_{i+1}, \ldots, x_n)}{\text{vol}(x_1, \ldots, x_n)} \\ \vdots \\ \frac{\text{vol}(x_1, \ldots, x_{n-1}, x)}{\text{vol}(x_1, \ldots, x_n)} \end{bmatrix}$$

is a linear map $V \to \mathbb{F}^n$. This means that we only need to check what happens when x is one of the vectors in the basis. If $x = x_i$, then

$$0 = \frac{\text{vol}(x_i, x_2, \ldots, x_n)}{\text{vol}(x_1, \ldots, x_n)},$$

$$\vdots \quad \vdots$$

$$1 = \frac{\text{vol}(x_1, \ldots, x_{i-1}, x_i, x_{i+1}, \ldots, x_n)}{\text{vol}(x_1, \ldots, x_n)},$$

$$\vdots \quad \vdots$$

$$0 = \frac{\text{vol}(x_1, \ldots, x_{n-1}, x_i)}{\text{vol}(x_1, \ldots, x_n)}.$$

Showing that $L(x_i) = e_i \in \mathbb{F}^n$. But this shows that L is the inverse to

$$[x_1 \cdots x_n] : \mathbb{F}^n \to V. \qquad \square$$

Cramer's rule is not necessarily very practical when solving equations, but it is often a useful abstract tool. It also comes in handy, as we shall see in Sect. 5.8 when solving inhomogeneous linear differential equations.

Cramer's rule can also be used to solve linear equations $L(x) = b$, as long as $L : V \rightarrow V$ is an isomorphism. In particular, it can be used to compute the inverse of L as is done in one of the exercises. To see how we can solve $L(x) = b$, we first select a basis $x_1, \ldots, x_n$ for V and then consider the problem of solving

$$
\begin{bmatrix} L(x_1) & \cdots & L(x_n) \end{bmatrix} \begin{bmatrix} \alpha_1 \\ \vdots \\ \alpha_n \end{bmatrix} = b.
$$

Since $L(x_1), \ldots, L(x_n)$ is also a basis, we know that this forces

$$
\alpha_1 = \frac{\mathrm{vol}(b, L(x_2), \ldots, L(x_n))}{\mathrm{vol}(L(x_1), \ldots, L(x_n))},
$$

$$
\vdots
$$

$$
\alpha_i = \frac{\mathrm{vol}(L(x_1), \ldots, L(x_{i-1}), b, L(x_{i+1}), \ldots, L(x_n))}{\mathrm{vol}(L(x_1), \ldots, L(x_n))},
$$

$$
\vdots
$$

$$
\alpha_n = \frac{\mathrm{vol}(L(x_1), \ldots, L(x_{n-1}), b)}{\mathrm{vol}(L(x_1), \ldots, L(x_n))}
$$

with $x = x_1\alpha_1 + \cdots + x_n\alpha_n$ being the solution. If we use $b = x_1, \ldots, x_n$, then we get the matrix representation for L^{-1} by finding the coordinates to the solutions of $L(x) = x_i$.

Example 5.6.2. As an example, let us see how we can solve

$$
\begin{bmatrix} 0 & 1 & \cdots & 0 \\ 0 & 0 & \ddots & \vdots \\ \vdots & \vdots & \ddots & 1 \\ 1 & 0 & \cdots & 0 \end{bmatrix} \begin{bmatrix} \xi_1 \\ \xi_2 \\ \vdots \\ \xi_n \end{bmatrix} = \begin{bmatrix} \beta_1 \\ \beta_2 \\ \vdots \\ \beta_n \end{bmatrix}.
$$

First, we see directly that

$$
\xi_2 = \beta_1,
$$

$$
\xi_3 = \beta_2,
$$

$$
\vdots
$$

$$
\xi_1 = \beta_n.
$$

From Cramer's rule, we get that

$$
\xi_1 = \frac{\begin{vmatrix} \beta_1 & 1 & \cdots & 0 \\ \beta_2 & 0 & \ddots & \vdots \\ \vdots & \vdots & \ddots & 1 \\ \beta_n & 0 & \cdots & 0 \end{vmatrix}}{\begin{vmatrix} 0 & 1 & \cdots & 0 \\ 0 & 0 & \ddots & \vdots \\ \vdots & \vdots & \ddots & 1 \\ 1 & 0 & \cdots & 0 \end{vmatrix}}
$$

A Laplace expansion along the first column tells us that

$$
\begin{vmatrix} \beta_1 & 1 & \cdots & 0 \\ \beta_2 & 0 & \ddots & \vdots \\ \vdots & \vdots & \ddots & 1 \\ \beta_n & 0 & \cdots & 0 \end{vmatrix} = \beta_1 \begin{vmatrix} 0 & 1 & \cdots & 0 \\ 0 & 0 & \ddots & \vdots \\ \vdots & \vdots & \ddots & 1 \\ 0 & 0 & \cdots & 0 \end{vmatrix} - \beta_2 \begin{vmatrix} 1 & 0 & \cdots & 0 \\ 0 & 0 & \ddots & \vdots \\ \vdots & \vdots & \ddots & 1 \\ 0 & 0 & \cdots & 0 \end{vmatrix}
$$

$$
\cdots + (-1)^{n+1} \beta_n \begin{vmatrix} 1 & 0 & \cdots & 0 \\ 0 & 1 & \ddots & \vdots \\ \vdots & \vdots & \ddots & 0 \\ 0 & 0 & \cdots & 1 \end{vmatrix}
$$

here all of the determinants are upper triangular and all but the last has zeros on the diagonal. Thus,

$$
\begin{vmatrix} \beta_1 & 1 & \cdots & 0 \\ \beta_2 & 0 & \ddots & \vdots \\ \vdots & \vdots & \ddots & 1 \\ \beta_n & 0 & \cdots & 0 \end{vmatrix} = (-1)^{n+1} \beta_n
$$

Similarly,

$$
\begin{vmatrix} 0 & 1 & \cdots & 0 \\ 0 & 0 & \ddots & \vdots \\ \vdots & \vdots & \ddots & 1 \\ 1 & 0 & \cdots & 0 \end{vmatrix} = (-1)^{n+1},
$$

so

$$
\xi_1 = \beta_n.
$$

Similar calculations will confirm our answers for $\xi_2, \ldots, \xi_n$. By using $b =$ $e_1, \ldots, e_n$, we can also find the inverse

$$
\begin{bmatrix}
0 & 1 & \cdots & 0 \\
0 & 0 & \ddots & \vdots \\
\vdots & \vdots & \ddots & 1 \\
1 & 0 & \cdots & 0
\end{bmatrix}^{-1}
=
\begin{bmatrix}
0 & 0 & \cdots & 1 \\
1 & 0 & \ddots & \vdots \\
\vdots & \ddots & \ddots & 0 \\
0 & \cdots & 1 & 0
\end{bmatrix}.
$$

Exercises

1. Let

$$
A_n =
\begin{bmatrix}
2 & -1 & 0 & \cdots & 0 \\
-1 & 2 & -1 & \cdots & 0 \\
0 & -1 & 2 & \ddots & \vdots \\
\vdots & \vdots & \ddots & \ddots & -1 \\
0 & 0 & \cdots & -1 & 2
\end{bmatrix}.
$$

   (a) Compute $\det A_n$ for $n = 1, 2, 3, 4$.
   (b) Compute A_n^{-1} for $n = 1, 2, 3, 4$.
   (c) Find $\det A_n$ and A_n^{-1} for general n.

2. Given a nontrivial volume form vol on an n-dimensional vector space V, a linear operator $L : V \to V$ and a basis $x_1, \ldots, x_n$ for V, define the *classical adjoint* $\mathrm{adj}\,(L) : V \to V$ by

$$
\begin{aligned}
\mathrm{adj}\,(L)\,(x) = &\ \mathrm{vol}\,(x, L\,(x_2), \ldots, L\,(x_n))\,x_1 \\
&+ \mathrm{vol}\,(L\,(x_1), x, L\,(x_3), \ldots, L\,(x_n))\,x_2 \\
&\quad\vdots \\
&+ \mathrm{vol}\,(L\,(x_1), \ldots, L\,(x_{n-1}), x)\,x_n.
\end{aligned}
$$

   (a) Show that $L \circ \mathrm{adj}\,(L) = \mathrm{adj}\,(L) \circ L = \det(L)\,1_V$.
   (b) Show that if L is an $n \times n$ matrix, then $\mathrm{adj}\,(L) = (\mathrm{cof}A)^t$, where $\mathrm{cof}A$ is the cofactor matrix whose ij entry is $(-1)^{i+j} \det A_{ij}$, where A_{ij} is the $(n-1) \times (n-1)$ matrix obtained from A by deleting the ith row and jth column (see Definition 5.4.2)
   (c) Show that $\mathrm{adj}\,(L)$ does not depend on the choice of basis $x_1, \ldots, x_n$ or volume form vol.

3. *(Lagrange Interpolation)* Use Cramer's rule and

$$p\,(t) = \det \begin{bmatrix} 1 & \cdots & 1 & 1 \\ \lambda_1 & \cdots & \lambda_n & t \\ \vdots & & \vdots & \vdots \\ \lambda_1^n & \cdots & \lambda_n^n & t^n \end{bmatrix}$$

to find $p \in P_n$ such that $p\,(t_0) = b_0, \ldots, p\,(t_n) = b_n$ where $t_0, \ldots, t_n \in \mathbb{C}$ are distinct.

4. Let $A \in \mathrm{Mat}_{n\times n}(\mathbb{F})$, where $\mathbb{F}$ is $\mathbb{R}$ or $\mathbb{C}$. Show that there is a constant C_n depending only on n such that if A is invertible, then

$$\left\|A^{-1}\right\| \le C_n \frac{\|A\|^{n-1}}{|\det\,(A)|}.$$

5. Let A be an $n \times n$ matrix whose entries are integers. If A is invertible show that A^{-1} has integer entries if and only if $\det\,(A) = \pm 1$.

6. Decide when the system

$$\begin{bmatrix} \alpha & -\beta \\ \beta & \alpha \end{bmatrix} \begin{bmatrix} \xi_1 \\ \xi_2 \end{bmatrix} = \begin{bmatrix} \beta_1 \\ \beta_2 \end{bmatrix}$$

can be solved for all β_1, β_2. Write down a formula for the solution.

7. For which α is the matrix invertible

$$\begin{bmatrix} \alpha & \alpha & 1 \\ \alpha & 1 & 1 \\ 1 & 1 & 1 \end{bmatrix}?$$

8. In this exercise, we will see how Cramer used his rule to study Leibniz's problem of when $Ax = b$ can be solved assuming that $A \in \mathrm{Mat}_{(n+1)\times n}(\mathbb{F})$ and $b \in \mathbb{F}^{n+1}$ (see Exercise 1 in Sect. 5.3). Assume in addition that $\mathrm{rank}\,(A) = n$. Then, delete one row from $[A|b]$ so that the resulting system $[A'|b']$ has a unique solution. Use Cramer's rule to solve $A'x = b'$ and then insert this solution in the equation that was deleted. Show that this equation is satisfied if and only if $\det[A|b] = 0$. Hint: The last equation is equivalent to a Laplace expansion of $\det[A|b] = 0$ along the deleted row.

9. For $a, b, c \in \mathbb{C}$ consider the real equation $a\xi + b\upsilon = c$, where $\xi, \upsilon \in \mathbb{R}$.

   (a) Write this as a system of the real equations.
   (b) Show that this system has a unique solution when $\mathrm{Im}\,(\bar{a}b) \neq 0$.
   (c) Use Cramer's rule to find a formula for ξ and υ that depends $\mathrm{Im}\,(\bar{a}b)$, $\mathrm{Im}\,(\bar{a}c)$, $\mathrm{Im}\,(\bar{b}c)$.

5.7 The Characteristic Polynomial

Now that we know that the determinant of a linear operator characterizes whether or not it is invertible, it would seem perfectly natural to define the characteristic polynomial as follows.

Definition 5.7.1. The characteristic polynomial of $L : V \to V$ is

$$\chi_L(t) = \det(t 1_V - L).$$

Clearly, a zero for the function $\chi_L(t)$ corresponds a value of t where $t 1_V - L$ is not invertible and thus $\ker(t 1_V - L) \neq \{0\}$, but this means that such a t is an eigenvalue. We now need to justify why this definition yields the same polynomial we constructed using Gauss elimination in Sect. 2.3.

Theorem 5.7.2. *Let $A \in \mathrm{Mat}_{n \times n}(\mathbb{F})$, then $\chi_A(t) = \det(t 1_{\mathbb{F}^n} - A)$ is a monic polynomial of degree n whose roots in $\mathbb{F}$ are the eigenvalues for $A : \mathbb{F}^n \to \mathbb{F}^n$. Moreover, this definition for the characteristic polynomial agrees with the one given using Gauss elimination.*

Proof. First we show that if $L : V \to V$ is a linear operator on an n-dimensional vector space, then $\chi_L(t) = \det(t 1_V - L)$ defines a monic polynomial of degree n. To see this, consider

$$\det(t 1_V - L) = \mathrm{vol}\left((t 1_V - L) e_1, \dots, (t 1_V - L) e_n\right)$$

and use linearity of vol to separate each of the terms $(t 1_V - L) e_k = t e_k - L(e_k)$. When doing this, we get to factor out t several times so it is easy to see that we get a polynomial in t. To check the degree, we group terms involving powers of t that are lower than n in the expression $O\left(t^{n-1}\right)$

$$
\begin{aligned}
\det(t 1_V - A) &= \mathrm{vol}\left((t 1_V - L) e_1, \dots, (t 1_V - L) e_n\right) \\
&= t \, \mathrm{vol}\left(e_1, (t 1_V - L) e_2, \dots, (t 1_V - L) e_n\right) \\
&\quad - \mathrm{vol}\left(L(e_1), (t 1_V - L) e_2, \dots, (t 1_V - L) e_n\right) \\
&= t \, \mathrm{vol}\left(e_1, (t 1_V - L) e_2, \dots, (t 1_V - L) e_n\right) + O\left(t^{n-1}\right) \\
&= t^2 \mathrm{vol}\left(e_1, e_2, \dots, (t 1_V - L) e_n\right) + O\left(t^{n-1}\right) \\
&\quad \vdots \\
&= t^n \mathrm{vol}\left(e_1, e_2, \dots, e_n\right) + O\left(t^{n-1}\right) \\
&= t^n + O\left(t^{n-1}\right).
\end{aligned}
$$

In Theorem 2.3.6, we proved that $(t1_{\mathbb{F}^n} - A) = PU$, where

$$U = \begin{bmatrix} r_1(t) & * & \cdots & * \\ 0 & r_2(t) & \cdots & * \\ \vdots & \vdots & \ddots & \vdots \\ 0 & 0 & \cdots & r_n(t) \end{bmatrix}$$

and P is the product of the elementary matrices: (1) I_{kl} interchanging rows, (2) $R_{kl}(r(t))$ which multiplies row l by a function $r(t)$ and adds it to row k, and (3) $M_k(\alpha)$ which simply multiplies row k by $\alpha \in \mathbb{F} - \{0\}$. For each fixed t, we have

$$\det(I_{kl}) = -1,$$
$$\det(R_{kl}(r(t))) = 1,$$
$$\det(M_k(\alpha)) = \alpha.$$

This means that

$$\det(t1_{\mathbb{F}^n} - A) = \det(PT)$$
$$= \det(P)\det(T)$$
$$= \det(P)\, r_1(t)\cdots r_n(t),$$

where $\det(P)$ is a nonzero scalar that does not depend on t and $r_1(t)\cdots r_n(t)$ is the function that we used to define the characteristic polynomial in Sect. 2.3. This shows that the two definitions have to agree. $\qquad\Box$

Remark 5.7.3. Recall that the Frobenius canonical form also lead us to a rigorous definition of the characteristic polynomial (see Sect. 2.7). Moreover, that definition definitely agrees with the definition from Sect. 2.3. It is also easy to see, using the above proof, that it agrees with the definition using determinants.

With this new definition of the characteristic polynomial, we can establish some further interesting properties.

Proposition 5.7.4. *Assume that* $L : V \to V$ *is a linear operator on an n-dimensional vector space with*

$$\chi_L(t) = t^n + \alpha_{n-1}t^{n-1} + \cdots + \alpha_1 t + \alpha_0.$$

Then,

$$\alpha_{n-1} = -\mathrm{tr}\,L,$$
$$\alpha_0 = (-1)^n \det L.$$

Proof. To show the last property, just note that

$$\alpha_0 = \chi_L(0)$$
$$= \det(-L)$$
$$= (-1)^n \det(L).$$

The first property takes a little more thinking. We use the calculation that lead to the formula

$$\det(t1_V - A) = \mathrm{vol}\left((t1_V - L)\, x_1, \ldots, (t1_V - L)\, x_n\right)$$
$$= t^n + O\left(t^{n-1}\right)$$

from the previous proof. Evidently, we have to calculate the coefficient in front of t^{n-1}. That term must look like

$$t^{n-1}\left(\mathrm{vol}\left(-L(e_1), e_2, \ldots, e_n\right) + \cdots + \mathrm{vol}\left(e_1, e_2, \ldots, -L(e_n)\right)\right).$$

Thus, we have to show

$$\mathrm{tr}(L) = \mathrm{vol}\left(L(e_1), e_2, \ldots, e_n\right) + \cdots + \mathrm{vol}\left(e_1, e_2, \ldots, L(e_n)\right).$$

To see this, expand

$$L(e_i) = \sum_{j=1}^{n} e_j \alpha_{ji}$$

so that $[\alpha_{ji}] = [L]$ and $\mathrm{tr}(L) = \alpha_{11} + \cdots + \alpha_{nn}$. Next note that if we insert that expansion in, say, $\mathrm{vol}\left(L(e_1), e_2, \ldots, e_n\right)$, then we have

$$\mathrm{vol}\left(L(e_1), e_2 \ldots, e_n\right) = \mathrm{vol}\left(\sum_{j=1}^{n} e_j \alpha_{j1}, e_2, \ldots, e_n\right)$$
$$= \mathrm{vol}\left(e_1 \alpha_{11}, e_2, \ldots, e_n\right)$$
$$= \alpha_{11} \mathrm{vol}\left(e_1, e_2, \ldots, e_n\right)$$
$$= \alpha_{11}.$$

This implies that

$$\mathrm{tr}(L) = \alpha_{11} + \cdots + \alpha_{nn}$$
$$= \mathrm{vol}\left(L(e_1), e_2, \ldots, e_n\right) +$$
$$\cdots + \mathrm{vol}\left(e_1, e_2, \ldots, L(e_n)\right).$$

$\square$

Proposition 5.7.5. *Assume that* $L : V \to V$ *is a linear operator on a finite-dimensional vector space. If* $M \subset V$ *is an* L*-invariant subspace, then* $\chi_{L|_M}(t)$ *divides* $\chi_L(t)$.

Proof. Select a basis $x_1, \ldots, x_n$ for V such that $x_1, \ldots, x_k$ form a basis for M. Then, the matrix representation for L in this basis looks like

$$[L] = \begin{bmatrix} A_{11} & A_{12} \\ 0 & A_{22} \end{bmatrix},$$

where $A_{11} \in \text{Mat}_{k \times k}(\mathbb{F})$, $A_{12} \in \text{Mat}_{k \times (n-k)}(\mathbb{F})$, and $A_{22} \in \text{Mat}_{(n-k) \times (n-k)}(\mathbb{F})$. This means that

$$t1_{\mathbb{F}^n} - [L] = \begin{bmatrix} t1_{\mathbb{F}^k} - A_{11} & A_{12} \\ 0 & t1_{\mathbb{F}^{n-k}} - A_{22} \end{bmatrix}.$$

Thus, we have

$$\chi_L(t) = \chi_{[L]}(t)$$
$$= \det(t1_{\mathbb{F}^n} - [L])$$
$$= \det(t1_{\mathbb{F}^k} - A_{11}) \det(t1_{\mathbb{F}^{n-k}} - A_{22}).$$

Now, A_{11} is the matrix representation for $L|_M$, so we have proven

$$\chi_L(t) = \chi_{L|_M}(t) \, p(t),$$

where $p(t)$ is some polynomial. $\qquad\qquad\qquad\qquad\qquad\qquad\qquad\qquad\square$

Exercises

1. Let $K, L : V \to V$ be linear operators on a finite-dimensional vector space.

   (a) Show that $\det(K - tL)$ is a polynomial in t.
   (b) If K or L is invertible show that $\det(tI - L \circ K) = \det(tI - K \circ L)$.
   (c) Show part b in general.

2. Let V be a finite-dimensional real vector space and $L : V \to V$ a linear operator.

   (a) Show that the number of complex roots of the characteristic polynomial is even. Hint: They come in conjugate pairs.
   (b) If $\dim_{\mathbb{R}} V$ is odd, then L has a real eigenvalue whose sign is the same as that of $\det L$.

(c) If $\dim_{\mathbb{R}} V$ is even and $\det L < 0$, then L has two real eigenvalues, one negative and one positive.

3. Let

$$A = \begin{bmatrix} \alpha & \gamma \\ \beta & \delta \end{bmatrix}.$$

Show that

$$\chi_A(t) = t^2 - (\operatorname{tr} A) t + \det A$$
$$= t^2 - (\alpha + \delta) t + (\alpha\delta - \beta\gamma).$$

4. Let $A \in \operatorname{Mat}_{3\times 3}(\mathbb{F})$ and $A = [\alpha_{ij}]$. Show that

$$\chi_A(t) = t^3 - (\operatorname{tr} A) t^2 + (|A_{11}| + |A_{22}| + |A_{33}|) t - \det A,$$

where A_{ii} is the companion matrix we get from eliminating the ith row and column in A.

5. Show that if L is invertible, then

$$\chi_{L^{-1}}(t) = \frac{(-t)^n}{\det L} \chi_L\left(t^{-1}\right).$$

6. Let $L : V \to V$ be a linear operator on a finite-dimensional inner product space with

$$\chi_L(t) = t^n + a_{n-1}t^{n-1} + \cdots + a_1 t + a_0.$$

Show that

$$\chi_{L^*}(t) = t^n + \bar{a}_{n-1}t^{n-1} + \cdots + \bar{a}_1 t + \bar{a}_0.$$

7. Let

$$\chi_L(t) = t^n + a_{n-1}t^{n-1} + \cdots + a_1 t + a_0$$

be the characteristic polynomial for $L : V \to V$. If vol is a volume form on V, show that

$$(-1)^k a_{n-k} \operatorname{vol}(x_1, \ldots, x_n)$$
$$= \sum_{i_1 < i_2 < \cdots < i_k} \operatorname{vol}\left(\ldots, x_{i_1-1}, L(x_{i_1}), x_{i_1+1}, \ldots, x_{i_k-1}, L(x_{i_k}), x_{i_k+1}, \ldots\right),$$

i.e., we are summing over all possible choices of $i_1 < i_2 < \cdots < i_k$ and in each summand replacing x_{i_j} by $L(x_{i_j})$.

8. Suppose we have a sequence $V_1 \xrightarrow{L_1} V_2 \xrightarrow{L_2} V_3$ of linear maps, where L_1 is one-to-one, L_2 is onto, and $\operatorname{im}(L_1) = \ker(L_2)$. Show that $\dim V_2 = \dim V_1 \dim V_3$.

Assume furthermore that we have linear operators $K_i : V_i \to V_i$ such that the diagram commutes

$$
\begin{array}{ccccc}
V_1 & \xrightarrow{L_1} & V_2 & \xrightarrow{L_2} & V_3 \\
K_1 \uparrow & & K_2 \uparrow & & K_3 \uparrow \\
V_1 & \xrightarrow{L_1} & V_2 & \xrightarrow{L_2} & V_3
\end{array}
$$

Show that

$$
\chi_{K_2}(t) = \chi_{K_1}(t)\,\chi_{K_3}(t).
$$

9. Using the definition

$$
\det A = \sum \operatorname{sign}(i_1, \ldots, i_n)\,\alpha_{i_1 1} \cdots \alpha_{i_n n}
$$

reprove the results from this section for matrices.

10. (*The Newton Identities*) In this exercise, we wish to generalize the formulae $\alpha_{n-1} = -\operatorname{tr}L$, $\alpha_0 = (-1)^n \det L$, for the characteristic polynomial

$$
t^n + \alpha_{n-1}t^{n-1} + \cdots + \alpha_1 t + \alpha_0 = (t - \lambda_1) \cdots (t - \lambda_n)
$$

of L.

(a) Prove that

$$
\alpha_k = (-1)^{n-k} \sum_{i_1 < \cdots < i_{n-k}} \lambda_{i_1} \cdots \lambda_{i_{n-k}}.
$$

(b) Prove that

$$
(\operatorname{tr}L)^k = (\lambda_1 + \cdots + \lambda_n)^k,
$$
$$
\operatorname{tr}(L^k) = \lambda_1^k + \cdots + \lambda_n^k.
$$

(c) Prove

$$
(\operatorname{tr}L)^2 = \operatorname{tr}(L^2) + 2\sum_{i<j} \lambda_i \lambda_j
$$

$$
= \operatorname{tr}(L^2) + 2\alpha_{n-2}.
$$

(d) Prove more generally that

$$
(\operatorname{tr}L)^k = k!\,(-1)^k \alpha_{n-k}
$$

$$
+ \binom{k}{2} (\operatorname{tr}L)^{k-2}\operatorname{tr}L^2
$$

$$
+ \left(\binom{k}{3} - \binom{k}{2} \right) (\operatorname{tr}L)^{k-3}\operatorname{tr}L^3
$$

$$+ \left(\binom{k}{4} - \binom{k}{3} + \binom{k}{2} \right) (\operatorname{tr} L)^{n-4} \operatorname{tr} L^4$$

$$\vdots$$

$$+ \left(\binom{k}{k} - \binom{k}{k-1} + \cdots + (-1)^k \binom{k}{2} \right) \operatorname{tr} L^k .$$

(e) If $\operatorname{tr} L = 0$, then

$$\left(\binom{n}{n} - \binom{n}{n-1} + \cdots + (-1)^n \binom{n}{2} \right) \operatorname{tr} L^n = n! \det L.$$

(f) If $\operatorname{tr} L = \operatorname{tr} L^2 = \cdots = \operatorname{tr} L^n = 0$, then $\chi_L(t) = t^n$.

5.8 Differential Equations*

We are now going to apply the theory of determinants to the study of linear differential equations. We start with the system $L(x) = \dot{x} - Ax = b$, where

$$x(t) \in \mathbb{C}^n,$$

$$b \in \mathbb{C}^n$$

$$A \in \operatorname{Mat}_{n \times n}(\mathbb{C})$$

and $x(t)$ is the vector-valued function we need to find. We know that the homogeneous problem $L(x) = 0$ has n linearly independent solutions $x_1, \ldots, x_n$. More generally, we can show something quite interesting about collections of solutions.

Lemma 5.8.1. *Let $x_1, \ldots, x_n$ be solutions to the homogeneous problem $L(x) = 0$; then,*

$$\frac{d}{dt} \left(\operatorname{vol}(x_1, \ldots, x_n) \right) = \operatorname{tr}(A) \operatorname{vol}(x_1, \ldots, x_n) .$$

In particular,

$$\operatorname{vol}(x_1, \ldots, x_n)(t) = \operatorname{vol}(x_1, \ldots, x_n)(t_0) \exp(\operatorname{tr}(A)(t - t_0)) .$$

Moreover, $x_1, \ldots, x_n$ are linearly independent solutions if and only if $x_1(t_0), \ldots, x_n(t_0) \in \mathbb{C}^n$ are linearly independent. Each of these two conditions in turn imply that $x_1(t), \ldots, x_n(t) \in \mathbb{C}^n$ are linearly independent for all t.

Proof. To compute the derivative, we find the Taylor expansion for

$$\mathrm{vol}\,(x_1,\ldots,x_n)\,(t+h)$$

in terms of h and then identify the term that is linear in h. This is done along the lines of our proof in the previous section that $\alpha_{n-1} = -\mathrm{tr}\,A$, where α_{n-1} is the coefficient in front of t^{n-1} in the characteristic polynomial.

$$
\begin{aligned}
&\mathrm{vol}\,(x_1,\ldots,x_n)\,(t+h) \\
&\quad = \mathrm{vol}\,(x_1\,(t+h),\ldots,x_n\,(t+h)) \\
&\quad = \mathrm{vol}\,(x_1\,(t) + Ax_1\,(t)\,h + o\,(h),\ldots,x_n\,(t) + Ax_n\,(t)\,h + o\,(h)) \\
&\quad = \mathrm{vol}\,(x_1\,(t),\ldots,x_n\,(t)) \\
&\qquad + h\mathrm{vol}\,(Ax_1\,(t),\ldots,x_n\,(t)) \\
&\qquad\vdots \\
&\qquad + h\mathrm{vol}\,(x_1\,(t),\ldots,Ax_n\,(t)) \\
&\qquad + o\,(h) \\
&\quad = \mathrm{vol}\,(x_1\,(t),\ldots,x_n\,(t)) + h\mathrm{tr}\,(A)\,\mathrm{vol}\,(x_1\,(t),\ldots,x_n\,(t)) + o\,(h)\,.
\end{aligned}
$$

Thus,

$$v\,(t) = \mathrm{vol}\,(x_1,\ldots,x_n)\,(t)$$

solves the differential equation

$$\dot{v} = \mathrm{tr}\,(A)\,v$$

implying that

$$v\,(t) = v\,(t_0)\exp\,(\mathrm{tr}\,(A)\,(t-t_0))\,.$$

In particular, we see that $v\,(t) \neq 0$ for all t provided $v\,(t_0) \neq 0$.

It remains to prove that $x_1,\ldots,x_n$ are linearly independent solutions if and only if $x_1\,(t_0),\ldots,x_n\,(t_0) \in \mathbb{C}^n$ are linearly independent. It is obvious that $x_1,\ldots,x_n$ are linearly independent if $x_1\,(t_0),\ldots,x_n\,(t_0) \in \mathbb{C}^n$ are linearly independent. Conversely, if we assume that $x_1\,(t_0),\ldots,x_n\,(t_0) \in \mathbb{C}^n$ are linearly dependent, then we can find $\alpha_1,\ldots,\alpha_n \in \mathbb{C}^n$ not all zero so that

$$\alpha_1 x_1\,(t_0) + \cdots + \alpha_n x_n\,(t_0) = 0.$$

Uniqueness of solutions to the initial value problem $L\,(x) = 0$, $x\,(t_0) = 0$, then implies that

$$x\,(t) = \alpha_1 x_1\,(t) + \cdots + \alpha_n x_n\,(t) \equiv 0$$

for all t.   □

We now claim that the inhomogeneous problem can be solved provided we have found a linearly independent set of solutions $x_1, \ldots, x_n$ to the homogeneous equation. The formula comes from Cramer's rule but is known as the *variations of constants method*. We assume that the solution x to

$$L(x) = \dot{x} - Ax = b,$$

$$x(t_0) = 0$$

looks like

$$x(t) = c_1(t) x_1(t) + \cdots + c_n(t) x_n(t),$$

where $c_1(t), \ldots, c_n(t) \in C^\infty(\mathbb{R}, \mathbb{C})$ are functions rather than constants. Then,

$$\dot{x} = c_1 \dot{x}_1 + \cdots + c_n \dot{x}_n + \dot{c}_1 x_1 + \cdots + \dot{c}_n x_n$$

$$= c_1 A x_1 + \cdots + c_n A x_n + \dot{c}_1 x_1 + \cdots + \dot{c}_n x_n$$

$$= A(x) + \dot{c}_1 x_1 + \cdots + \dot{c}_n x_n.$$

In other words,

$$L(x) = \dot{c}_1 x_1 + \cdots + \dot{c}_n x_n.$$

This means that for each t, the values $\dot{c}_1(t), \ldots, \dot{c}_n(t)$ should solve the linear equation

$$\dot{c}_1 x_1 + \cdots + \dot{c}_n x_n = b.$$

Cramer's rule for solutions to linear systems (see Sect. 5.6) then tells us that

$$\dot{c}_1(t) = \frac{\text{vol}(b, \ldots, x_n)(t)}{\text{vol}(x_1, \ldots, x_n)(t)},$$

$$\vdots \quad \vdots$$

$$\dot{c}_n(t) = \frac{\text{vol}(x_1, \ldots, b)(t)}{\text{vol}(x_1, \ldots, x_n)(t)},$$

implying that

$$c_1(t) = \int_{t_0}^t \frac{\text{vol}(b, \ldots, x_n)(s)}{\text{vol}(x_1, \ldots, x_n)(s)} ds,$$

$$\vdots \quad \vdots$$

$$c_n(t) = \int_{t_0}^t \frac{\text{vol}(x_1, \ldots, b)(s)}{\text{vol}(x_1, \ldots, x_n)(s)} ds.$$

In practice, there are more efficient methods that can be used when we know something about b. These methods also use linear algebra.

Having dealt with systems, we next turn to higher order equations: $L(x) = p(D)(x) = f$, where

$$p(D) = D^n + \alpha_{n-1} D^{n-1} + \cdots + \alpha_1 D + \alpha_0$$

is a polynomial with complex or real coefficients and $f(t) \in C^\infty (\mathbb{R}, \mathbb{C})$. This can be translated into a system $\dot{z} - Az = b$, or

$$\dot{z} - \begin{bmatrix} 0 & 1 & \cdots & & 0 \\ \vdots & \ddots & \vdots & & \vdots \\ 0 & \cdots & 0 & 1 \\ -\alpha_0 & \cdots & -\alpha_{n-2} & -\alpha_{n-1} \end{bmatrix} z = \begin{bmatrix} 0 \\ \vdots \\ 0 \\ f \end{bmatrix},$$

by using

$$z = \begin{bmatrix} x \\ Dx \\ \vdots \\ D^{n-1}x \end{bmatrix}.$$

If we have n functions $x_1, \ldots, x_n \in C^\infty (\mathbb{R}, \mathbb{C})$, then the *Wronskian* is defined as

$$W(x_1, \ldots, x_n)(t) = \text{vol}(z_1, \ldots, z_n)(t)$$

$$= \det \begin{bmatrix} x_1(t) & \cdots & x_n(t) \\ (Dx_1)(t) & \cdots & (Dx_n)(t) \\ \vdots & \ddots & \vdots \\ (D^{k-1}x_1)(t) & \cdots & (D^{k-1}x_n)(t) \end{bmatrix}.$$

In the case where $x_1, \ldots, x_n$ solve $L(x) = p(D)(x) = 0$, this tells us that

$$W(x_1, \ldots, x_n)(t) = W(x_1, \ldots, x_n)(t_0) \exp(-\alpha_{n-1}(t - t_0)).$$

Finally, we can again try the variation of constants method to solve the inhomogeneous equation. It is slightly tricky to do this directly by assuming that

$$x(t) = c_1(t) x_1(t) + \cdots + c_n(t) x_n(t).$$

Instead, we use the system $\dot{z} - Az = b$, and guess that

$$z = c_1(t) z_1(t) + \cdots + c_n(t) z_n(t).$$

This certainly implies that

$$x(t) = c_1(t) x_1(t) + \cdots + c_n(t) x_n(t),$$

but the converse is not true. As above, we get

$$c_1(t) = \int_{t_0}^t \frac{\text{vol}(b, \ldots, z_n)(s)}{\text{vol}(z_1, \ldots, z_n)(s)} ds,$$

$$\vdots \quad \vdots$$

$$c_n(t) = \int_{t_0}^t \frac{\text{vol}(z_1, \ldots, b)(s)}{\text{vol}(z_1, \ldots, z_n)(s)} ds.$$

Here

$$\text{vol}(z_1, \ldots, z_n) = W(x_1, \ldots, x_n).$$

The numerator can also be simplified by using a Laplace expansion along the column vector b. This gives us

$$\text{vol}(b, z_2, \ldots, z_n) = \begin{vmatrix} 0 & x_1 & \cdots & x_n \\ \vdots & \vdots & \cdots & \vdots \\ 0 & D^{n-2}x_2 & \cdots & D^{n-2}x_n \\ b & D^{n-1}x_2 & \cdots & D^{n-1}x_n \end{vmatrix}$$

$$= (-1)^{n+1} b \begin{vmatrix} x_1 & \cdots & x_n \\ \vdots & \cdots & \vdots \\ D^{n-2}x_2 & \cdots & D^{n-2}x_n \end{vmatrix}$$

$$= (-1)^{n+1} b W(x_2, \ldots, x_n).$$

Thus,

$$c_1(t) = (-1)^{n+1} \int_{t_0}^t \frac{b(s) W(x_2, \ldots, x_n)(s)}{W(x_1, \ldots, x_n)(s)} ds,$$

$$\vdots \quad \vdots$$

$$c_n(t) = (-1)^{n+n} \int_{t_0}^t \frac{b(s) W(x_1, \ldots, x_{n-1})(s)}{W(x_1, \ldots, x_n)(s)} ds,$$

and therefore, a solution to the inhomogeneous equation is given by

$$x(t) = \left((-1)^{n+1} \int_{t_0}^t \frac{b(s) W(x_2, \ldots, x_n)(s)}{W(x_1, \ldots, x_n)(s)} ds \right) x_1(t) + \cdots$$

$$+ \left((-1)^{n+n} \int_{t_0}^t \frac{b(s) W(x_1, \ldots, x_{n-1})(s)}{W(x_1, \ldots, x_n)(s)} ds \right) x_n(t)$$

$$= \sum_{k=1}^{n} (-1)^{n+k} x_k (t) \int_{t_0}^{t} \frac{b(s) W(x_1, \ldots, \hat{x}_k, \ldots, x_n)(s)}{W(x_1, \ldots, x_n)(s)} ds.$$

Let us try to solve a concrete problem using these methods.

Example 5.8.2. Find the complete set of solutions to $\ddot{x} - 2\dot{x} + x = \exp(t)$. We see that $\ddot{x} - 2\dot{x} + x = (D - 1)^2 x$, thus the characteristic equation is $(\lambda - 1)^2 = 1$. This means that we only get one solution $x_1 = \exp(t)$ from the eigenvalue $\lambda = 1$. The other solution is then given by $x_2(t) = t \exp(t)$. We now compute the Wronskian to check that they are linearly independent:

$$W(x_1, x_2) = \begin{vmatrix} \exp(t) & t \exp(t) \\ \exp(t) & (1 + t) \exp(t) \end{vmatrix}$$

$$= \exp(2t) \begin{vmatrix} 1 & t \\ 1 & (1 + t) \end{vmatrix}$$

$$= ((1 + t) - t) \exp(2t)$$

$$= \exp(2t).$$

Note, we could also have found x_2 from our knowledge that

$$W(x_1, x_2)(t) = W(x_1, x_2)(t_0) \exp(2(t - t_0)).$$

Assuming that $t_0 = 0$ and we want $W(x_1, x_2)(t_0) = 1$, we simply need to solve

$$W(x_1, x_2)(t) = x_1 \dot{x}_2 - \dot{x}_1 x_2 = \exp(2t).$$

Since $x_1 = \exp(t)$, this implies that

$$\dot{x}_2 - x_2 = \exp(t).$$

Hence,

$$x_2(t) = \exp(t) \int_0^t \exp(-s) \exp(t) ds$$

$$= t \exp(t)$$

as expected.

The variation of constants formula now tells us to compute

$$c_1(t) = (-1)^{2+1} \int_0^t \frac{f(s) x_2(s)}{W(x_1, x_2)(s)} ds$$

$$= -\int_0^t \frac{\exp(s) (s \exp(s))}{\exp(2s)} ds$$

$$= -\int_0^t s\,ds$$

$$= -\frac{1}{2}t^2$$

and

$$c_2(t) = (-1)^{2+2} \int_0^t \frac{f(s)\,x_1(s)}{W(x_1, x_2)(s)}\,ds$$

$$= \int_0^t 1\,ds$$

$$= t.$$

Thus,

$$x = -\frac{1}{2}t^2 x_1(t) + t x_2(t)$$

$$= -\frac{1}{2}t^2 \exp(t) + t\,(t\exp(t))$$

$$= \frac{1}{2}t^2 \exp(t)$$

solves the inhomogeneous problem and $x = \alpha_1 \exp(t) + \alpha_2 t \exp(t) + \frac{1}{2}t^2 \exp(t)$ represents the complete set of solutions.

Exercises

(1) Let $p_0(t), \ldots, p_n(t) \in \mathbb{C}[t]$ and assume that $t \in \mathbb{R}$. If

$$p_i(t) = \alpha_{ni}t^n + \cdots + \alpha_{1i}t + \alpha_{0i},$$

show that

$$W(p_0, \ldots, p_n) = \det \begin{bmatrix} p_0(t) & \cdots & p_n(t) \\ (Dp_0)(t) & \cdots & (Dp_n)(t) \\ \vdots & & \vdots \\ (D^n p_0)(t) & \cdots & (D^n p_n)(t) \end{bmatrix}$$

$$= \det \begin{bmatrix} \alpha_{00} & \cdots & \alpha_{0n} \\ \alpha_{10} & \cdots & \alpha_{1n} \\ 2\alpha_{20} & \cdots & 2\alpha_{2n} \\ \vdots & & \vdots \\ n!\alpha_{n0} & \cdots & n!\alpha_{nn} \end{bmatrix}$$

$$= n! \cdot (n-1)! \cdot \cdots \cdot 2 \cdot 1 \det \begin{bmatrix} \alpha_{00} & \cdots & \alpha_{0n} \\ \alpha_{10} & \cdots & \alpha_{1n} \\ \alpha_{20} & \cdots & \alpha_{2n} \\ \vdots & & \vdots \\ \alpha_{n0} & \cdots & \alpha_{nn} \end{bmatrix}.$$

(2) Let $x_1, \ldots, x_n$ be linearly independent solutions to

$$p(D)(x) = \left(D^n + \alpha_{n-1} D^{n-1} + \cdots + \alpha_0 \right)(x) = 0.$$

Attempt the following questions without using what we know about existence
and uniqueness of solutions to differential equations.

(a) Show that

$$p(D)(x) = \frac{W(x_1 \ldots, x_n, x)}{W(x_1 \ldots, x_n)}.$$

(b) Conclude that $p(D)(x) = 0$ if and only if $W(x, x_1 \ldots, x_n) = 0$.
(c) If $W(x, x_1 \ldots, x_n) = 0$, then x is a linear combination of $x_1, \ldots, x_n$.
(d) If x, y are solutions with the same initial values: $x(0) = y(0)$, $Dx(0) = Dy(0), \ldots, D^{n-1}x(0) = D^{n-1}y(0)$, then $x = y$.

(3) Assume two monic polynomials $p, q \in \mathbb{C}[t]$ have the property that
$p(D)(x) = 0$ and $q(D)(x) = 0$ have the same solutions. Is it true that
$p = q$? Hint: If $p(D)(x) = 0 = q(D)(x)$, then $\gcd(p, q)(D)(x) = 0$.
(4) Assume that x is a solution to $p(D)(x) = 0$, where $p(D) = D^n + \cdots + \alpha_1 D + \alpha_0$.

(a) Show that the phase shifts $x_\omega(t) = x(t + \omega)$ are also solutions.
(b) If the vectors

$$\begin{bmatrix} x(\omega_1) \\ Dx(\omega_1) \\ \vdots \\ D^{n-1}x(\omega_1) \end{bmatrix}, \ldots, \begin{bmatrix} x(\omega_n) \\ Dx(\omega_n) \\ \vdots \\ D^{n-1}x(\omega_n) \end{bmatrix}$$

form a basis for $\mathbb{C}^n$ for some choice of $\omega_1, \ldots, \omega_n \in \mathbb{R}$, then all solutions
to $p(D)(x) = 0$ are linear combinations of the phase-shifted solutions
$x_{\omega_1}, \ldots, x_{\omega_n}$.
(c) If the vectors

$$\begin{bmatrix} x(\omega_1) \\ Dx(\omega_1) \\ \vdots \\ D^{n-1}x(\omega_1) \end{bmatrix}, \ldots, \begin{bmatrix} x(\omega_n) \\ Dx(\omega_n) \\ \vdots \\ D^{n-1}x(\omega_n) \end{bmatrix}$$

never form a basis for $\mathbb{C}^n$ for any choice of $\omega_1, \ldots, \omega_n \in \mathbb{R}$, then x is a solution to a kth equation for $k < n$. Hint: If x is not a solution to a lower order equation, the $x, Dx, \ldots, D^{n-1}x$ is a (cyclic) basis for the solution space.

(5) Find a formula for the real solutions to the system

$$\begin{bmatrix} \dot{x}_1 \\ \dot{x}_2 \end{bmatrix} - \begin{bmatrix} a & -b \\ b & a \end{bmatrix} \begin{bmatrix} x_1 \\ x_2 \end{bmatrix} = \begin{bmatrix} b_1 \\ b_2 \end{bmatrix},$$

where $a, b \in \mathbb{R}$ and $b_1, b_2 \in C^\infty(\mathbb{R}, \mathbb{R})$.

(6) Find a formula for the real solutions to the equation

$$\ddot{x} + a\dot{x} + bx = f,$$

where $a, b \in \mathbb{R}$ and $f \in C^\infty(\mathbb{R}, \mathbb{R})$.

References

[Axler]   Axler, S.: Linear Algebra Done Right. Springer-Verlag, New York (1997)
[Bretscher]   Bretscher, O.: Linear Algebra with Applications, 2nd edn. Prentice-Hall, Upper Saddle River (2001)
[Curtis]   Curtis, C.W.: Linear Algebra: An Introductory Approach. Springer-Verlag, New York (1984)
[Greub]   Greub, W.: Linear Algebra, 4th edn. Springer-Verlag, New York (1981)
[Halmos]   Halmos, P.R.: Finite-Dimensional Vector Spaces. Springer-Verlag, New York (1987)
[Hoffman-Kunze]   Hoffman, K., Kunze, R.: Linear Algebra. Prentice-Hall, Upper Saddle River (1961)
[Lang]   Lang, S.: Linear Algebra, 3rd edn. Springer-Verlag, New York (1987)
[Roman]   Roman, S.: Advanced Linear Algebra, 2nd edn. Springer-Verlag, New York (2005)
[Serre]   Serre, D.: Matrices, Theory and Applications. Springer-Verlag, New York (2002)

P. Petersen, *Linear Algebra*, Undergraduate Texts in Mathematics,
DOI 10.1007/978-1-4614-3612-6, © Springer Science+Business Media New York 2012

Index

A
Adjoint, 242
Affine subspace, 62
Algebraic multiplicity, 169
Algebraic numbers, 18
Annihilator, 99
Augmented matrix, 84
Axiom of choice, 18
Axioms
  Field, 6
  Vector Space, 8

B
Basis, 14
Bessel Inequality, 240
Bianchi's identity, 300

C
Canonical form
  Frobenius, 185
  Jordan, 193
  rational, 188
Cauchy-Schwarz inequality, 211, 222, 223
Cayley-Hamilton theorem, 178, 192
Change of basis, 48
Characteristic
  Equation, 134
  Polynomial, 123, 136, 140, 188, 367
  Value, 123
Column equivalence, 94
Column operations, 94, 343
Column rank, 75, 92
Companion matrix, 176, 350
Complement
  existence of, 61, 73
  uniqueness of, 65

Complementary subspaces, 57
Completely reducible, 268
Complexification, 13
Conjugate vector space, 13
Cramer's rule, 361
Cyclic subspace, 174
Cyclic subspace decomposition, 178

D
Darboux vector, 300, 308
Degree, 114
Descartes Rule of Signs, 328
Determinant, 355
Diagonalizable, 158
Differential, 29
Differentiation operator, 26
Dimension, 15, 44, 90
Dimension formula, 66, 90, 111
  for subspaces, 69
Direct sum, 58
Dual basis, 99, 234
Dual map, 102
Dual space, 98

E
Eigenspace, 133
Eigenvalue, 124, 127, 133
Eigenvector, 127, 133
Elementary
  Column operations, 94, 343
  Matrices, 86
  Row operations, 84
Elementary divisors, 188
Equivalence, 95
Expansion, 15
Exponential function, 255

P. Petersen, *Linear Algebra*, Undergraduate Texts in Mathematics,
DOI 10.1007/978-1-4614-3612-6, © Springer Science+Business Media New York 2012